Mechanics Modeling of Sheet Metal Forming

Sing C. Tang
Jwo Pan

SAE International™
Warrendale, Pa.

For permission and licensing requests, contact:

SAE Permissions
400 Commonwealth Drive
Warrendale, PA 15096-0001 USA
E-mail: permissions@sae.org
Tel: 724-772-4028
Fax: 724-772-4891

Library of Congress Cataloging-in-Publication Data

Tang, Sing C.
Mechanics modeling of sheet metal forming / Sing C. Tang, Jwo Pan.
p. cm.
Includes bibliographical references and index.
ISBN 978-0-7680-0896-8
1. Sheet-metal work. 2. Continuum mechanics. I. Pan, J. (Jwo).
II. Title.

TS250.T335 2007
671.8'23011--dc22 2006039364

SAE International
400 Commonwealth Drive
Warrendale, PA 15096-0001 USA
E-mail: CustomerService@sae.org
Tel: 877-606-7323 (inside USA and Canada)
724-776-4970 (outside USA)
Fax: 724-776-1615

ISBN 978-0-7680-0896-8

SAE Order No. R-321

Printed in the United States of America.

green press INITIATIVE

SAE International is committed to preserving ancient forests and natural resources. We elected to print *R-321 Mechanics Modeling Of Sheet Metal Forming* on 30% post consumer recycled paper, processed chlorine free. As a result, for this printing, we have saved:

6 Trees (40' tall and 6-8" diameter)
2,511 Gallons of Wastewater
1,010 Kilowatt Hours of Electricity
277 Pounds of Solid Waste
544 Pounds of Greenhouse Gases

SAE International made this paper choice because our printer, Thomson-Shore, Inc., is a member of Green Press Initiative, a nonprofit program dedicated to supporting authors, publishers, and suppliers in their efforts to reduce their use of fiber obtained from endangered forests.

For more information, visit www.greenpressinitiative.org

Thanks to our families for their support and patience.

To my wife Kin Ling

—Sing C. Tang

To my mom Mei-Chin and my wife Michelle

—Jwo Pan

Contents

Preface

Beverage cans and many parts in aircraft, appliances, and automobiles are made of thin sheet metals formed by stamping operations at room temperature. Thus, sheet metal forming processes play an important role in mass production. Conventionally, the forming process and tool designs are based on the trial-and-error method or the pure geometric method of surface fitting that requires an actual hardware tryout that is called a die tryout. This design process often is expensive and time consuming because forming tools must be built for each trial. Significant savings are possible if a designer can use simulation tools based on the principles of mechanics to predict formability before building forming tools for tryout. Due to the geometric complexity of sheet metal parts, especially automotive body panels, development of an analytical method based on the mechanics principles to predict formability is difficult, if not impossible. Because of modern computer technology, the numerical finite element method at the present time is feasible for such a highly nonlinear analysis using a digital computer, especially one equipped with vector and parallel processors.

Although simulation of sheet metal forming processes using a modern digital computer is an important technology, a comprehensive book on this subject seems to be lacking in the literature. Fundamental principles are discussed in some books for forming sheet metal parts with simple geometry such as plane strain or axisymmetry. In contrast, detailed theoretically sound formulations based on the principles of continuum mechanics for finite or large deformation are presented in this book for implementation into simulation codes. The contents of this book represent proof of the usefulness of advanced continuum mechanics, plasticity theories, and shell theories to practicing engineers. The governing equations are presented with specified boundary and initial conditions, and these equations are solved using a modern digital computer (engineering workstation) via finite element methods. Therefore, the forming of any complex part such as an automotive inner panel can be simulated. We hope that simulation engineers who read this book will then be able to use simulation software wisely and better understand the output of the simulation software. Therefore, this book is not only a textbook but also a reference book for practicing engineers. Because advanced topics are discussed in the book, readers should have some basic knowledge of mechanics, constitutive laws, finite element methods, and matrix and tensor analyses.

Chapter 1 gives a brief introduction to typical automotive sheet metal forming processes. Basic mechanics, vectors and tensors, and constitutive laws for elastic and plastic materials are reviewed in Chapters 2 and 3, based on course material taught at the University of Michigan by Dr. Jwo Pan. The remaining chapters are drawn from the experience of Dr. Sing C. Tang, who had been working on simulations of real automotive sheet metal parts at Ford Motor Company for more than 15 years. Chapter 2 presents the fundamental concepts of tensors, stress, and strain. The definitions of the stresses and strains in tensile tests then are discussed. Readers should pay special attention to the kinematics of finite deformation and the definitions of different stress tensors due to finite deformation because extremely large deformation occurs in sheet metal forming processes.

Chapter 3 reviews the linear elastic constitutive laws for small or infinitesimal deformation. Hooke's law for isotropic linear elastic materials, which is widely used in many mechanics analyses, is discussed first. Anisotropic linear elastic behavior also is discussed in detail. Then, deviatoric stresses and deviatoric strains are introduced. These concepts are used as the basis for development of pressure-independent incompressible anisotropic plasticity theory. Chapter 3 also discusses fundamentals of mathematical plasticity theories. In sheet metal forming processes, most of the deformation is plastic. Therefore, knowledge of plasticity is essential in using simulation software and in understanding simulation results. Different mathematical models for uniaxial tensile stress-strain relations are introduced first. Then the yield conditions for isotropic incompressible materials under multiaxial stress states are presented. Because sheet metals generally are plastically anisotropic, the anisotropic yield conditions are discussed in detail. The basic concepts of the formation of constitutive laws with consideration of plastic hardening behavior of materials also are presented. Finally, the principles of plastic localization and modeling of failure processes based on void mechanics are summarized.

Chapter 4 introduces formulations for analyses of sheet metal forming processes, including binder closing, stretching/drawing, trimming, flanging, and hemming. More attention is paid to the most basic analysis of the stretching/drawing process, which then can be extended to analyses of all other processes. The formulations include equations of motion, constitutive equations, tool surface modeling, surface contact forces, and draw-bead modeling.

Chapter 5 discusses thin shell theories. Tensors with reference to the curvilinear coordinate system are used. Most sheet metal parts are made of thin sheets and can be modeled by thin shells for numerical efficiency and accuracy. Engineers may be tempted to use three-dimensional (3-D) solid elements, which are more general, to model a metal sheet under plastic deformation. However, the solid element model contains too many degrees of freedom to be solved using the current generation of digital computers. Even for the explicit time integration method, we cannot handle a finite element model with too many degrees of freedom for reasonable computation accuracy and time. The reason is that the dimension in the thickness direction of the sheet is very small compared to other dimensions. To satisfy the stability requirement for a numerical solution using the explicit time integration method, an extremely small time increment for a three-dimensional mesh must be used. However, it still is not practical at the present time, and the shell model is emphasized in this book.

Chapter 6 presents formulations of two shell elements for finite element models appropriate for use in computation. The interpolation (shape) function for the C^1 continuous shell element is complex but accurate, and it provides good convergence for the implicit integration method. The interpolation function for the C^0 continuous element is simple, but it might have a shear locking problem for thin sheets.

Chapter 7 presents solution methods for the equations of motion by the explicit time integration and implicit time integration methods. The contact forces are computed by the direct, Lagrangian multiplier, or penalty methods. If the dynamic effects are neglected, the equations of motion are reduced to the equations of equilibrium that are solved by the quasi-static method. Although the quasi-static method is more appropriate for analyses of sheet metal forming processes, it has convergence problems. Also, it would break down for a singular stiffness matrix when structural instability occurs. Structural stability problems also are discussed in Chapter 7. The radial return method is discussed to compute the stress increment from a given strain increment for more accurate numerical results. Computation of springback also is discussed briefly. For more efficient computations, adaptive meshing is introduced. Finally, various numerical examples for forming, springback, and flanging operations are given.

Chapter 8 on buckling and wrinkling analyses briefly introduces Rik's approach to the solution of snap-through and bifurcation buckling. This type of instability may occur when the global stiffness matrix in the quasi-static method becomes singular. Because analyses of sheet metal forming processes mainly involve surface contact with friction, Rik's method cannot be applied directly without modification. Some methods are suggested to compute sheet deformation continuously to the post-buckling and wrinkling region. Numerical examples for buckling and wrinkling in production automotive panels are demonstrated at the end of Chapter 8.

Recently, hydroforming processes have become popular in manufacturing automotive body panels and structural members. Although we do not specifically include simulations of hydroforming processes in this book, the principles and solution methods presented in this book can be applied to the simulation of hydroforming processes. In fact, one specifies the hydropressure instead of a punch movement in simulations of hydroforming processes. Therefore, the methods proposed in this book are ready to be applied to simulations of hydroforming processes with slight modifications.

We would like to thank Professor Pai-Chen Lin of the National Chung-Cheng University for preparing most of the figures in this book. We also want to thank Ms. Selina Pan of the University of Michigan for preparing some figures in this book.

Sing C. Tang
Jwo Pan
Ann Arbor, Michigan
June, 2006

CHAPTER

1

Introduction to Typical Automotive Sheet Metal Forming Processes

Beverage cans and many parts in aircraft, appliances, cars, and trucks are made of thin sheet metals formed by stamping operations at room temperature. Thus, sheet metal forming processes play an important role in mass production. The conventional method used for forming processes and tool design is based on the trial-and-error method or the pure geometric method of surface fitting that requires actual hardware tryout or the so-called die tryout. This design process often is expensive and time consuming because forming tools must be built for each trial. Significant savings are possible if designers can use computer simulation tools to predict formability and the final part dimensions, based on the principles of mechanics, before building forming tools for tryout. Due to the geometric complexity of sheet metal parts, especially automotive body panels, development of an analytical method based on the mechanics and mathematical principles to predict formability is difficult, if not impossible. The finite element method is feasible at the present time for such a highly nonlinear analysis.

Based on nonlinear thin shell theories with consideration of elastic-plastic finite deformation, sheet metal forming processes can be analyzed by using the present generation of high-speed digital computers. Because neither displacement nor force boundary conditions are specified exactly in the analysis, a surface contact problem with consideration of friction is solved in the analysis in order to predict failure due to (1) necking or splitting, (2) buckling or wrinkling, (3) loose metal without enough stretch, and (4) excessive shape distortion from the springback. The speed of the tool hitting a sheet for an automotive sheet metal part usually is low (about 0.2 m/sec). Therefore, for the size of typical automotive parts, the inertia of the sheet metal can be neglected, and a quasi-static analysis can be used to avoid undesirable oscillation in the corresponding dynamic analysis.

Based on our past research experience, a quasi-static analysis is possibly the most reliable method for designers to use, provided that a dependable quasi-static commercial code is available. Based on a quasi-static analysis, the speed of tool travel and the damping coefficient do not have to be specified among the input parameters, whereas these parameters must be adjusted artificially when a dynamic explicit code is used. (This will be explained in Chapter 7.) A quasi-static computational method can be used to determine the shape of a binder wrap (to be defined later in this chapter), the stress and strain distributions during the punch/die contact with the sheet, and the springback after the sheet is released from the tool and trimmed. However, a quasi-static analysis poses numerical difficulties caused by the surface contact with friction and the presence of buckling and wrinkling in the sheet. Researchers have resolved most of these numerical problems.

Reliability and accuracy of the solutions were demonstrated in the NUMISHEET conferences of the past decade [Makinouchi *et al.*, 1993; Lee *et al.*, 1996; Gelin and Picart, 1999; Yang *et al.*, 2002; Smith *et al.*, 2005; Cao *et al.*, 2005]. Here, NUMISHEET is the abbreviation of "Numerical Simulation of 3-D Sheet Metal Forming Processes." The use of computational simulations of forming processes is critical in the effort to develop virtual manufacturing capability in the automotive industry. Three-dimensional sheet forming simulation using supercomputers and high-speed workstations is one of the world's most exciting and challenging subjects, involving many different disciplines such as numerical methods, mechanics, materials, tribology, industrial practice, and process experiments. The first international NUMISHEET conference was held in Switzerland in 1991, and the second one was held in Japan in 1993. The more recent ones were held in Korea in 2002 and in the United States in 2005. The conference usually consisted of two parts. The first part covered research and engineering advances in the field, including theory and numerical modeling, material modeling, instability prediction, contact modeling, computer-aided design/computer-aided engineering (CAD/CAE) systems, and experimental verification. The second part covered benchmark results, experiments, and simulations.

In this chapter, we first will introduce typical forming processes used in the automotive industry. Figure 1.1 shows four stages of a typical forming process for an automotive body panel: (1) binder closing, (2) stretching and drawing, (3) trimming, and (4) flanging. As shown in the top two parts of this figure, the upper and lower binder rings first close to clamp the sheet, and then the upper punch moves to stretch or draw the sheet. As shown in the lower two parts of the figure, a formed part then will be trimmed and finally flanged. We will establish the necessary mechanics and mathematical principles for simulations of these four forming stages. The mechanics and mathematical principles presented in this book are valid for all of these forming processes. These typical automotive forming processes are explained in the following sections.

1.1 Stretching and Drawing

The first two stages—(1) binder closing and (2) stretching and drawing, which are the so-called draw die operations—are the fundamental operations of sheet metal forming. The mechanics and mathematical models for simulation of the draw die operations can be extended to the operations of trimming, flanging, and hemming. (Hemming is a method to join two sheet metal parts and will be discussed later in this chapter.) Chapter 4 will explain in more detail the modeling of draw die operations. Figures 1.2(a) and 1.2(b) show the conventional draw process. In the first stage as shown in Figure 1.2(a), the binder ring (generally on a curved surface) sets the draw-beads (shown as the grooves on the binder surface). The binder ring holds the perimeter of the sheet and pre-forms the sheet to optimize the strain distribution in the subsequent stretching and drawing operation. The deformed sheet in this stage is called the binder wrap. The deformation of the sheet inside the die cavity between the upper die punch and the lower die can be computed by using the thin shell theory. In the second stage as shown in Figure 1.2(b),

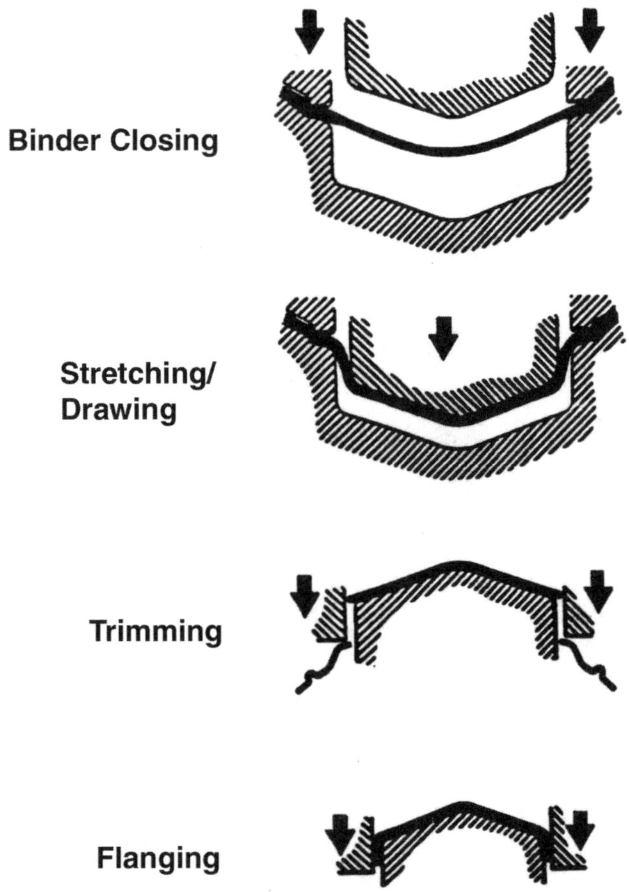

Figure 1.1 Various forming stages.

the punch travels down through the binder ring into the cavity and presses the sheet until the final part shape of the part is formed. The key role of a mechanics and mathematical model for the simulation of the draw die operation is to establish the surface contact with friction that occurs between the sheet and the tool surfaces.

There are two commonly used draw die operations that vary from the conventional draw described in the previous paragraph. The first variation is the inverted stretching and drawing operation with a three-piece tool, as shown in Figure 1.3. The tool used in this process contains three pieces: a lower binder ring, an upper die, and a lower punch. In the first stage, which is called the binder wrap stage as shown in Figure 1.3(a), the sheet is laid on the surface of the lower binder ring that is supported by a set of pressure cylinders. The upper die first is lowered to contact the sheet as shown in Figure 1.3(a) to complete the binder wrap stage. In the second stage, which is called the die closure stage as shown in Figure 1.3(b), the upper die and lower binder ring travel down together until reaching the home position of the forming tools, which is the final position before a sheet metal part is released.

The second variation is the inverted drawing operation with a four-piece tool, as shown in Figure 1.4. The tool used in this process contains four pieces: an upper binder ring, a lower binder ring, an upper die, and a lower die punch. The additional tool for this case comes from the separation of the upper binder ring from the upper die. The first stage of this process, the binder wrap stage as shown in Figure 1.4(a), is similar to that of the inverted stretching and drawing process with a three-piece tool, except the upper

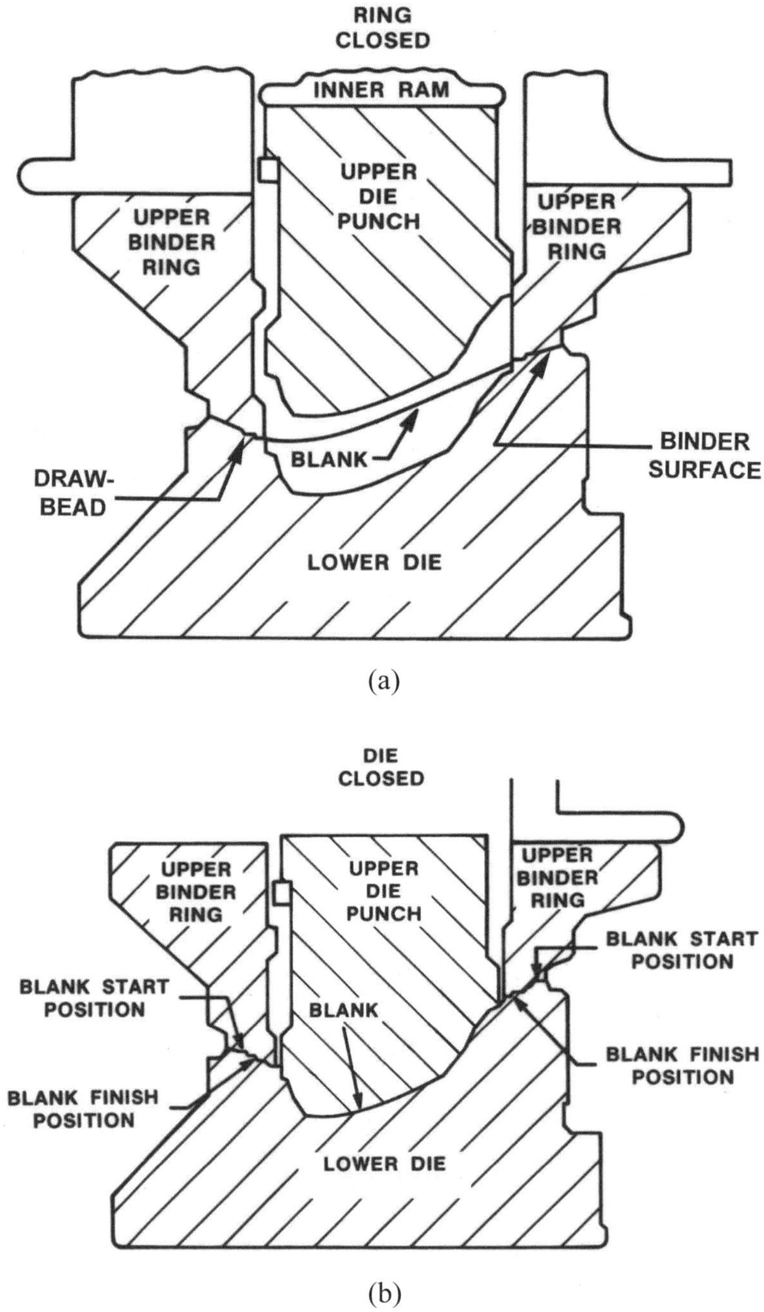

Figure 1.2 Conventional draw operation: (a) binder wrap, and (b) die closure.

die is replaced by the upper binder ring. In the second stage of this process, which is the punch wrap stage as shown in Figure 1.4(b), both the upper and lower binder rings travel downward together and thereby stretch the sheet over the lower die punch. In the third stage, which is the upper die closure stage as shown in Figure 1.4(c), the upper die travels downward until the cavity between the upper die and the lower die punch closes, and then the final shape of the part is formed. Although the inverted drawing process with a four-piece tool is more expensive because of a separate upper binder ring from the upper die, it offers at least two advantages over the conventional or the inverted stretching and drawing process with a three-piece tool. The first advantage is to save

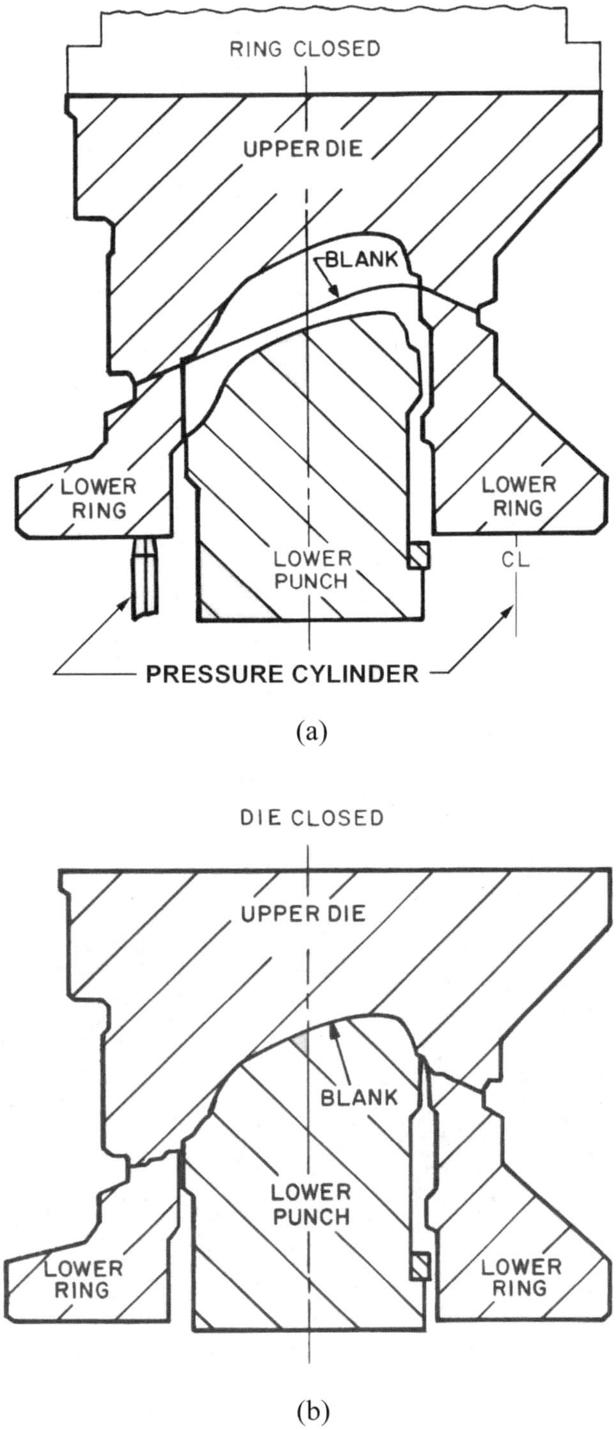

Figure 1.3 Inverted draw operation with a three-piece tool: (a) binder wrap, and (b) die closure.

blank material by 15 to 20%, and the second advantage is to form a panel with more uniform stretching over the entire surface to avoid splitting and loose metal (where the sheet metal is not stretched enough).

After the die closure, the sheet metal part is formed and then is released from the draw die. The shape of the part more or less deviates from the desired one that conforms to

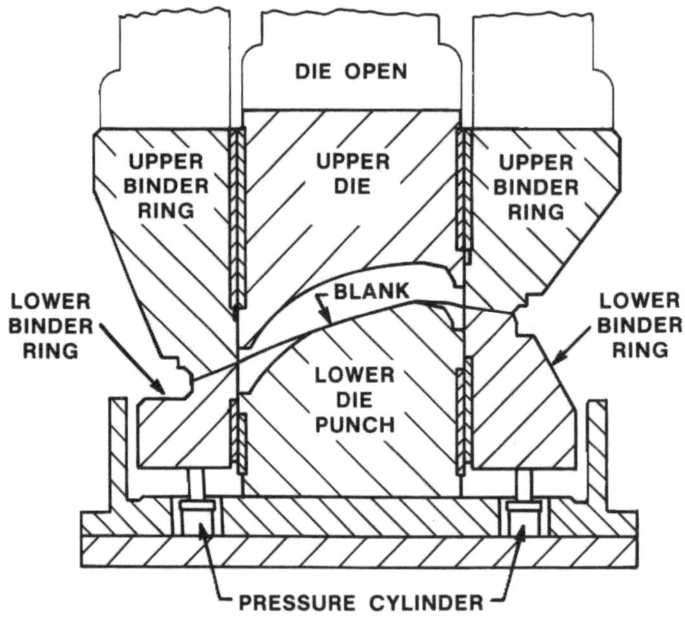

(a)

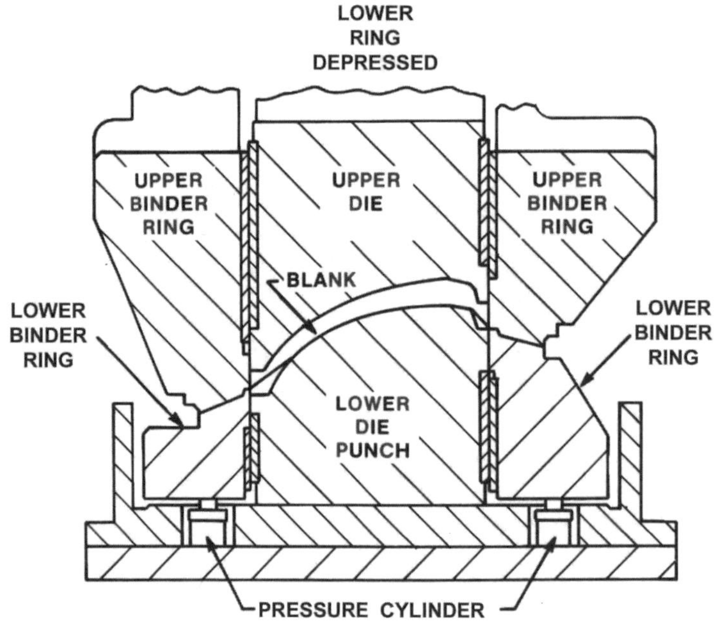

(b)

Figure 1.4 Inverted draw operation with a four-piece tool: (a) binder wrap, and (b) punch wrap.

the die surface because of the springback that results from a small amount of elastic unloading and/or plastic reloading. Because of the complex geometry of the draw-beads that are shown as the grooves on the binder surface in Figure 1.2(a), design engineers usually compute the springback after the trimming operation.

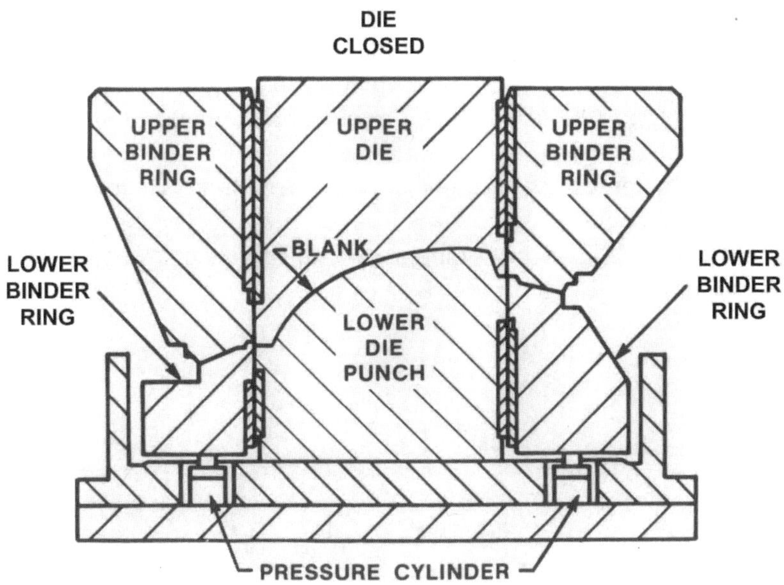

(c)

Figure 1.4 *(Continued)*
(c) Die closure.

1.2 Trimming

The finite element model of a sheet metal part for the draw die operation is used as the basic model for a trimming simulation. The elements inside the trim line are kept, and the elements for the extra metal outside the trim line are deleted or trimmed off. The internal forces acting at those nodes on the trim line are released by superimposing the external forces of equal magnitudes but opposite signs to the internal nodal forces. With application of these forces to the trimmed part and with the appropriate support to eliminate any rigid body motion, the additional deformation of the sheet metal part due to the springback can be computed.

We consider the geometric nonlinearity due to large deformation and the material nonlinearity due to elastic unloading and reversed plastic loading in the springback analysis. The commonly used isotropic hardening rule generally is good for monotonically increasing loading conditions. However, it usually predicts unreasonably high magnitudes of the stresses, for example, under uniaxial cyclic straining conditions as discussed in Tang [1990]. Therefore, an anisotropic hardening rule to establish cyclic stress-strain relations such as that discussed in Section 3.7 of Chapter 3 should be used.

1.3 Flanging and Hemming

The purpose of a flange in an automotive sheet metal part is to add strength to the part or to connect the part to other parts. There are three types of flanges: (1) flat, (2) stretch, and (3) compression flanges, as shown in Figure 1.5. For a flat flange as shown in Figure 1.5(a), it is bent at a die corner, and there is no severe stretch or compression at the flange tip. Two cylindrical panels with stretch and compression flanges are shown in Figures 1.5(b) and 1.5(c), respectively. If we make a flange outward as shown in

Figure 1.5(b), it is a stretch flange that has high tensile strains at the flange tip. If we make a flange inward as shown in Figure 1.5(c), it is a compression flange that has compressive strains at the flange tip. In a flanging operation, we are interested not only in determining whether a flange can be formed but also in the shape distortion after the operation. The failure mode for stretch flanging is due to fracture, whereas the failure mode for compression flanging is due to wrinkling. We may extend the simulation for a draw die operation to a flanging operation. Releasing the external loads from the forming tools allows the computation of the shape distortion after a flanging operation. This is essentially a springback analysis.

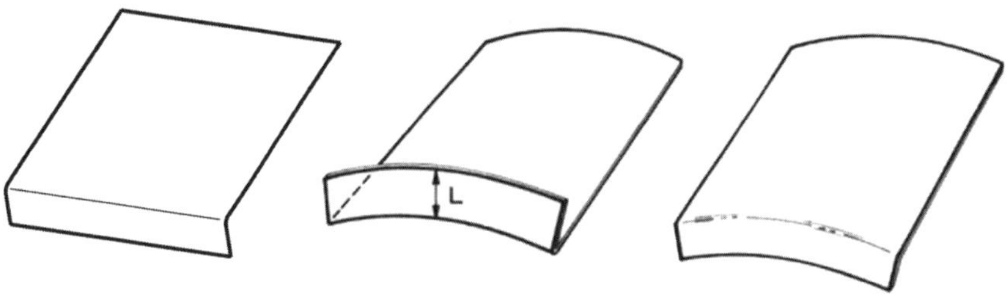

(a) (b) (c)

Figure 1.5 Flanging operation: (a) straight flange, (b) stretch flange, and (c) compression flange.

For a stretch flange, the strain component parallel to the boundary along the tip of a flange can be very high without fracture. There is no difficulty in extending the method used to analyze a stretching and drawing operation to simulate a stretch flange forming. Likewise, there is no difficulty in extending the method used to simulate a compression flange forming. However, when the final shape of a flange is close to vertical (almost 90°) and the strain component in the direction perpendicular to the boundary along the tip of a flange is small, the contact logic used in a stretching and drawing simulation may fail. Because of the small strain component, the contact with the tool surfaces of the punch and flanging die could be unstable because the contact points can be on and off contact with the tool surfaces during the computation. Therefore, appropriate modification of the contact logic to take care of the almost 90° flange wall sometimes is necessary.

As mentioned, hemming is a method used to join two sheet metal parts. For example, the flange of the outer panel of an automotive hood is bent over the inner panel. This joining method is neat and compact. However, the joint strength is less than that of welding. Therefore, adhesives can be used in addition to the hemming to increase the strength and tightness of the joint. Hemming usually is performed in two steps, as shown in Figure 1.6(a). First, the flange on the outer panel is bent to an angle that is called the pre-hemming, as shown in the middle diagram of Figure 1.6(a). In the second step, the flange of the outer panel is bent down to the final position, as shown in the right diagram of Figure 1.6(a).

During a hemming operation, the dimension of the parts can be changed and either reduced or enlarged. For example, the outer panel can decrease in size, as shown in Figure 1.6(b). This decrease in size is called flange inrolling [Svensson, 1998]. This reduction in the size of the outer panel must be compensated in a flange die for the

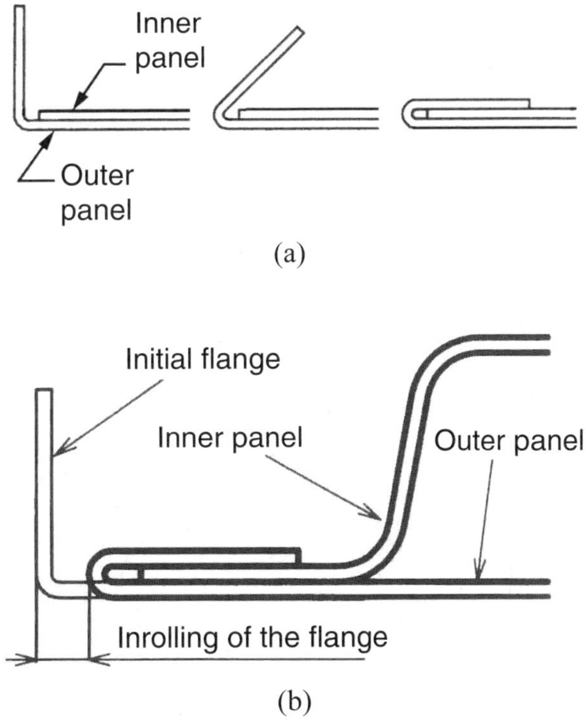

Figure 1.6 Hemming operation: (a) (from left to right) initial, pre-hemming, and final hemming stages; and (b) initial flange and final hemming stages. (Courtesy of Swets & Zeitlinger Publishers)

outer panel to make an assembled part with a precisely correct dimension. Computer simulations may be used to predict the amount of inrolling. A springback analysis can be used for this prediction. However, three-dimensional solid element models should be applied for accuracy.

1.4 References

Cao, J., Shi, M.F., Stoughton, T.B., Wang, C.-T., and Zhang, L. (eds.), The NUMISHEET 2005 Benchmark Study, Detroit, MI, 2005.

Gelin, J.C. and Picart, P. (eds.), Proceedings of the 4th International Conference—"NUMISHEET'99," Besancon, France, 1999.

Lee, J.K., Kinzel, G.L., and Wagoner, R.H. (eds.), Proceedings of the 3rd International Conference—"NUMISHEET'96," Dearborn, MI, 1996.

Makinouchi, A., Nakamachi, E., Oñate, E., and Wagoner, R.H. (eds.), Proceedings of the 2nd International Conference—"NUMISHEET'93," Isehara, Japan, 1993.

Smith, L., Pourboghrat, F., Yoon, J.-W., and Stoughton, T.B. (eds.), Proceedings of the 6th International Conference—"NUMISHEET'05," Detroit, MI, 2005.

Svensson, M., "Hemming Simulation," Proceedings of the 6th International Conference on Numerical Methods in Industrial Forming Processes—UMIFORM'98, Enschede, The Netherlands, 1998, p. 925.

Tang, S.C., "An Anisotropic Hardening Rule for the Analysis of Sheet Metal Forming Operations," Advanced Technology of Plasticity 1990, Proceedings of the Third International Conference on the Technology of Plasticity (3ICTP), Vol. 3, Kyoto, Japan, 1990, p. 1149.

Yang, D.-Y., Soo, I.O., Huh, H., and Kim, Y.H. (eds.), Proceedings of the 5th International Conference—"NUMISHEET'02," Jeju Island, Korea, 2002.

CHAPTER

2

Tensor, Stress, and Strain

Because we will adopt finite deformation kinematics to develop a finite element formulation based on thin shell theories, some background on tensor operations based on a general coordinate system is needed and therefore is introduced briefly in this chapter. For this chapter, we assume the reader has some background in vector and tensor analysis. First, the concepts of stress and equilibrium are reviewed. The principal stresses and stress invariants then are introduced. Because sheet metal forming involves large plastic deformation, we then discuss the concepts of finite deformation kinematics. We also discuss the small strain theory, which is widely used to describe deformation. Different stress measures for finite element formulations are discussed. Finally, the stresses and strains for tensile tests of sheet specimens are discussed.

2.1 Transformation of Vectors and Tensors in Cartesian Coordinate Systems

Figure 2.1 shows a set of rectangular Cartesian coordinates (x, y, z). We now change the conventional coordinate notation from (x, y, z) to (x_1, x_2, x_3) to take advantage of the indicial notation for vector and tensor operations. The base vectors for the coordinate system (x, y, z) are denoted as $\mathbf{e}_x$, $\mathbf{e}_y$, and $\mathbf{e}_z$. Now, the base vectors for the coordinate system (x_1, x_2, x_3) are denoted as $\mathbf{e}_1$, $\mathbf{e}_2$, and $\mathbf{e}_3$. Note that these base vectors are orthogonal to each other, and the magnitudes of these base vectors are unity. We can express the position vector $\mathbf{x}$ for point P as

$$\mathbf{x} = x\mathbf{e}_x + y\mathbf{e}_y + z\mathbf{e}_z \tag{2.1.1}$$

based on the Cartesian coordinates (x, y, z). Alternatively, we can express the position vector $\mathbf{x}$ for point P as

$$\mathbf{x} = x_1\mathbf{e}_1 + x_2\mathbf{e}_2 + x_3\mathbf{e}_3 = \sum_{i=1}^{3} x_i\mathbf{e}_i \tag{2.1.2}$$

based on the Cartesian coordinates (x_1, x_2, x_3). We can omit the summation sign, and the position vector $\mathbf{x}$ can now be written as

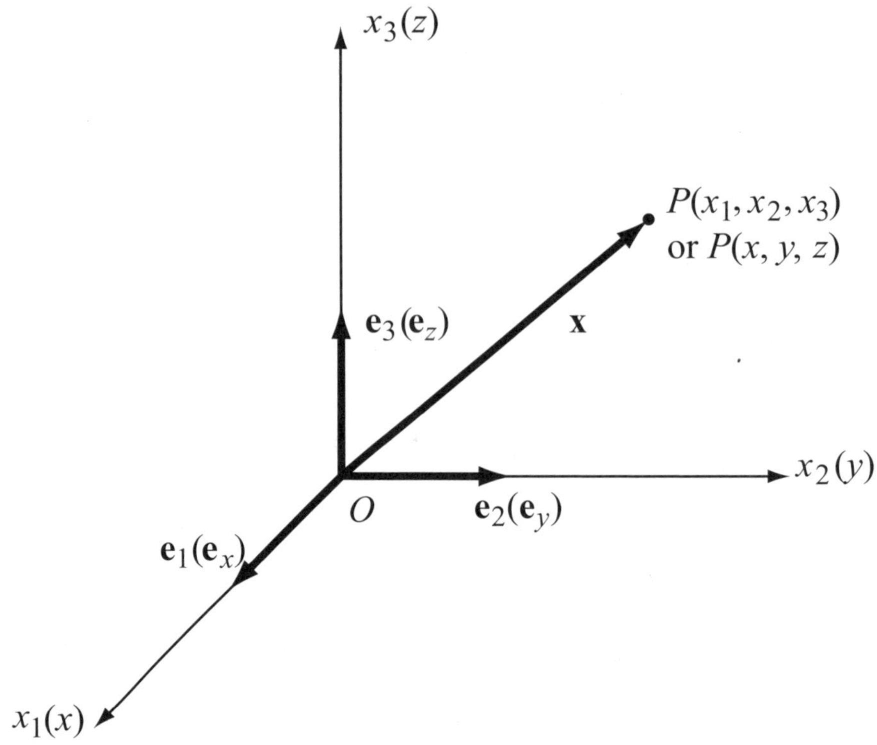

Figure 2.1 A Cartesian coordinate system.

$$\mathbf{x} = x_i \mathbf{e}_i \tag{2.1.3}$$

The repeated subscript i is called the dummy index. The summation convention is adopted for repeated indices. Usually, Latin indices such as i and j have the range from 1 to 3 for three-dimensional problems, and Greek indices such as α and β have the range of 1 to 2 for two-dimensional problems. When we mention the components of the position vector **x**, we can use x_i where the subscript i can have the value of 1, 2, or 3.

Figure 2.2 shows two sets of rectangular Cartesian coordinates (x_1, x_2, x_3) and (x_1', x_2', x_3') with the same origin O. The base vectors for the two coordinate systems are denoted as $\mathbf{e}_i$ and $\mathbf{e}_j'$ where the subscripts i and j can have the values of 1, 2, or 3. The position vector **x** can be expressed in terms of either of the two sets of the coordinates and the corresponding base vectors as

$$\mathbf{x} = x_i \mathbf{e}_i = x_j' \mathbf{e}_j' \tag{2.1.4}$$

To determine the component of **x** in the $\mathbf{e}_j$ direction, we can use the dot product as

$$x_j = \mathbf{e}_j \cdot \mathbf{x} = \mathbf{e}_j \cdot x_i \mathbf{e}_i = \delta_{ji} x_i \tag{2.1.5}$$

Here, δ_{ji} represent the Kronecker delta, which are defined as

$$\delta_{ji} = 1 \quad \text{when} \quad j = i \tag{2.1.6}$$

$$\delta_{ji} = 0 \quad \text{when} \quad j \neq i \tag{2.1.7}$$

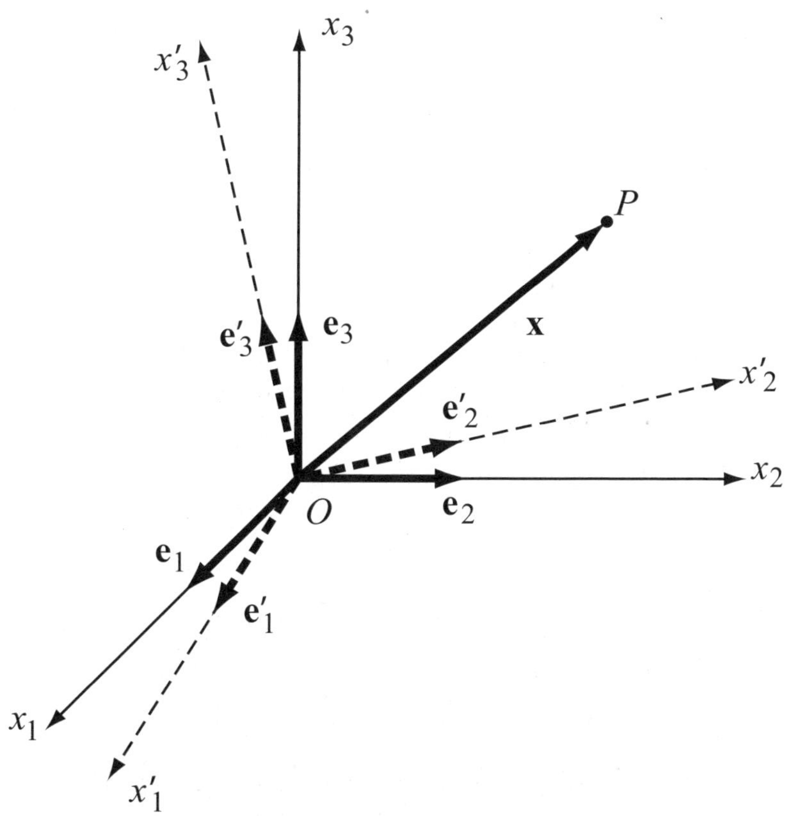

Figure 2.2 Two Cartesian coordinate systems.

To determine the component of $\mathbf{x}$ in the $\mathbf{e}'_j$ direction, we can also use the dot product as

$$x'_j = \mathbf{e}'_j \cdot \mathbf{x} = \mathbf{e}'_j \cdot x_i \mathbf{e}_i = \alpha_{ji} x_i \tag{2.1.8}$$

or

$$x'_j = \alpha_{ji} x_i \tag{2.1.9}$$

Here, $\alpha_{ji} (= \mathbf{e}'_j \cdot \mathbf{e}_i)$ represent the direction cosines between the base vectors $\mathbf{e}'_j$ and $\mathbf{e}_i$. In Eq. 2.1.9, the index j is called the free index where the subscript j can have the value of 1, 2, or 3. The index i again is called the dummy index because the summation convention is implied for the repeated index i. The transformation of the coordinates in Eq. 2.1.9 can be expressed in the matrix form as

$$\begin{bmatrix} x'_1 \\ x'_2 \\ x'_3 \end{bmatrix} = \begin{bmatrix} \alpha_{11} & \alpha_{12} & \alpha_{13} \\ \alpha_{21} & \alpha_{22} & \alpha_{23} \\ \alpha_{31} & \alpha_{32} & \alpha_{33} \end{bmatrix} \begin{bmatrix} x_1 \\ x_2 \\ x_3 \end{bmatrix} \tag{2.1.10}$$

Similarly,

$$x_i = \mathbf{e}_i \cdot \mathbf{x} = \mathbf{e}_i \cdot \mathbf{x}' = \mathbf{e}_i \cdot x'_j \mathbf{e}'_j = \alpha_{ji} x'_j \tag{2.1.11}$$

or

$$x_i = \alpha_{ji} x'_j \tag{2.1.12}$$

Equation 2.1.12 can be expressed in the matrix form as

$$\begin{bmatrix} x_1 \\ x_2 \\ x_3 \end{bmatrix} = \begin{bmatrix} \alpha_{11} & \alpha_{12} & \alpha_{13} \\ \alpha_{21} & \alpha_{22} & \alpha_{23} \\ \alpha_{31} & \alpha_{32} & \alpha_{33} \end{bmatrix}^T \begin{bmatrix} x_1' \\ x_2' \\ x_3' \end{bmatrix} \tag{2.1.13}$$

where the superscript T means the transpose of the matrix. Note that occasionally repeated indices do not mean summation if we specify "no sum on" (nso). For example, the ith diagonal element of a matrix $\mathbf{C}$ is represented by "C_{ii} nso i." For a rectangular Cartesian coordinate system, the partial derivatives of Eqs. 2.1.9 and 2.1.12 can be expressed in terms of α_{ij} as

$$\frac{\partial x_i'}{\partial x_j} = \frac{\partial x_j}{\partial x_i'} = \alpha_{ij} \tag{2.1.14}$$

Similarly, we consider a vector $\mathbf{V}$, which can be expressed as

$$\mathbf{V} \equiv V_i \mathbf{e}_i = V_j' \mathbf{e}_j' \tag{2.1.15}$$

where V_i are the components of vector $\mathbf{V}$ in the $\mathbf{e}_i$ direction, and V_j' are the components of vector $\mathbf{V}$ in the $\mathbf{e}_j'$ direction. Similar to the discussion for the position vector $\mathbf{x}$, we can write

$$V_j' = \alpha_{ji} V_i \tag{2.1.16}$$

and

$$V_i = \alpha_{ji} V_j' \tag{2.1.17}$$

We may generalize the transformation rule to a tensor of the second order or higher. For example, we consider a second order tensor $\mathbf{W}$. The transformation rule is

$$W_{ij}' = \alpha_{ik} \alpha_{jl} W_{kl} \tag{2.1.18}$$

or

$$W_{ij} = \alpha_{ki} \alpha_{lj} W_{kl}' \tag{2.1.19}$$

Examples of second order tensors are the strain and stress tensors. Elastic modulus or compliance tensors generally are expressed as fourth order tensors. In a rectangular Cartesian coordinate system in this section, we may use either a subscript or a superscript to represent the components of a vector, a tensor, and a base vector, such as $V^i \equiv V_i$, $W^{ij} \equiv W^{ij}$, and $\mathbf{e}^i \equiv \mathbf{e}_i$. However, when a general coordinate system is used, a subscript or superscript denotes the covariant or contravariant component of the field quantity of interest.

2.2 Transformation of Vectors and Tensors in General Coordinate Systems

A general coordinate system often is used to describe the geometry of a boundary value problem in order to simplify the mathematical equations. For example, cylindrical and spherical coordinate systems commonly are used. Transformation of the components of a vector or tensor in one coordinate system to those in another system often is needed. In sheet metal forming simulation, the computational formulation is based on the finite strain thin shell theory, which is more conveniently expressed in terms of the curvilinear coordinate system. However, when we obtain the computational results, we need to express the results of the stress and strain measures in terms of the familiar Cartesian coordinate system with a set of orthonormal base vectors. Therefore, in this section, we present a summary of the transformation rule for tensor components between a general coordinate system and a Cartesian coordinate system. Note again that it does not matter whether a subscript or a superscript is used for the coordinates and the base vectors in a Cartesian coordinate system. However, in a general coordinate system, a subscript or superscript for the coordinates and the base vectors represents the covariant or contravariant components [Spain, 1956]. For example, we can express a position vector **x** in a Cartesian coordinate system as

$$\mathbf{x} = x^i \mathbf{e}_i = x_i \mathbf{e}^i \tag{2.2.1}$$

For this case, $\mathbf{e}^i = \mathbf{e}_i$ and $x^i = x_i$ for the Cartesian coordinate system. Now we express the position vector **x** in terms of a general coordinate system with a set of three base vectors $\mathbf{g}_1$, $\mathbf{g}_2$, and $\mathbf{g}_3$, as shown in Figure 2.3. The general coordinate system has

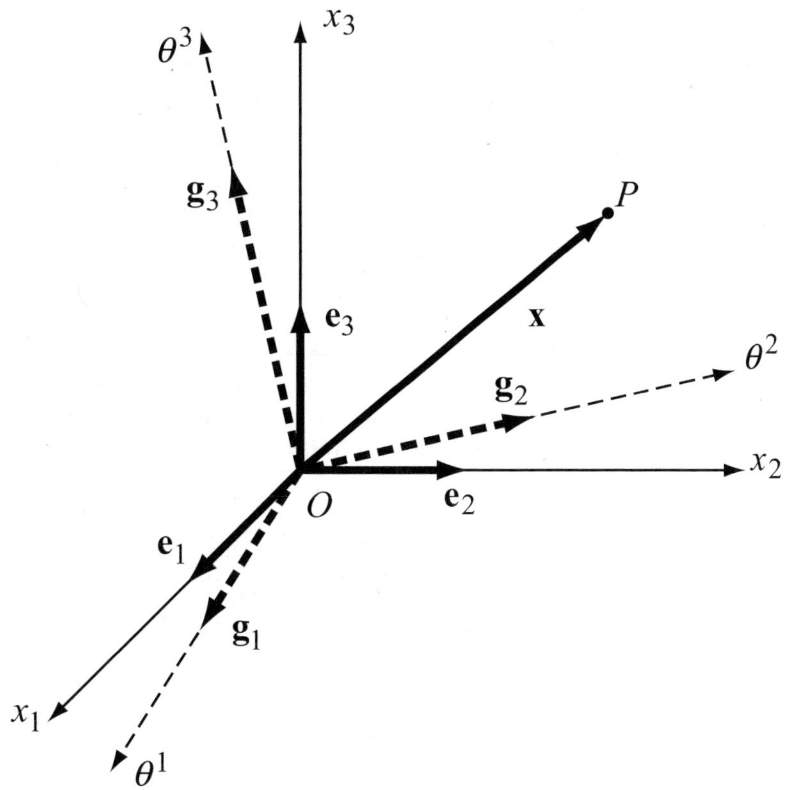

Figure 2.3 A Cartesian coordinate system and a general coordinate system.

three coordinates θ^1, θ^2, and θ^3, corresponding to the three base vectors $\mathbf{g}_1$, $\mathbf{g}_2$, and $\mathbf{g}_3$, respectively. We call $\mathbf{g}_1$, $\mathbf{g}_2$, and $\mathbf{g}_3$ the covariant base vectors. Note that we use the subscripts for the base vectors. The covariant base vectors $\mathbf{g}_1$, $\mathbf{g}_2$, and $\mathbf{g}_3$ do not have to possess a magnitude of unity, and they do not have to be orthogonal to each other. The position vector $\mathbf{x}$ can be expressed in terms of the two coordinate systems as

$$\mathbf{x} = x^i \mathbf{e}_i = \theta^r \mathbf{g}_r \tag{2.2.2}$$

The covariant base vectors $\mathbf{g}_r$ can be expressed as

$$\mathbf{g}_r = \frac{\partial x^i}{\partial \theta^r} \mathbf{e}_i \tag{2.2.3}$$

We also can define the contravariant or the reciprocal base vectors $\mathbf{g}^r$

$$\mathbf{g}^r = \frac{\partial \theta^r}{\partial x^i} \mathbf{e}^i \tag{2.2.4}$$

Note that unlike the base vectors $\mathbf{e}_i$ and $\mathbf{e}^i$ in the Cartesian coordinate system, the covariant and contravariant base vectors $\mathbf{g}_r$ and $\mathbf{g}^r$ for the general coordinate system are different. We will demonstrate this for a two-dimensional case later. When we use the thin shell theory to describe the kinematics of curved sheet metals for sheet forming simulation and modeling, the base vectors $\mathbf{g}_r$ and $\mathbf{g}^r$ in general are different point by point and are functions of the location or the coordinates of the material point of interest. We can easily show that the dot product of $\mathbf{g}_r$ and $\mathbf{g}^s$ is

$$\mathbf{g}_r \cdot \mathbf{g}^s = \delta_r^s \tag{2.2.5}$$

where δ_r^s represent the Kronecker delta ($\delta_r^s = 1$ when $r = s$, and $\delta_r^s = 0$ when $r \neq s$). Therefore, $\mathbf{g}_r$ and $\mathbf{g}^s$ are orthogonal to each other when $r \neq s$.

Now we can express a vector $\mathbf{V}$ referred to the covariant and contravariant base vectors $\mathbf{g}_r$ and $\mathbf{g}^s$ in a general coordinate system as

$$\mathbf{V} = \bar{V}^s \mathbf{g}_s = \bar{V}_r \mathbf{g}^r \tag{2.2.6}$$

where $\bar{V}^s$ and $\bar{V}_r$ are the contravariant and covariant components of vector $\mathbf{V}$, respectively. Note that a bar is used to represent the component of the vector in the general coordinate system. Figure 2.4 shows the vector $\mathbf{V}$ expressed in terms of the covariant and contravariant base vectors in a two-dimensional plane. The figure shows three sets of base vectors as $(\mathbf{e}_1, \mathbf{e}_2)$, $(\mathbf{g}_1, \mathbf{g}_2)$, and $(\mathbf{g}^1, \mathbf{g}^2)$. Note that $\mathbf{e}_1$ is perpendicular to $\mathbf{e}_2$, $\mathbf{g}_1$ is perpendicular to $\mathbf{g}^2$, and $\mathbf{g}^1$ is perpendicular to $\mathbf{g}_2$. Note that the magnitudes of $\mathbf{g}_1$, $\mathbf{g}_2$, $\mathbf{g}^1$, and $\mathbf{g}^2$are not equal to each other.

For a second order tensor $\mathbf{W}$, we can express $\mathbf{W}$ in terms of the covariant and contravariant base vectors $\mathbf{g}_r$ and $\mathbf{g}^s$ as

$$\mathbf{W} = \bar{W}^{rs} \mathbf{g}_r \mathbf{g}_s = \bar{W}_{rs} \mathbf{g}^r \mathbf{g}^s \tag{2.2.7}$$

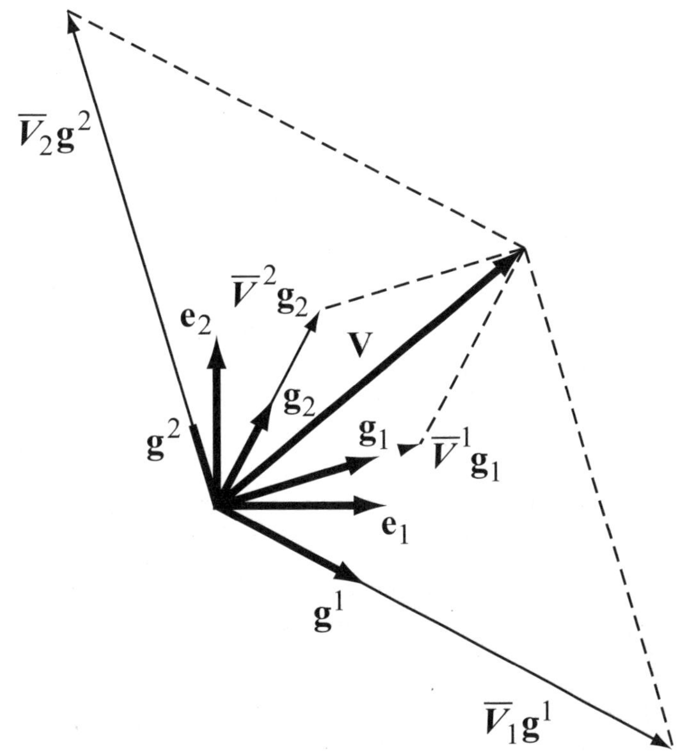

Figure 2.4 A vector **V** with respect to the Cartesian, covariant, and contravariant coordinate systems as shown.

where $\overline{W}^{rs}$ and $\overline{W}_{rs}$ are the contravariant and covariant components of **W**, respectively. We usually use the covariant components to express a Lagrangian strain tensor and the contravariant components to express a Cauchy stress tensor in a curvilinear coordinate system when the sheets are under large deformation.

To examine the magnitudes of a vector and a tensor in a general coordinate system, we need to use metric tensors. The metric tensor g_{rs} is defined as

$$g_{rs} = \mathbf{g}_r \cdot \mathbf{g}_s \tag{2.2.8}$$

Here, g_{rs} are symmetrical ($g_{rs} = g_{sr}$) due to the commutation of the dot product. In general, all the terms of g_{rs} are nonzeros for the set of the general base vectors $\mathbf{g}_1$, $\mathbf{g}_2$, and $\mathbf{g}_3$. When $\mathbf{g}_1$, $\mathbf{g}_2$, and $\mathbf{g}_3$ are orthogonal to each other, only the diagonal terms survive when the metric tensor g_{rs} is expressed in the matrix form. When $\mathbf{g}_1$, $\mathbf{g}_2$, and $\mathbf{g}_3$ are orthonormal (i.e., have the magnitudes of unity and are orthogonal to each other), the metric tensor g_{rs} reduces to δ_{rs}. The metric tensor g^{rs} is defined as

$$g^{rs} = \mathbf{g}^r \cdot \mathbf{g}^s \tag{2.2.9}$$

Here, g^{rs} also are symmetrical due to the commutation of the dot product. Based on the definitions of the base vectors in Eqs. 2.2.3 and 2.2.4, we can easily show that

$$g_{rq}g^{qs} = \delta_r^s \tag{2.2.10}$$

Using the metric tensors, we can easily establish the relationship between the covariant components $\overline{V}_r$ and the contravariant components $\overline{V}^s$ of vector **V** by the dot products as

$$\bar{V}_r = \mathbf{V} \cdot \mathbf{g}_r = g_{rs}\bar{V}^s \tag{2.2.11}$$

$$\bar{V}^r = \mathbf{V} \cdot \mathbf{g}^r = g^{rs}\bar{V}_s \tag{2.2.12}$$

Similarly, we can establish the relationship between the covariant components $\bar{W}_{rs}$ and the contravariant components $\bar{W}^{rs}$ of a tensor $\mathbf{W}$ as

$$\bar{W}_{rs} = g_{rq}g_{st}\bar{W}^{qt} \tag{2.2.13}$$

$$\bar{W}^{rs} = g^{rq}g^{st}\bar{W}_{qt} \tag{2.2.14}$$

We can obtain the mixed components of a second order tensor that can be converted from the covariant or the contravariant components using the metric tensors as

$$\bar{W}_r^s = g^{st}\bar{W}_{rt} = g_{rt}\bar{W}^{st} \tag{2.2.15}$$

The magnitudes of the covariant and contravariant base vectors can be obtained from the components of the metric tensors as

$$|\mathbf{g}_r| = (\mathbf{g}_r \cdot \mathbf{g}_r)^{1/2} = (g_{rr})^{1/2} \quad \text{nso r} \tag{2.2.16}$$

$$|\mathbf{g}^r| = (\mathbf{g}^r \cdot \mathbf{g}^r)^{1/2} = (g^{rr})^{1/2} \quad \text{nso r} \tag{2.2.17}$$

We can convert the components of a vector $\mathbf{V}$ in a rectangular coordinate system to those in a general coordinate system or vice versa. Taking the dot product of the base vector $\mathbf{e}^i$, $\mathbf{e}_i$, $\mathbf{g}^r$, or $\mathbf{g}_r$ and $\mathbf{V}$ in Eq. 2.2.6 and using Eqs. 2.2.3 and 2.2.4, we can derive the transformation laws for the components for the vector $\mathbf{V}$ as

$$V^i = \frac{\partial x^i}{\partial \theta^r}\bar{V}^r \tag{2.2.18}$$

$$V_i = \frac{\partial \theta^r}{\partial x^i}\bar{V}_r \tag{2.2.19}$$

$$\bar{V}^r = \frac{\partial \theta^r}{\partial x^i}V^i \tag{2.2.20}$$

$$\bar{V}_r = \frac{\partial x^i}{\partial \theta^r}V_i \tag{2.2.21}$$

Note that $V_i = V^i$ for the rectangular Cartesian coordinate system. Based on Eqs. 2.2.7, 2.2.3, and 2.2.4, we can derive the transformation laws for the components of the second order tensor $\mathbf{W}$ as

$$W^{ij} = \frac{\partial x^i}{\partial \theta^r}\frac{\partial x^j}{\partial \theta^s}\overline{W}^{rs} \tag{2.2.22}$$

$$W_{ij} = \frac{\partial \theta^r}{\partial x^i}\frac{\partial \theta^s}{\partial x^j}\overline{W}_{rs} \tag{2.2.23}$$

$$\overline{W}^{rs} = \frac{\partial \theta^r}{\partial x^i}\frac{\partial \theta^s}{\partial x^j}W^{ij} \tag{2.2.24}$$

$$\overline{W}_{rs} = \frac{\partial x^i}{\partial \theta^r}\frac{\partial x^j}{\partial \theta^s}W_{ij} \tag{2.2.25}$$

In general, we use thin shell finite elements for sheet metal forming simulations. For a thin shell, we use the middle surface as the reference. Although it initially is flat before forming, the reference surface becomes curved during a forming process. We use a set of curvilinear coordinates θ^α ($\alpha = 1, 2$) to represent the reference surface. This set of coordinates also is called the Gaussian in geometric terms or convected coordinates in mechanics terms. If the set is defined in a two-dimensional rectangular coordinate system on an initially flat sheet, the metric tensor is the Kronecker delta. After deformation, the metric tensor on the deformed reference surface becomes $a_{\alpha\beta}$ ($\alpha, \beta = 1, 2$), of which the base vectors are not necessarily orthogonal and their magnitudes are not necessarily unity.

2.3 Stress and Equilibrium

We consider a region of a solid with reference to a Cartesian coordinate system as shown in Figure 2.5. The region has a volume V and a surface S. As shown in this

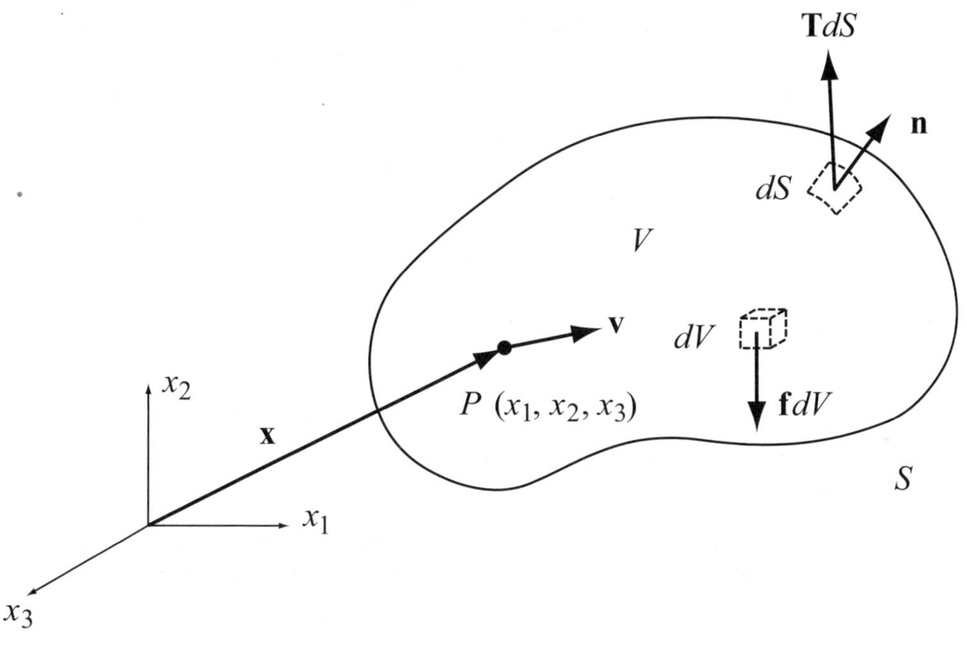

Figure 2.5 A region of a solid with a surface S and a volume V.

figure, $\mathbf{x}$ represents the position vector for point P, and $\mathbf{v}$ represents the velocity of point P. Also, dS represents a differential element of the surface, and dV represents a differential element of the volume. Here, $\mathbf{n}$ represents the outward unit normal to the differential surface element dS, and $\mathbf{T}$ represents the force vector per unit area acting on the differential surface element dS. Therefore, $\mathbf{T}dS$ represents the differential force acting on the differential surface element dS. Likewise, $\mathbf{f}$ represents the body force per unit volume, and $\mathbf{f}dV$ represents the differential force of the differential volume element dV. The linear momentum principle requires

$$\int_S \mathbf{T}dA + \int_V \mathbf{f}dV = \int_V \rho\mathbf{a}dV \tag{2.3.1}$$

where ρ represents the density and $\mathbf{a}$ represents the acceleration. Note that the momentum flux across the surface is not considered here.

The stress vector $\mathbf{T}$ represents the force per unit area on any surface in an average sense from a continuum mechanics viewpoint. The orientation expressed in terms of the normal of the surface element of interest is important. Consider a differential volume element with respect to the Cartesian coordinate system, as shown in Figure 2.6. As shown in this figure, we denote $\mathbf{T}_1$, $\mathbf{T}_2$, and $\mathbf{T}_3$ as the tractions on the differential surfaces dS_1, dS_2, and dS_3, which have the outward normal in the x_1, x_2, and x_3 directions, respectively. Because $\mathbf{T}_1$, $\mathbf{T}_2$, and $\mathbf{T}_3$ are vectors, they can be decomposed into the components in the x_1, x_2, and x_3 directions, as shown in Figure 2.6, as

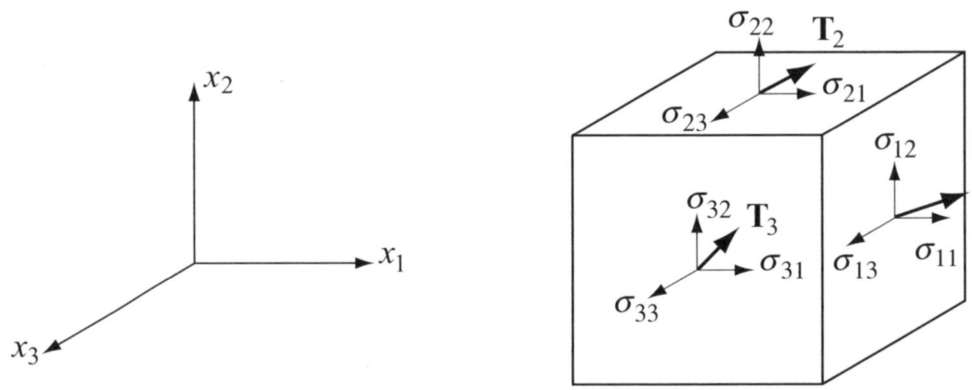

Figure 2.6 A differential volume element with respect to the Cartesian coordinate system as shown. The tractions and the stress components are shown.

$$\mathbf{T}_1 = \sigma_{11}\mathbf{e}_1 + \sigma_{12}\mathbf{e}_2 + \sigma_{13}\mathbf{e}_3 \tag{2.3.2}$$

$$\mathbf{T}_2 = \sigma_{21}\mathbf{e}_1 + \sigma_{22}\mathbf{e}_2 + \sigma_{23}\mathbf{e}_3 \tag{2.3.3}$$

$$\mathbf{T}_3 = \sigma_{31}\mathbf{e}_1 + \sigma_{32}\mathbf{e}_2 + \sigma_{33}\mathbf{e}_3 \tag{2.3.4}$$

where $\mathbf{e}_1$, $\mathbf{e}_2$, and $\mathbf{e}_3$ represent the unit vectors in the x_1, x_2, and x_3 directions, respectively. When we use the indicial notation, Eqs. 2.3.2 through 2.3.4 can be written as

$$\mathbf{T}_i = \sigma_{ij}\mathbf{e}_j \tag{2.3.5}$$

where i represents the free index, which means that i can be assigned a value of 1, 2, or 3 when the range of i is from 1 to 3. Here, j represents the dummy index where the summation convention must be applied. For σ_{ij}, the first subscript i indicates the direction of the normal to the differential area, and the second subscript j indicates the direction of the force acting on the differential area.

When the differential area dS is not perpendicular to the x_1, x_2, and x_3 directions, we can consider the equilibrium of the forces on a Cauchy tetrahedron as shown in Figure 2.7 to obtain the traction **T** on the inclined surface with the unit outward normal vector **n**. In Figure 2.7, the tetrahedron has four differential surfaces: dS_1, dS_2, dS_3, and dS. Because the unit outward normal vectors for dS_1, dS_2, and dS_3 are in the negative x_1, x_2, and x_3 directions, the tractions on these surfaces are denoted as $-\mathbf{T}_1$, $-\mathbf{T}_2$, and $-\mathbf{T}_3$, based on the principle of action–reaction. With consideration of the linear momentum equation for the tetrahedron as the size of the tetrahedron approaches zero, the volume integrals approach zero faster than the surface integrals. Therefore, only the surface integrals of traction over the four faces survive. Summing all the tractions over the four faces leads to the traction vector **T** expressed in terms of the inner product of the unit outward normal vector **n** and the stress tensor $\boldsymbol{\sigma}$ as

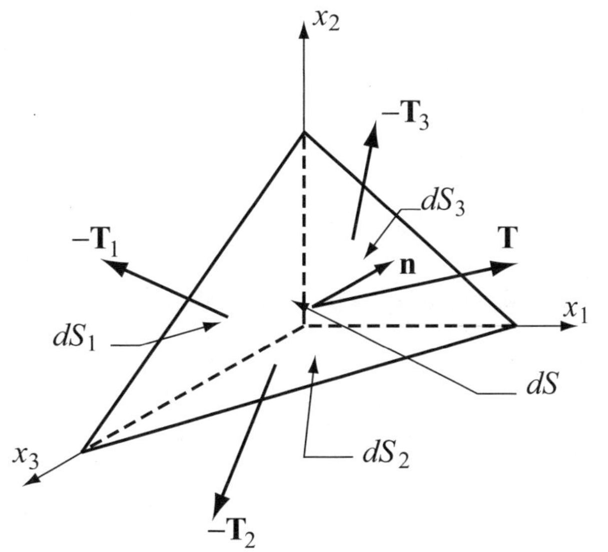

Figure 2.7 A tetrahedron with the tractions $\mathbf{T}$, $-\mathbf{T}_1$, $-\mathbf{T}_2$, and $-\mathbf{T}_3$ on the differential surfaces dS, dS_1, dS_2, and dS_3, respectively, with the normals in the $\mathbf{n}$, $-\mathbf{e}_1$, $-\mathbf{e}_2$, and $-\mathbf{e}_3$ directions.

$$\mathbf{T} = \mathbf{n} \cdot \boldsymbol{\sigma} \tag{2.3.6}$$

where the unit outward normal vector **n** is expressed as

$$\mathbf{n} = n_i\mathbf{e}_i \tag{2.3.7}$$

and the stress tensor $\boldsymbol{\sigma}$ is expressed in the dyadic form as

$$\boldsymbol{\sigma} = \sigma_{jk}\mathbf{e}_j\mathbf{e}_k \tag{2.3.8}$$

In indicial notation, the component T_k of the traction vector $\mathbf{T}$ can be related to the components σ_{ij} of the stress tensor $\boldsymbol{\sigma}$ as

$$T_k = n_j \sigma_{jk} \tag{2.3.9}$$

The linear momentum equation can be written in the component form as

$$\int_S T_i dS + \int_V f_i dV = \int_V \rho a_i dV \tag{2.3.10}$$

Substituting Eq. 2.3.9 into Eq. 2.3.10 and using the divergence theorem, we can rewrite Eq. 2.3.10 as

$$\int_V \left(\frac{\partial \sigma_{ji}}{\partial x_j} + f_i - \rho a_i \right) dV = 0 \tag{2.3.11}$$

Because Eq. 2.3.11 must hold for any volume, the local equations of motion become

$$\frac{\partial \sigma_{ji}}{\partial x_j} + f_i - \rho a_i = 0 \tag{2.3.12}$$

When we consider the angular moment equation, the stress tensor $\boldsymbol{\sigma}$ can be shown to be symmetric as

$$\sigma_{ij} = \sigma_{ji} \tag{2.3.13}$$

Therefore, for three-dimensional solids where the indices range from 1 to 3, the stress tensor $\boldsymbol{\sigma}$ has six components. When the contribution of the acceleration term ρa_i is small compared to the other terms, the local equations of motion are reduced to the local equilibrium equations as

$$\frac{\partial \sigma_{ji}}{\partial x_j} + f_i = 0 \tag{2.3.14}$$

For a typical sheet metal forming process, the contribution of the acceleration term is quite small; therefore, quasi-static modeling based on the local equilibrium equations is reasonable. When the magnitudes of the body force f_i are much smaller than those involving the stress components σ_{ij}, Eq. 2.3.14 becomes

$$\frac{\partial \sigma_{ji}}{\partial x_j} = 0 \tag{2.3.15}$$

2.4 Principal Stresses and Stress Invariants

As shown in Figure 2.8, for a differential area dS with the unit normal vector $\mathbf{n}$, the direction of the traction vector $\mathbf{T}$ may not be in the same direction of the unit normal vector $\mathbf{n}$. We can express this in mathematical form as

$$\mathbf{T}/\|\mathbf{T}\| \neq \mathbf{n} \tag{2.4.1}$$

where the symbol $\|\ \|$ represents the magnitude of the vector quantity enclosed. However, we can always find a normal vector $\mathbf{n}$ for a differential area dS such that the direction of the traction vector $\mathbf{T}$ is the same as that of the unit normal vector $\mathbf{n}$. Mathematically, we can write

$$\mathbf{T}/\|\mathbf{T}\| = \mathbf{n} \tag{2.4.2}$$

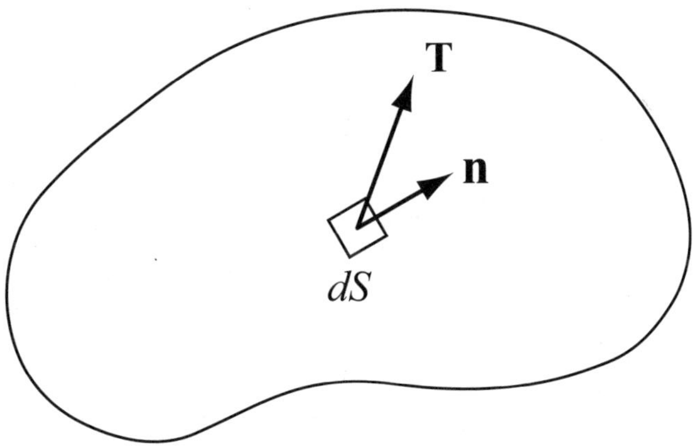

Figure 2.8 A traction vector **T** on a differential area dS with the unit normal vector **n**. The direction of **T** may not be the same as the direction of the unit normal vector **n**.

This means that in this direction, only the normal stress exists, and no shear stress exists. Because the traction vector $\mathbf{T}$ is in the same direction of the unit normal vector $\mathbf{n}$, we can write

$$\mathbf{T} = \mathbf{n} \cdot \boldsymbol{\sigma} = \sigma_p \mathbf{n} \tag{2.4.3}$$

Here, σ_p represents the principal stress or the eigenvalue of the stress tensor $\boldsymbol{\sigma}$. In the component form,

$$n_j \sigma_{jk} = \sigma_p n_k = \sigma_p \delta_{jk} n_j \tag{2.4.4}$$

where δ_{jk} represent the Kronecker delta. Equation 2.4.4 can be written as

$$\left(\sigma_{jk} - \sigma_p \delta_{jk}\right) n_j = 0 \tag{2.4.5}$$

Equation 2.4.5 represents three equations for the three components n_j. Equation 2.4.5 can be expressed in the matrix form as

$$\begin{bmatrix} \sigma_{11}-\sigma_p & \sigma_{12} & \sigma_{13} \\ \sigma_{21} & \sigma_{22}-\sigma_p & \sigma_{23} \\ \sigma_{31} & \sigma_{32} & \sigma_{33}-\sigma_p \end{bmatrix} \begin{Bmatrix} n_1 \\ n_2 \\ n_3 \end{Bmatrix} = 0 \tag{2.4.6}$$

Because the components of the stress tensor $\boldsymbol{\sigma}$ are real and symmetrical, nontrivial solutions for n_j exist when the determinant is

$$\det\left|\sigma_{jk} - \sigma_p \delta_{jk}\right| = 0 \tag{2.4.7}$$

or in the matrix form as

$$\begin{vmatrix} \sigma_{11}-\sigma_p & \sigma_{12} & \sigma_{13} \\ \sigma_{21} & \sigma_{22}-\sigma_p & \sigma_{23} \\ \sigma_{31} & \sigma_{32} & \sigma_{33}-\sigma_p \end{vmatrix} = 0 \tag{2.4.8}$$

Equation 2.4.8 represents a cubic equation for σ_p

$$-\sigma_p^3 + I_1\sigma_p^2 - I_2\sigma_p + I_3 = 0 \tag{2.4.9}$$

Here, I_1, I_2, and I_3 are independent of the coordinate system chosen. They are called the invariants of the stress tensor $\boldsymbol{\sigma}$. I_1, I_2, and I_3 are expressed as

$$I_1 = \sigma_{kk} = \sigma_{11} + \sigma_{22} + \sigma_{33} \tag{2.4.10}$$

$$I_2 = \frac{1}{2}\left(\sigma_{ii}\sigma_{jj} - \sigma_{ij}\sigma_{ij}\right) \tag{2.4.11}$$

$$I_3 = \left|\boldsymbol{\sigma}\right| = \frac{1}{6}\varepsilon_{ijk}\varepsilon_{pqr}\sigma_{ip}\sigma_{jq}\sigma_{kr} \tag{2.4.12}$$

where ε_{ijk} is the permutation symbol, which is defined as

$$\varepsilon_{ijk} = \begin{Bmatrix} 1 & \text{when } i,j,k \text{ are even permutations of } 1,2,3 \\ -1 & \text{when } i,j,k \text{ are odd permutations of } 1,2,3 \\ 0 & \text{when any subscripts are repeated} \end{Bmatrix} \tag{2.4.13}$$

Because $\boldsymbol{\sigma}$ is real and symmetric ($\sigma_{ij} = \sigma_{ji}$), we will have three real roots (eigenvalues) that are called the principal stresses. The corresponding directions of the eigenvectors are the principal directions. In general, these directions are mutually orthogonal to each other. Figure 2.9 shows a differential material element with the principal stresses σ_{I}, σ_{II}, and σ_{III} corresponding to the principal directions $\mathbf{n}_{\mathrm{I}}$, $\mathbf{n}_{\mathrm{II}}$, and $\mathbf{n}_{\mathrm{III}}$, respectively.

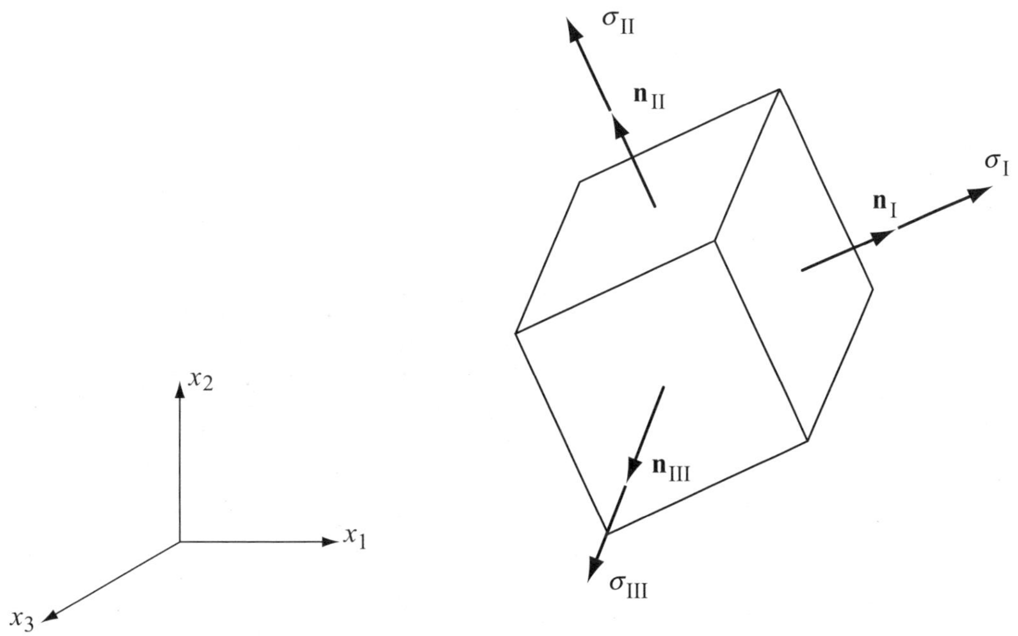

Figure 2.9 A differential material element with the principal stresses σ_{I}, σ_{II}, and σ_{III} corresponding to the principal directions $\mathbf{n}_{\mathrm{I}}$, $\mathbf{n}_{\mathrm{II}}$, and $\mathbf{n}_{\mathrm{III}}$, respectively.

2.5 Finite Deformation Kinematics

We consider a solid in motion, as shown in Figure 2.10. The reference configuration represents the solid occupied at time $t = 0$. The current configuration represents the solid occupied at time t. We consider a material point P as shown in the figure. The material point P at time 0 has a position vector $\mathbf{X}$, which defines the material coordinates X_1, X_2, and X_3 for point P. The material point P moves to a new position P' at time t. Now the material point P has a new position vector $\mathbf{x}$, which defines the spatial coordinates x_1, x_2, and x_3 for the material point P. The spatial position vector $\mathbf{x}$ for the material point P at the current time t can be expressed in terms of the reference position vector $\mathbf{X}$ as

$$\mathbf{x} = \mathbf{x}(\mathbf{X}, t) \tag{2.5.1}$$

The displacement vector $\mathbf{u}$ for the material point P at the current time t is defined as

$$\mathbf{u} = \mathbf{x} - \mathbf{X} = \mathbf{x}(\mathbf{X}, t) - \mathbf{X} \tag{2.5.2}$$

As shown in Figure 2.11, we now consider a differential material vector $d\mathbf{X}$ in the neighborhood of point P at $t = 0$. At the current time t, the differential material vector $d\mathbf{X}$ is

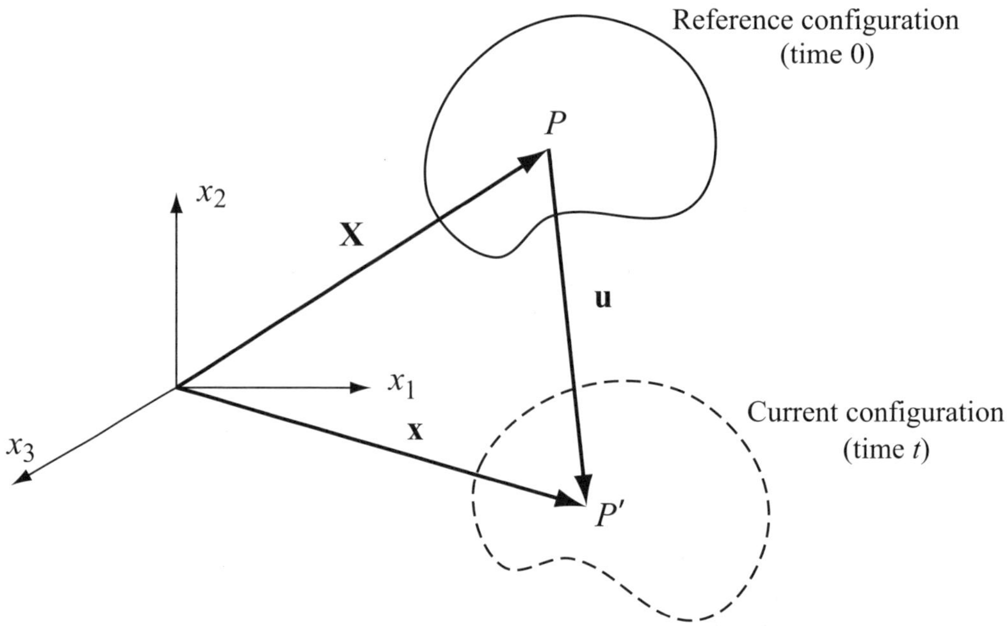

Figure 2.10 A material point P in the reference configuration at time 0 and the current configuration at time t.

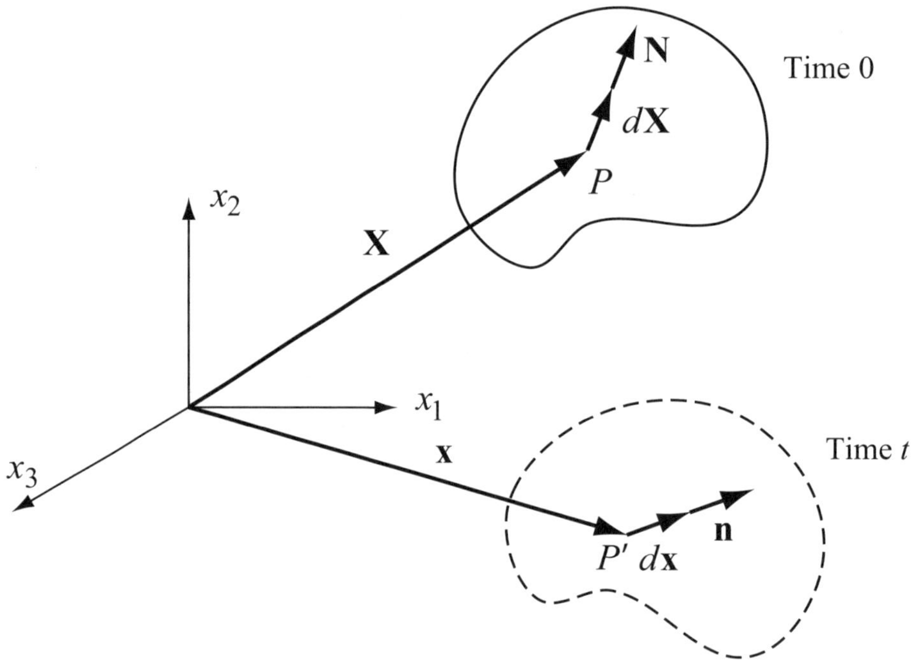

Figure 2.11 A differential material vector $d\mathbf{X}$ and the differential deformed material vector $d\mathbf{x}$.

deformed into $d\mathbf{x}$. The magnitudes of $d\mathbf{X}$ and $d\mathbf{x}$ are defined as dS and ds, respectively, as

$$dS = \sqrt{d\mathbf{X} \cdot d\mathbf{X}} = \sqrt{dX_i dX_i} \tag{2.5.3}$$

$$ds = \sqrt{d\mathbf{x} \cdot d\mathbf{x}} = \sqrt{dx_i dx_i} \tag{2.5.4}$$

The stretch ratio λ of the differential material vector $d\mathbf{X}$ is defined as

$$\lambda = ds/dS \tag{2.5.5}$$

The orientations of the material vectors $d\mathbf{X}$ and $d\mathbf{x}$ are defined as $\mathbf{N}$ and $\mathbf{n}$, respectively, as

$$\mathbf{N} = d\mathbf{X}/\|d\mathbf{X}\| \tag{2.5.6}$$

$$\mathbf{n} = d\mathbf{x}/\|d\mathbf{x}\| \tag{2.5.7}$$

The rotation of the material element $d\mathbf{X}$ can be obtained from a comparison of $\mathbf{N}$ and $\mathbf{n}$.

The deformation gradient tensor $\mathbf{F}$ is defined as

$$d\mathbf{x} = \mathbf{F} \cdot d\mathbf{X} \tag{2.5.8}$$

This equation can be written in the component form as

$$dx_i = F_{ij} dX_j \tag{2.5.9}$$

where

$$F_{ij} = \frac{\partial x_i(\mathbf{X},t)}{\partial X_j} \tag{2.5.10}$$

In this way, the deformation gradient tensor can be used to characterize the general deformation near the material point P, for example, the stretch and rotation of the differential material vector $d\mathbf{X}$.

The Lagrangian strain tensor $\mathbf{E}$ can be defined by

$$(ds)^2 - (dS)^2 = 2d\mathbf{X} \cdot \mathbf{E} \cdot d\mathbf{X} \tag{2.5.11}$$

Substituting Eqs. 2.5.3 and 2.5.4 into Eq. 2.5.11 with use of Eq. 2.5.8 gives

$$\mathbf{E} = \frac{1}{2}\left(\mathbf{F}^{\mathbf{T}} \cdot \mathbf{F} - \mathbf{I}\right) \tag{2.5.12}$$

where $\mathbf{F}^T$ represents the transposition of $\mathbf{F}$, and $\mathbf{I}$ represents the unit tensor. In the component form, Eq. 2.5.12 can be expressed as

$$E_{ij} = \frac{1}{2}\left(F_{ki}F_{kj} - \delta_{ij}\right) \tag{2.5.13}$$

Equation 2.5.13 can be written as

$$E_{ij} = \frac{1}{2}\left(\frac{\partial x_k}{\partial X_i}\frac{\partial x_k}{\partial X_j} - \delta_{ij}\right) \tag{2.5.14}$$

The spatial coordinates x_i can be related to the material coordinates X_i as

$$x_i = X_i + u_i \tag{2.5.15}$$

Substituting Eq. 2.5.15 into Eq. 2.5.14 gives

$$E_{ij} = \frac{1}{2}\left(\frac{\partial u_i}{\partial X_j} + \frac{\partial u_j}{\partial X_i} + \frac{\partial u_k}{\partial X_i}\frac{\partial u_k}{\partial X_j}\right) \tag{2.5.16}$$

When the magnitudes of the displacement gradients $\partial u_i/\partial X_j$ are small, the Lagrangian strains E_{ij} are reduced to the small or infinitesimal strains ε_{ij} as

$$\varepsilon_{ij} = \frac{1}{2}\left(\frac{\partial u_i}{\partial X_j} + \frac{\partial u_j}{\partial X_i}\right) \tag{2.5.17}$$

2.6 Small Strain Theory

Elementary definitions of strains can be illustrated based on simple deformation modes of a differential rectangular material element with respect to the Cartesian coordinate, as shown in Figure 2.12. In this figure, the rectangular elements have two sides ΔX_1 and ΔX_2 parallel to the x_1 and x_2 directions. As shown in Figure 2.12(a), the rectangular element is elongated in the x_1 direction. The elongation of ΔX_1 in the x_1 direction is Δu_1. The strain ε_{11} can be estimated as

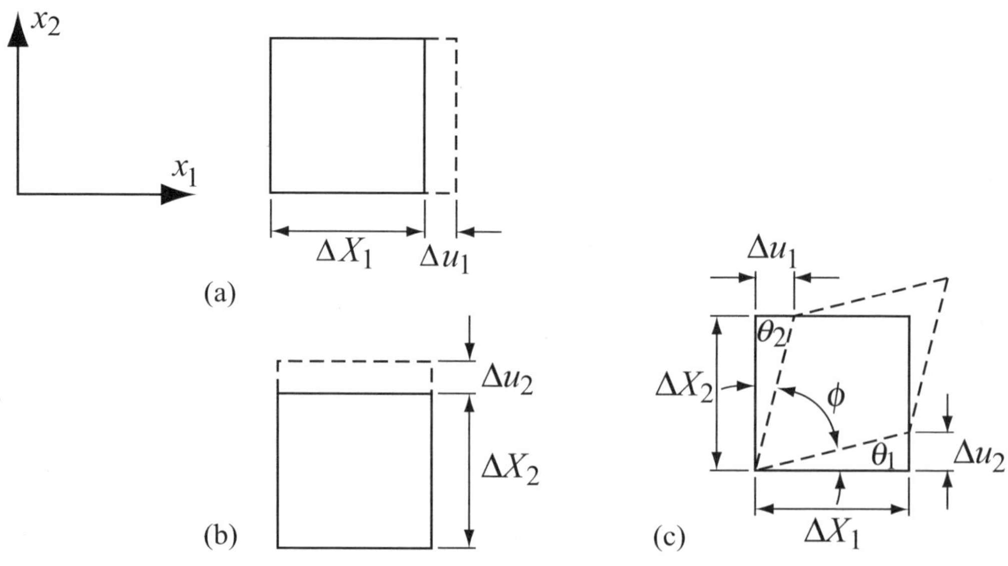

Figure 2.12 Elementary definitions of the strains based on simple deformation modes of a differential rectangular material element. (a) Tensile deformation in the x_1 direction, (b) tensile deformation in the x_2 direction, and (c) shear deformation with respect to the x_1 and x_2 directions.

$$\varepsilon_{11} = \frac{\Delta u_1}{\Delta X_1} \tag{2.6.1}$$

As shown in Figure 2.12(b), the rectangular element is elongated in the x_2 direction. The elongation of ΔX_2 in the x_2 direction is Δu_2. The strain ε_{22} can be estimated as

$$\varepsilon_{22} = \frac{\Delta u_2}{\Delta X_2} \tag{2.6.2}$$

As shown in Figure 2.12(c), the rectangular material element is changed into a parallelogram. As shown in this figure, the rotation of ΔX_1 gives Δu_2, and the rotation of ΔX_2 gives Δu_1. The orientation changes for ΔX_1 and ΔX_2 are denoted as θ_1 and θ_2 as

$$\theta_1 = \frac{\Delta u_2}{\Delta X_1} \tag{2.6.3}$$

$$\theta_2 = \frac{\Delta u_1}{\Delta X_2} \tag{2.6.4}$$

In this figure, the angle between ΔX_1 and ΔX_2 decreases from $\pi/2$ to ϕ. The engineering shear strain γ_{12} is defined as the decrease of the angle from $\pi/2$ to ϕ. Therefore, the engineering shear strain γ_{12} can be estimated as

$$\gamma_{12} = \theta_1 + \theta_2 = \frac{\Delta u_2}{\Delta X_1} + \frac{\Delta u_1}{\Delta X_2} \tag{2.6.5}$$

As ΔX_1 and $\Delta X_2 \to 0$, the definitions of ε_{11}, ε_{22}, and γ_{12} are

$$\varepsilon_{11} = \frac{\partial u_1}{\partial X_1} \tag{2.6.6}$$

$$\varepsilon_{22} = \frac{\partial u_2}{\partial X_2} \tag{2.6.7}$$

$$\gamma_{12} = \frac{\partial u_2}{\partial X_1} + \frac{\partial u_1}{\partial X_2} \tag{2.6.8}$$

The shear strain for convenience of tensorial and mathematical operations is defined as

$$\varepsilon_{12} = \frac{1}{2}\left(\frac{\partial u_2}{\partial X_1} + \frac{\partial u_1}{\partial X_2}\right) = \frac{1}{2}\gamma_{12} \tag{2.6.9}$$

Figure 2.13 shows two neighboring material points P and Q and their displacements. We now consider the small strain tensor from the viewpoint of the displacement gradient tensor. Points P and Q have the displacement vectors $\mathbf{u}^P$ and $\mathbf{u}^Q$, respectively, as

$$\mathbf{u}^P = u_i^P \mathbf{e}_i \tag{2.6.10}$$

$$\mathbf{u}^Q = u_j^Q \mathbf{e}_j \tag{2.6.11}$$

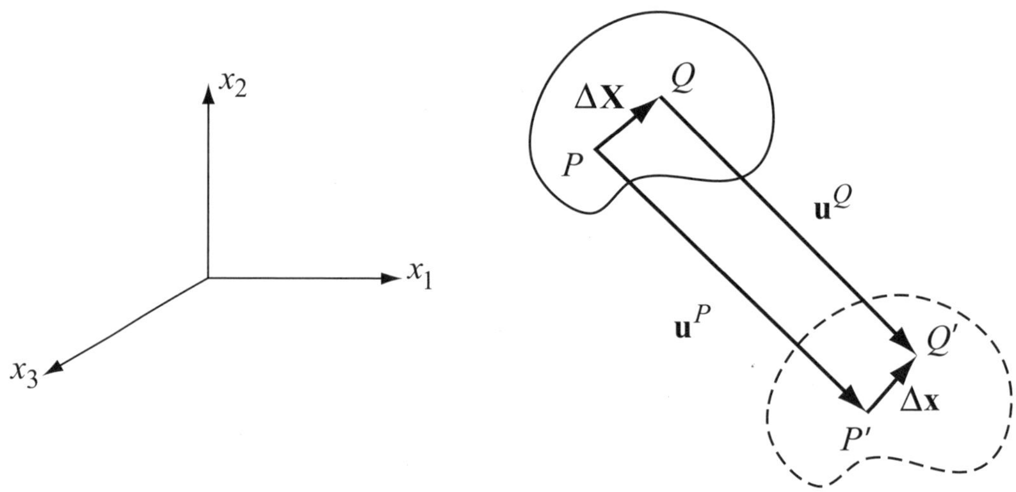

Figure 2.13 Two neighboring material points P and Q and their displacements.

The relative displacement $\Delta\mathbf{u}$ is defined as

$$\Delta\mathbf{u} = \mathbf{u}^Q - \mathbf{u}^P \tag{2.6.12}$$

In the component form,

$$\Delta u_i = u_i^Q - u_i^P \tag{2.6.13}$$

The vector $\Delta\mathbf{X}$ from P to Q can be written as

$$\Delta\mathbf{X} = \mathbf{X}^Q - \mathbf{X}^P \tag{2.6.14}$$

In the component form,

$$\Delta X_j = X_j^Q - X_j^P \tag{2.6.15}$$

The components of the displacement gradient tensor $\nabla\mathbf{u}$ can be written as

$$u_{i,j} = \frac{\partial u_i}{\partial X_j} \tag{2.6.16}$$

To the first order approximation, we can write

$$u_i^Q - u_i^P = u_{i,j}^P\left(X_j^Q - X_j^P\right) \tag{2.6.17}$$

where $u_{i,j}^P$ represent the components of the displacement gradient tensor at point P.

As the magnitude of $\Delta\mathbf{X}$ decreases to zero, Eq. 2.6.17 can be written as

$$du_i = u_{i,j}dX_j \tag{2.6.18}$$

We can decompose the displacement gradient tensor into a symmetric part and an anti-symmetric part as

$$u_{i,j} = \frac{1}{2}\left(u_{i,j} + u_{j,i}\right) + \frac{1}{2}\left(u_{i,j} - u_{j,i}\right) \tag{2.6.19}$$

The first part or the symmetric part is defined as the infinitesimal strain tensor ε_{ij} as

$$\varepsilon_{ij} = \frac{1}{2}\left(u_{i,j} + u_{j,i}\right) = \varepsilon_{ji} \tag{2.6.20}$$

The infinitesimal strain tensor $\boldsymbol{\varepsilon}$ has six components. The infinitesimal strain tensor defined here is consistent with the elementary definitions of the small strains from the elongation and change of angle for two line elements in the two-dimensional cases, as discussed earlier. The definition also is consistent with that of the Lagrangian strain when the displacement gradients are small. The second part or the anti-symmetric part is defined as the rotational tensor ω_{ij} as

$$\omega_{ij} = \frac{1}{2}\left(u_{i,j} - u_{j,i}\right) = -\omega_{ji} \tag{2.6.21}$$

The rotational tensor $\boldsymbol{\omega}$ has three components. The rotational tensor $\boldsymbol{\omega}$ can be shown to relate to the rotation of a line element $d\mathbf{X}$ under the displacement field.

2.7 Different Stress Tensors

When a solid body is under finite deformation, different stress tensors are needed for convenience. The Cauchy stress tensor $\boldsymbol{\sigma}$ is based on the force per unit deformed area. Other stress tensors can be defined from the work rate per unit reference volume. The work rate per unit reference volume can be written as

$$\begin{aligned} \dot{W} &= |\mathbf{F}|\boldsymbol{\sigma}:\mathbf{D} \\ &= \boldsymbol{\tau}:\mathbf{D} \\ &= \mathbf{t}:\dot{\mathbf{F}} \\ &= \mathbf{S}:\dot{\mathbf{E}} \end{aligned} \tag{2.7.1}$$

where a colon ":" denotes the dyadic product. In Eq. 2.7.1, $|\mathbf{F}|$ represents the determinant of the deformation gradient tensor $\mathbf{F}$, and $\mathbf{D}$ represents the rate of the deformation tensor, which is the symmetric part of the velocity gradient tensor $\nabla\mathbf{u}$. Here, $\boldsymbol{\tau}$ represents the Kirchhoff stress tensor, which can be defined as

$$\boldsymbol{\tau} = |\mathbf{F}|\boldsymbol{\sigma} \tag{2.7.2}$$

In Eq. 2.7.1, $\mathbf{t}$ represents the nominal or the first Piola-Kirchhoff stress tensor, which is defined as

$$\mathbf{t} = |\mathbf{F}|\mathbf{F}^{-1}\cdot\boldsymbol{\sigma} \tag{2.7.3}$$

where $\mathbf{F}^{-1}$ represents the inverse of the deformation gradient tensor $\mathbf{F}$. In Eq. 2.7.1, $\dot{\mathbf{F}}$ is the deformation gradient rate tensor, which represents the work-conjugate strain rate tensor to the nominal stress tensor $\mathbf{t}$. Note that the nominal stress tensor $\mathbf{t}$ is not symmetrical. In Eq. 2.7.1, $\mathbf{S}$ represents the second Piola-Kirchhoff stress tensor, which is defined as

$$\mathbf{S} = |\mathbf{F}|\mathbf{F}^{-1}\cdot\boldsymbol{\sigma}\cdot\left(\mathbf{F}^{-1}\right)^{\mathrm{T}} \tag{2.7.4}$$

where $\left(\mathbf{F}^{-1}\right)^{\mathrm{T}}$ represents the transpose of the inverse of the deformation gradient tensor $\mathbf{F}^{-1}$. In Eq. 2.7.1, $\dot{\mathbf{E}}$ is the Lagrangian strain rate tensor, which represents the work-conjugate strain rate tensor to the second Piola-Kirchhoff stress tensor $\mathbf{S}$. Note that the second Piola-Kirchhoff stress tensor $\mathbf{S}$ is symmetric. These stresses are convenient to use when a convected coordinate system is used to describe the kinematics.

Now we consider a solid body with reference to a Cartesian coordinate system, as shown in Figure 2.14. Here, $\mathbf{X}$ represents the position vector for the material point P in the reference or undeformed configuration, $\mathbf{x}$ represents the position vector for the material point P in the current or deformed configuration, and $\mathbf{u}$ represents the displacement vector for the material point P. The body has a volume V_0 and a surface S_0 in the

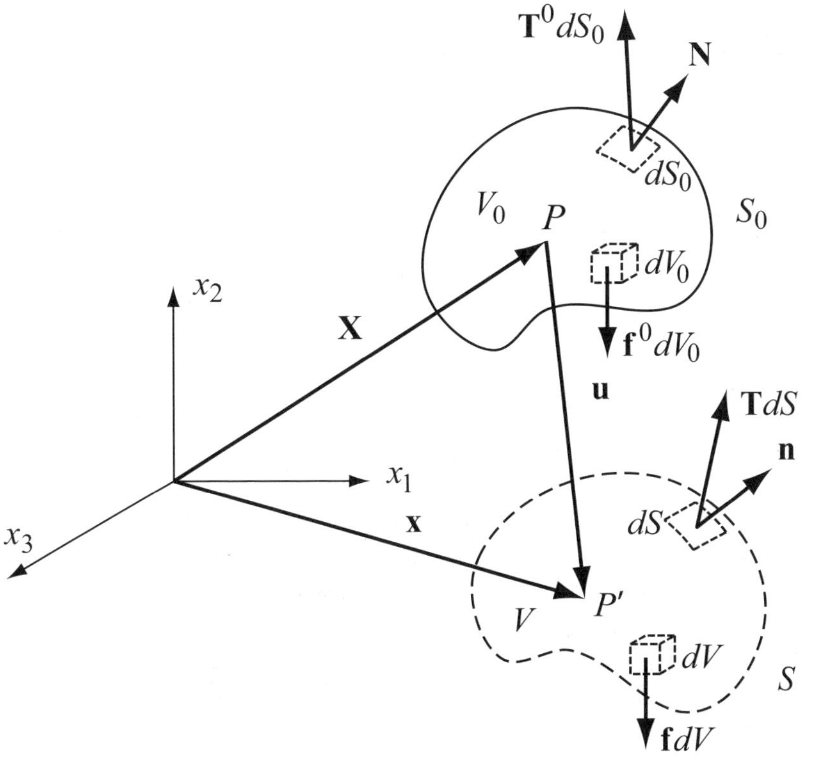

Figure 2.14 A solid with a material point P in the undeformed and deformed configurations.

undeformed configuration. The body has a volume V and a surface S in the deformed configuration. The classical principle of virtual work with respect to the deformed configuration can written as

$$\int_S T_i \delta u_i dS + \int_V (f_i - \rho a_i)\delta u_i dV = \int_V \sigma_{ij}\delta\varepsilon_{ij} dV \tag{2.7.5}$$

where δu_i represent the components of the virtual displacement $\delta\mathbf{u}$. Here, $\delta\varepsilon_{ij}$ are

$$\delta\varepsilon_{ij} = \frac{1}{2}\left(\frac{\partial \delta u_i}{\partial x_j} + \frac{\partial \delta u_j}{\partial x_i}\right) \tag{2.7.6}$$

associated with the virtual displacement $\delta\mathbf{u}$. The principle of virtual work with respect to the undeformed configuration can be written as

$$\int_{S_0} T_i^0 \delta u_i dS_0 + \int_{V_0}\left(f_i^0 - \rho_0 a_i\right)\delta u_i dV_0 = \int_{V_0} t_{ij}\frac{\partial \delta u_j}{\partial X_j} dV_0 \tag{2.7.7}$$

where X_j represent the components of the position vector $\mathbf{X}$, and t_{ij} represent the components of the first Piola-Kirchhoff stress tensor $\mathbf{t}$. A subscript or superscript zero represents a field quantity for a material point in the undeformed configuration. In this case, the equations of motion based on the undeformed configuration can be written as

$$\frac{\partial t_{ji}}{\partial X_j} + f_i^0 - \rho_0 a_i = 0 \tag{2.7.8}$$

The traction $\mathbf{T}^0$ referred to the undeformed surface with a unit normal $\mathbf{N}$ is related to the first Piola-Kirchhoff stress tensor $\mathbf{t}$ as

$$T_j^0 = N_i t_{ij} \tag{2.7.9}$$

When a solid body is under large deformation, a convected coordinate system can be adopted to describe the kinematics. In a convected coordinate system, a set of coordinates (ξ_1, ξ_2, ξ_3) is used to identify a material point in a solid in the undeformed and deformed configurations. The position vectors of the material point of interest in the undeformed and deformed configurations are denoted as

$${}_0\mathbf{x} = {}_0\mathbf{x}(\xi_1, \xi_2, \xi_3) \tag{2.7.10}$$

$$\mathbf{x} = \mathbf{x}(\xi_1, \xi_2, \xi_3) \tag{2.7.11}$$

The displacement $\mathbf{u}$ is defined as

$$\mathbf{u} = \mathbf{x}(\xi_1, \xi_2, \xi_3) - {}_0\mathbf{x}(\xi_1, \xi_2, \xi_3) \tag{2.7.12}$$

The displacement $\mathbf{u}$ usually is defined with respect to the base vectors associated with the undeformed configuration. Because the sheet usually is flat before forming, a Cartesian coordinate system typically is used to describe the undeformed configuration. The covariant base vectors in the undeformed and deformed configurations are defined as

$${}_0\mathbf{g}_i = \frac{\partial\, {}_0\mathbf{x}}{\partial \xi^i} \tag{2.7.13}$$

$$\mathbf{g}_i = \frac{\partial \mathbf{x}}{\partial \xi^i} \tag{2.7.14}$$

Usually, when a Cartesian coordinate system is used to describe the undeformed configuration, the base vectors ${}_0\mathbf{g}_i$ can be thought of as the usual Cartesian base vectors that have a magnitude of unity and are mutually orthogonal to each other. Note that

$$\mathbf{g}_i = \mathbf{F} \cdot {}_0\mathbf{g}_i \tag{2.7.15}$$

The metric tensors associated with the undeformed and deformed bodies are

$$_0g_{ij} = {_0\mathbf{g}_i} \cdot {_0\mathbf{g}_j} \tag{2.7.16}$$

$$g_{ij} = \mathbf{g}_i \cdot \mathbf{g}_j \tag{2.7.17}$$

The covariant components of the Lagrangian strain **E** with respect to the base vectors in the undeformed configuration are

$$E_{ij} = \frac{1}{2}\left(g_{ij} - {_0g_{ij}}\right) \tag{2.7.18}$$

The Lagrangian strain tensor **E** can be written in terms of the covariant derivatives of the displacement **u** as

$$E_{ij} = \frac{1}{2}\left(u_{i,j} + u_{j,i} + {_0g_{kl}}u^k_{,i}u^l_{,j}\right) \tag{2.7.19}$$

The stress tensors usually are defined with respect to the base vectors associated with the deformed configuration, especially when plasticity theories are used to describe the constitutive laws. Because

$$\mathbf{S} = \mathbf{F}^{-1} \cdot \boldsymbol{\tau} \cdot \left(\mathbf{F}^{-1}\right)^{\mathrm{T}} \tag{2.7.20}$$

It easily can be shown that the contravariant components of the second Piola-Kirchhoff stress **S** with respect to the undeformed base vectors are equal to the contravariant components of the Kirchhoff stress $\boldsymbol{\tau}$ with respect to the deformed base vectors

$$S^{ij} = {_0\mathbf{g}_i} \cdot \mathbf{S} \cdot {_0\mathbf{g}_j} = \mathbf{g}_i \cdot \boldsymbol{\tau} \cdot \mathbf{g}_j = \tau^{ij} \tag{2.7.21}$$

Finally, the classical principle of virtual work can be written alternatively with respect to the undeformed configuration as

$$\int_{S_0} {_0T^i}\delta u_i dS_0 + \int_{V_0} (f_i^0 - \rho_0 a_i)\delta u_i dV_0 = \int_{V_0} \tau^{ij}\delta E_{ij} dV_0 \tag{2.7.22}$$

where

$$\delta E_{ij} = \frac{1}{2}\left(\delta u_{i,j} + \delta u_{j,i} + {_0g_{kl}}\left(u^k_{,i}\delta u^l_{,j} + u^k_{,j}\delta u^l_{,i}\right)\right) \tag{2.7.23}$$

Note that the principle of virtual work is integrated over the original known configuration of the solid body. This equation is the starting point for our finite element formulations. Then the following step is to specify the constitutive relations between the Kirchhoff stress tensor $\boldsymbol{\tau}$ and the Lagrangian strain tensor **E**. Usually, a rate form of the principle of virtual work is needed. Therefore, a rate form of the constitutive relation between $\boldsymbol{\tau}$ and **E** is needed for setting up the finite element formulation. Commonly used constitutive laws in the rate form for plastically deformed materials are discussed in Chapter 3.

2.8 Stresses and Strains from Tensile Tests

Tensile tests typically are used to obtain the stress-strain relations for sheet metals. The tensile stress-strain relation is used to extract the elastic modulus and the plastic strain hardening relation. Figure 2.15 shows a typical sheet specimen. The middle portion of the sheet specimen has the uniform cross-sectional area A. As shown in this figure, the middle portion of the specimen is designated as the gage section with the gage length L. The tensile stress-strain relation is extracted from the relation of the force and the elongation or extension of the gage length. When the reference gage length is marked as L_0 and the deformed gage length is marked as L, the engineering strain e is defined as

$$e = \frac{L - L_0}{L_0} \tag{2.8.1}$$

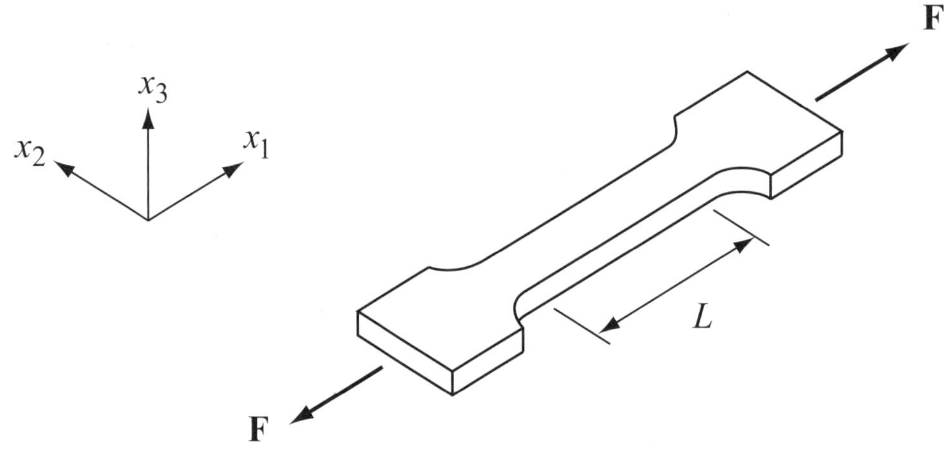

Figure 2.15 A typical sheet specimen.

We can define the differential true strain $d\varepsilon$ as

$$d\varepsilon = \frac{dl}{l} \tag{2.8.2}$$

where l is the current gage length, and dl represents the differential gage length. The true strain or the logarithmic strain ε is defined as the differential strain integrated over the deformation history from the reference gage length L_0 to the current or deformed gage length L as

$$\varepsilon = \int_0^{\varepsilon} d\varepsilon = \int_{L_0}^{L} \frac{dl}{l} \tag{2.8.3}$$

Carrying out the integration gives the true strain ε as

$$\varepsilon = \ln\left(\frac{L}{L_0}\right) \tag{2.8.4}$$

As the sheet specimen is extended, the cross-sectional area of the gage section decreases. When the extension of the specimen is small due only to elastic deformation of the materials, the decrease in the cross-sectional area of the gage section is small and negligible. The engineering stress S is defined as the force F applied to the specimen divided by the original cross-sectional area A_0 as

$$S = \frac{F}{A_0} \tag{2.8.5}$$

When the extension of the bar becomes large due to plastic deformation, the decrease of the cross-sectional area of the gage length becomes large and not negligible. The true stress σ is then defined as the force F divided by the deformed or current cross-sectional area A as

$$\sigma = \frac{F}{A} \tag{2.8.6}$$

When the large plastic strains are considered in the deformation history, the true stress and true strain relation usually is used as the input of the materials. The engineering stress-strain relation and the true stress-strain relation are plotted schematically in Figure 2.16. In this figure, the initial part of the curve is linear, which represents the linear elastic response of the material. The elastic portion is exaggerated in this figure. As shown in Figure 2.16, the true stress-strain relation gives a higher stress for a given strain when compared to the engineering stress-strain curve. For ductile sheet metals, the engineering stress-strain curve usually shows a maximum, as shown in Figure 2.16.

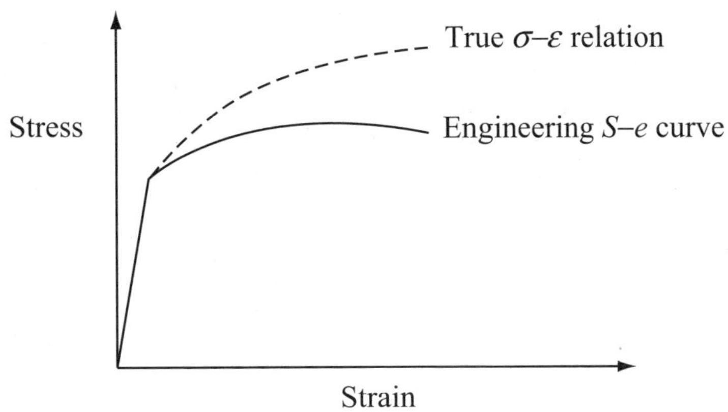

Figure 2.16 An engineering stress-strain curve and the corresponding true stress-strain curve.

2.9 Reference

Spain, B., *Tensor Calculus*, Oliver and Boyd, Edinburgh and London, U.K., 1956.

Chapter 3

Constitutive Laws

In uniaxial tensile tests of ductile sheet metals, stress-strain curves usually become nonlinear when the strains are large. Figure 3.1 shows a schematic of the stress-strain curve from a tensile test. In this figure, σ represents the tensile stress, and ε represents the tensile strain. Also in this figure, the initial part of the curve is linear. As the strain ε continues to increase, the stress-strain curve becomes nonlinear. Sheet metals usually have both elastic and plastic anisotropy due to rolling processes. Therefore, material anisotropy is discussed in detail in this chapter. Isotropic and anisotropic linear elastic responses of materials are discussed first. Deviatoric stresses and strains then are introduced for the development of plasticity theories. Then, we concentrate on the nonlinear or strain hardening response of materials. We begin with simple mathematical models for uniaxial tensile stress-strain relations. Then we concentrate on the yield conditions under general stress states for isotropic incompressible materials. We also discuss the yield conditions for plastic anisotropic materials. In addition, we discuss the basic concepts of constitutive laws with consideration of isotropic and anisotropic hardening behavior of materials. Detailed constitutive laws are discussed based on the von Mises yield condition and the quadratic anisotropic yield condition of Hill [1948], with consideration of isotropic and anisotropic hardening laws. Finally, we summarize

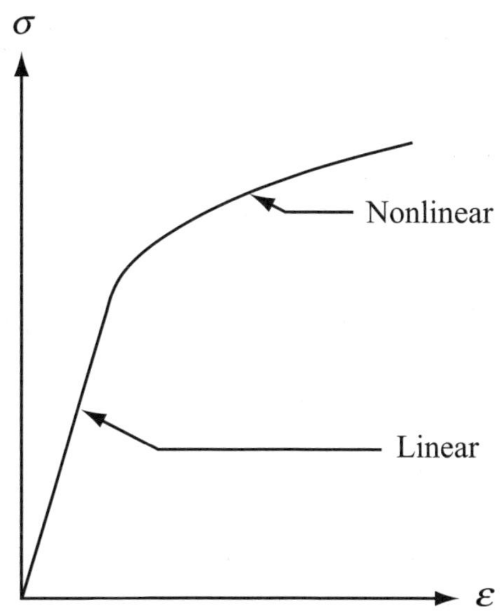

Figure 3.1 A schematic of the stress-strain curve from a tensile test.

the principles of plastic localization, the forming limit diagram, and modeling of failure processes based on void mechanics.

3.1 Linear Elastic Isotropic Materials

Figure 3.2 shows a sheet specimen with respect to a Cartesian coordinate system and a schematic of the tensile stress-strain relation obtained from a sheet specimen. When the tensile axis is in the x_1 direction, the stress is denoted as σ_{11} and the strain is denoted as ε_{11}. When we discuss the linear elastic behavior of sheet metals, the stress is thought of as the Cauchy stress. The strain is thought of as the small or infinitesimal strain because the elastic strains usually are very small, and the geometry change is small. The slope of the initial linear part of the stress-strain curve is defined as the Young's modulus E as

$$E = \frac{\sigma_{11}}{\varepsilon_{11}} \tag{3.1.1}$$

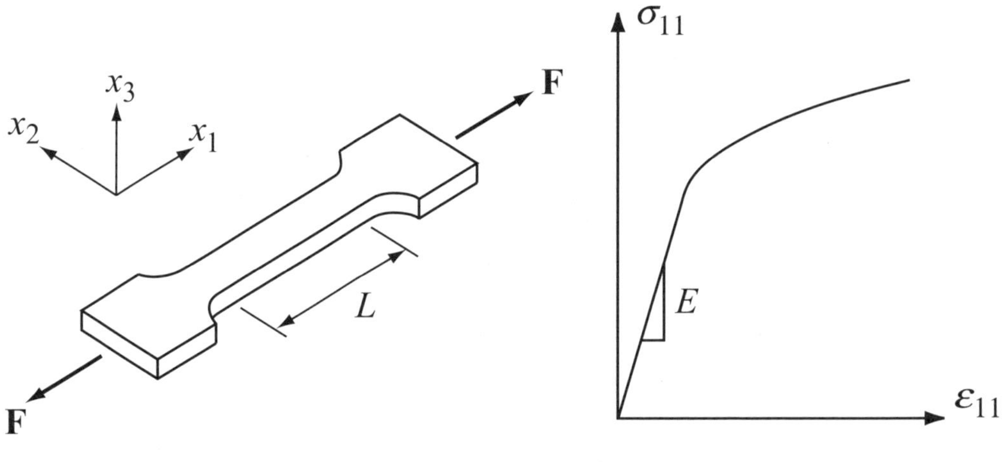

Figure 3.2 A sheet specimen with respect to a Cartesian coordinate system and a schematic of the tensile stress-strain relation obtained from a sheet specimen.

The dimensions of the width w and the thickness t of a sheet metal specimen decrease as the tensile strain ε_{11} increases. Therefore, the specimen has the strains ε_{22} and ε_{33} in the x_2 and x_3 directions, respectively, according to the Cartesian coordinates shown in Figure 3.2. For isotropic elastic materials, the normal strains ε_{22} and ε_{33} in the x_2 and x_3 directions are equal to each other. The Poisson's ratio ν is defined as

$$\nu = -\frac{\varepsilon_{22}}{\varepsilon_{11}} = -\frac{\varepsilon_{33}}{\varepsilon_{11}} \tag{3.1.2}$$

When we consider a material element under pure shear with respect to in the x_1 and x_2 axes as shown in Figure 3.3, the shear modulus G is defined from σ_{12} and $\gamma_{12}(= 2\varepsilon_{12})$ as

$$G = \frac{\sigma_{12}}{\gamma_{12}} = \frac{\sigma_{12}}{2\varepsilon_{12}} \tag{3.1.3}$$

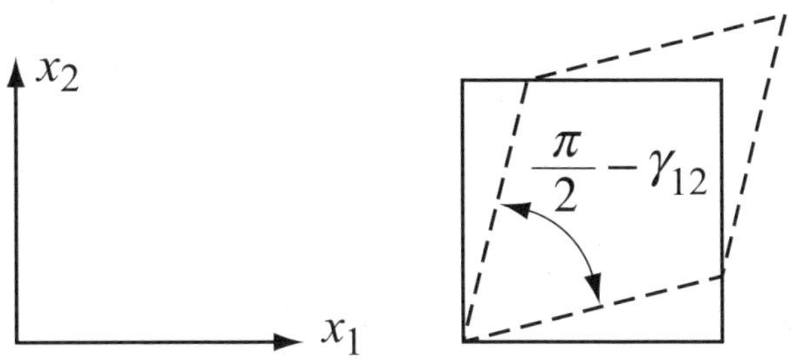

Figure 3.3 Pure shear with respect to the x_1 and x_2 axes.

For isotropic materials, only two elastic constants are needed to describe the elastic stress-strain relation by consideration of isotropy. The shear modulus G can be related to the Young's modulus E and Poisson's ratio ν as

$$G = \frac{E}{2(1+\nu)} \tag{3.1.4}$$

A generalized Hooke's law for isotropic linear elastic materials can be written as

$$\varepsilon_{ij} = \frac{1+\nu}{E}\sigma_{ij} - \frac{\nu}{E}\sigma_{kk}\delta_{ij} \tag{3.1.5}$$

where δ_{ij} represent the Kronecker delta. The elastic stress-strain relation can be expressed for each strain component as

$$\varepsilon_{11} = \frac{\sigma_{11}}{E} - \nu\frac{\sigma_{22}}{E} - \nu\frac{\sigma_{33}}{E} \tag{3.1.6}$$

$$\varepsilon_{22} = -\nu\frac{\sigma_{11}}{E} + \frac{\sigma_{22}}{E} - \nu\frac{\sigma_{33}}{E} \tag{3.1.7}$$

$$\varepsilon_{33} = -\nu\frac{\sigma_{11}}{E} - \nu\frac{\sigma_{22}}{E} + \frac{\sigma_{33}}{E} \tag{3.1.8}$$

$$\varepsilon_{23} = \frac{1}{2G}\sigma_{23} = \frac{1+\nu}{E}\sigma_{23} \tag{3.1.9}$$

$$\varepsilon_{31} = \frac{1}{2G}\sigma_{31} = \frac{1+\nu}{E}\sigma_{31} \tag{3.1.10}$$

$$\varepsilon_{12} = \frac{1}{2G}\sigma_{12} = \frac{1+\nu}{E}\sigma_{12} \tag{3.1.11}$$

The deviatoric stresses are important stress measures for the shear parts of the stresses. The deviatoric stress tensor s_{ij} is defined as

$$s_{ij} = \sigma_{ij} - \frac{1}{3}\sigma_{kk}\delta_{ij} \tag{3.1.12}$$

Note that

$$s_{kk} = 0 \tag{3.1.13}$$

The deviatoric stress tensor s_{ij} can be expressed in the matrix form as

$$[s_{ij}] = \begin{bmatrix} \frac{2\sigma_{11} - \sigma_{22} - \sigma_{33}}{3} & \sigma_{12} & \sigma_{13} \\ \sigma_{21} & \frac{2\sigma_{22} - \sigma_{11} - \sigma_{33}}{3} & \sigma_{23} \\ \sigma_{31} & \sigma_{32} & \frac{2\sigma_{33} - \sigma_{11} - \sigma_{22}}{3} \end{bmatrix} \tag{3.1.14}$$

The corresponding deviatoric strain tensor ε'_{ij} is defined as

$$\varepsilon'_{ij} = \varepsilon_{ij} - \frac{1}{3}\varepsilon_{kk}\delta_{ij} \tag{3.1.15}$$

Note that

$$\varepsilon'_{kk} = 0 \tag{3.1.16}$$

The deviatoric strain tensor ε'_{ij} can be expressed in the matrix form as

$$[\varepsilon'_{ij}] = \begin{bmatrix} \frac{2\varepsilon_{11} - \varepsilon_{22} - \varepsilon_{33}}{3} & \varepsilon_{12} & \varepsilon_{13} \\ \varepsilon_{21} & \frac{2\varepsilon_{22} - \varepsilon_{11} - \varepsilon_{33}}{3} & \varepsilon_{23} \\ \varepsilon_{31} & \varepsilon_{32} & \frac{2\varepsilon_{33} - \varepsilon_{11} - \varepsilon_{22}}{3} \end{bmatrix} \tag{3.1.17}$$

The strain energy ϕ for linear elastic isotropic materials can be expressed as

$$\phi = \frac{1}{2}\sigma_{ij}\varepsilon_{ij} \tag{3.1.18}$$

Substituting Eqs. 3.1.12 and 3.1.15 into Eq. 3.1.18 gives

$$\phi = \frac{1}{2} s_{ij} \varepsilon'_{ij} + \frac{1}{6} \sigma_{kk} \varepsilon_{ll} \tag{3.1.19}$$

Combining Eqs. 3.1.12, 3.1.15, and 3.1.5 gives

$$s_{ij} = 2G\varepsilon'_{ij} \tag{3.1.20}$$

The hydrostatic tension $\sigma_{kk}/3$ can be related to the volumetric strain ε_{kk} by the bulk modulus B as

$$\frac{\sigma_{kk}}{3} = B\varepsilon_{kk} \tag{3.1.21}$$

Substituting Eqs. 3.1.20 and 3.1.21 into Eq. 3.1.19 gives

$$\phi = G\varepsilon'_{ij}\varepsilon'_{ij} + \frac{1}{2} B\varepsilon_{kk}\varepsilon_{ll} \tag{3.1.22}$$

We can define the first part and the second part of the strain energy in Eq. 3.1.22 as

$$\phi_d = G\varepsilon'_{ij}\varepsilon'_{ij} \tag{3.1.23}$$

$$\phi_v = \frac{1}{2} B\varepsilon_{kk}\varepsilon_{ll} \tag{3.1.24}$$

Here, ϕ_d represents the deviatoric part of the strain energy, and ϕ_v represents the volumetric part of the strain energy. Note that the bulk modulus B can be related to the Young's modulus E and Poisson's ratio v by

$$B = \frac{E}{3(1-2v)} \tag{3.1.25}$$

For elastic incompressible materials, ε_{kk} equals 0 for any given amount of σ_{kk}. Therefore, the bulk modulus B should be infinity. Consequently, the Poisson's ratio v is

$$v = \frac{1}{2} \tag{3.1.26}$$

Finally, the inverse of the stress-strain relation in Eq. 3.1.5 can be written as

$$\sigma_{ij} = D_{ijkl}\varepsilon_{kl} \tag{3.1.27}$$

where

$$D_{ijkl} = \frac{E}{1+\nu}\left(\frac{1}{2}\left(\delta_{ik}\delta_{jl} + \delta_{il}\delta_{jk}\right) + \frac{\nu}{1-2\nu}\delta_{ij}\delta_{kl}\right) \tag{3.1.28}$$

3.2 Linear Elastic Anisotropic Materials

For general anisotropic materials, the linear stress-strain relation usually is written as

$$\sigma_{ij} = C_{ijkl}\varepsilon_{kl} \tag{3.2.1}$$

where C_{ijkl} represent the elastic moduli. Note that σ_{ij} and ε_{kl} are symmetric. Therefore, C_{ijkl} should have the symmetries of

$$C_{ijkl} = C_{jikl} \tag{3.2.2}$$

and

$$C_{ijkl} = C_{ijlk} \tag{3.2.3}$$

Due to the existence of the strain energy function, C_{ijkl} also should have the symmetry of

$$C_{ijkl} = C_{klij} \tag{3.2.4}$$

Therefore, the total number of elastic constants is reduced from 81 to 21 by considering the symmetries as discussed above.

Because C_{ijkl} is a fourth order tensor, it is hard to visualize the material symmetry in the four-dimensional matrix form. Therefore, the stresses and strains can be written as

$$\{\sigma_i\} = \begin{bmatrix} \sigma_1 \\ \sigma_2 \\ \sigma_3 \\ \sigma_4 \\ \sigma_5 \\ \sigma_6 \end{bmatrix} = \begin{bmatrix} \sigma_{11} \\ \sigma_{22} \\ \sigma_{33} \\ \sigma_{23} \\ \sigma_{31} \\ \sigma_{12} \end{bmatrix} \tag{3.2.5}$$

$$\{e_i\} = \begin{bmatrix} e_1 \\ e_2 \\ e_3 \\ e_4 \\ e_5 \\ e_6 \end{bmatrix} = \begin{bmatrix} \varepsilon_{11} \\ \varepsilon_{22} \\ \varepsilon_{33} \\ \gamma_{23} \\ \gamma_{31} \\ \gamma_{12} \end{bmatrix} = \begin{bmatrix} \varepsilon_{11} \\ \varepsilon_{22} \\ \varepsilon_{33} \\ 2\varepsilon_{23} \\ 2\varepsilon_{31} \\ 2\varepsilon_{12} \end{bmatrix} \tag{3.2.6}$$

Then the stress-strain relation can be expressed as

$$\begin{bmatrix}\sigma_1\\\sigma_2\\\sigma_3\\\sigma_4\\\sigma_5\\\sigma_6\end{bmatrix}=\begin{bmatrix}C_{11}&C_{12}&C_{13}&C_{14}&C_{15}&C_{16}\\C_{21}&C_{22}&C_{23}&C_{24}&C_{25}&C_{26}\\C_{31}&C_{32}&C_{33}&C_{34}&C_{35}&C_{36}\\C_{41}&C_{42}&C_{43}&C_{44}&C_{45}&C_{46}\\C_{51}&C_{52}&C_{53}&C_{54}&C_{55}&C_{56}\\C_{61}&C_{62}&C_{63}&C_{64}&C_{65}&C_{66}\end{bmatrix}\begin{bmatrix}e_1\\e_2\\e_3\\e_4\\e_5\\e_6\end{bmatrix} \tag{3.2.7}$$

Due to the existence of the strain energy density function, C_{ij} must be symmetric as

$$C_{ij}=C_{ji} \tag{3.2.8}$$

Now it is easy to count the number of elastic constants from the matrix shown in Eq. 3.2.7 as 21.

For discussion of symmetry, the stress-strain relation can be written back in the form as

$$\begin{bmatrix}\sigma_{11}\\\sigma_{22}\\\sigma_{33}\\\tau_{23}\\\tau_{31}\\\tau_{12}\end{bmatrix}=\begin{bmatrix}C_{11}&C_{12}&C_{13}&C_{14}&C_{15}&C_{16}\\C_{21}&C_{22}&C_{23}&C_{24}&C_{25}&C_{26}\\C_{31}&C_{32}&C_{33}&C_{34}&C_{35}&C_{36}\\C_{41}&C_{42}&C_{43}&C_{44}&C_{45}&C_{46}\\C_{51}&C_{52}&C_{53}&C_{54}&C_{55}&C_{56}\\C_{61}&C_{62}&C_{63}&C_{64}&C_{65}&C_{66}\end{bmatrix}\begin{bmatrix}\varepsilon_{11}\\\varepsilon_{22}\\\varepsilon_{33}\\\gamma_{23}\\\gamma_{31}\\\gamma_{12}\end{bmatrix} \tag{3.2.9}$$

For example, symmetry usually exists in single crystals. When one plane of symmetry exists in the material, the material is said to have a monoclinic symmetry. For example, when the Cartesian coordinate system is chosen such that the $x_1 - x_2$ plane is a symmetry plane, the requirement of symmetry gives

$$\begin{bmatrix}\sigma_{11}\\\sigma_{22}\\\sigma_{33}\\\tau_{23}\\\tau_{31}\\\tau_{12}\end{bmatrix}=\begin{bmatrix}C_{11}&C_{12}&C_{13}&0&0&C_{16}\\C_{21}&C_{22}&C_{23}&0&0&C_{26}\\C_{31}&C_{32}&C_{33}&0&0&C_{36}\\0&0&0&C_{44}&C_{45}&0\\0&0&0&C_{54}&C_{55}&0\\C_{61}&C_{62}&C_{63}&0&0&C_{66}\end{bmatrix}\begin{bmatrix}\varepsilon_{11}\\\varepsilon_{22}\\\varepsilon_{33}\\\gamma_{23}\\\gamma_{31}\\\gamma_{12}\end{bmatrix} \tag{3.2.10}$$

Therefore, the number of elastic constants is reduced to 13 with the use of $C_{ij}=C_{ji}$.

When three mutually orthogonal planes of symmetry exist in the material, the material is said to have orthotropic symmetry. When the Cartesian coordinate system is defined such that the $x_1 - x_2$, $x_2 - x_3$, and $x_3 - x_1$ planes are symmetry planes, the requirement of symmetry gives

$$\begin{bmatrix} \sigma_{11} \\ \sigma_{22} \\ \sigma_{33} \\ \tau_{23} \\ \tau_{31} \\ \tau_{12} \end{bmatrix} = \begin{bmatrix} C_{11} & C_{12} & C_{13} & 0 & 0 & 0 \\ C_{21} & C_{22} & C_{23} & 0 & 0 & 0 \\ C_{31} & C_{32} & C_{33} & 0 & 0 & 0 \\ 0 & 0 & 0 & C_{44} & 0 & 0 \\ 0 & 0 & 0 & 0 & C_{55} & 0 \\ 0 & 0 & 0 & 0 & 0 & C_{66} \end{bmatrix} \begin{bmatrix} \varepsilon_{11} \\ \varepsilon_{22} \\ \varepsilon_{33} \\ \gamma_{23} \\ \gamma_{31} \\ \gamma_{12} \end{bmatrix} \tag{3.2.11}$$

The number of elastic constants reduces to 9 with the use of $C_{ij} = C_{ji}$.

Further, when the isotropy exists with respect to, for example, the $x_1 - x_2$ plane, the material is said to have transversely isotropic symmetry or normal anisotropy. The requirement of the symmetry gives

$$C_{11} = C_{22} \tag{3.2.12}$$

$$C_{13} = C_{23} \tag{3.2.13}$$

$$C_{44} = C_{55} \tag{3.2.14}$$

$$C_{66} = \frac{C_{11} - C_{12}}{2} \tag{3.2.15}$$

Therefore, the number of elastic constants is reduced to 5 for transversely isotropic materials. Finally, when the material becomes isotropic, the symmetry requirement gives

$$C_{11} = C_{33} \tag{3.2.16}$$

$$C_{44} = C_{66} \tag{3.2.17}$$

$$C_{13} = C_{23} \tag{3.2.18}$$

Therefore, the number of elastic constants is reduced to 2.

Back to the fourth order tensor C_{ijkl} for isotropic materials, C_{ijkl} can be written as

$$C_{ijkl} = \lambda\delta_{ij}\delta_{kl} + \mu\left(\delta_{ik}\delta_{jl} + \delta_{il}\delta_{jk}\right) \tag{3.2.19}$$

where λ and μ are Lame elastic constants. Therefore, the elastic stress-strain relation for isotropic materials can be written as

$$\sigma_{ij} = 2\mu\varepsilon_{ij} + \lambda\varepsilon_{kk}\delta_{ij} \tag{3.2.20}$$

The Lame constants λ and μ can be related to the Young's modulus E, shear modulus G, and Poisson ratio ν as

$$\mu = G = \frac{E}{2(1+\nu)} \tag{3.2.21}$$

and

$$\lambda = \frac{2G\nu}{1-2\nu} = \frac{E\nu}{(1+\nu)(1-2\nu)} \tag{3.2.22}$$

3.3 Different Models for Uniaxial Stress-Strain Curves

When the tensile stress-strain relation becomes nonlinear for strain hardening materials, the stress-strain curves are quite different for different metals, such as those of aluminum and low-carbon steel near the yield point. As shown in Figure 3.4, the 0.2% strain offset usually is used to determine the yield stress σ_0. The yield stress is called the 0.2% offset yield stress. As shown in Figure 3.4, as the strain continues to increase, the stress increases nonlinearly. When the strain decreases, the stress decreases, and the stress-strain curve usually follows the curve with the slope of the elastic modulus E. When the stress decreases to 0, the strain decreases by the amount of the elastic strain ε_e. As shown in Figure 3.4, when the stress decreases to 0, some nonrecoverable strain ε_p remains, which is defined as the plastic strain. Therefore, within the context of the small strain approach, the total strain ε can be decomposed into the elastic strain ε_e and the plastic strain ε_p as

$$\varepsilon = \varepsilon_e + \varepsilon_p \tag{3.3.1}$$

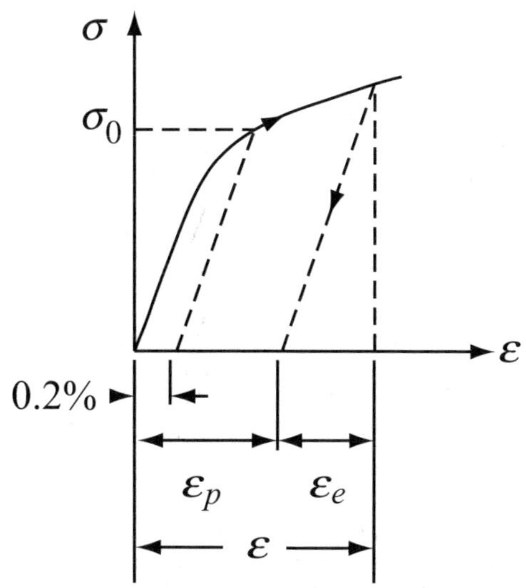

Figure 3.4 Definitions of the 0.2% offset yield stress, the elastic strain, the plastic strain, and the total strain.

Simple mathematical models for tensile stress-strain curves are used in general. The simplest one is a rigid perfectly plastic idealization, as shown in Figure 3.5(a). Here, σ_0 represents the yield stress. As shown in Figure 3.5(a), there is no elastic strain and no strain hardening. This model usually is taken for examining the plastic behavior where the elastic strains are small compared to the plastic strains. For rigid perfectly plastic materials, the total strain ε is the plastic strain ε_p

$$\varepsilon = \varepsilon_p \tag{3.3.2}$$

Figure 3.5(b) shows an elastic perfectly plastic model where the elastic behavior must be considered. This model usually is taken for examining the elastic-plastic response of a region where a part of the region is in elastic conditions and a part of the region is in plastic conditions. For elastic perfectly plastic materials, the total strain ε is the sum of the elastic strain ε_e (= σ_0/E) and the plastic strain ε_p as

$$\varepsilon = \frac{\sigma_0}{E} + \varepsilon_p \tag{3.3.3}$$

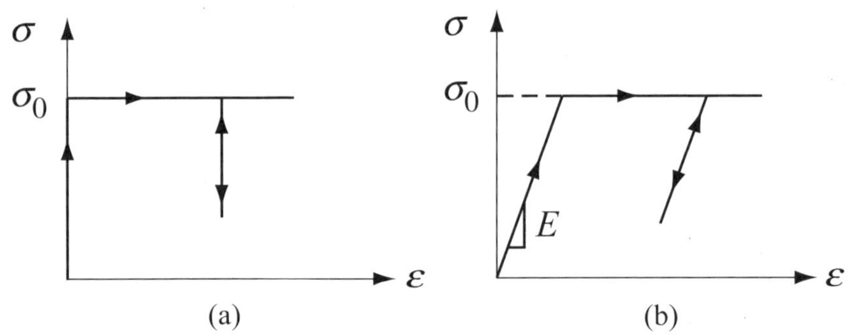

Figure 3.5 (a) Rigid perfectly plastic and (b) elastic perfectly plastic models.

Figure 3.6(a) shows a rigid linear strain hardening model. The total strain ε (= ε_p) for rigid linear strain hardening materials is

$$\varepsilon = \frac{\sigma - \sigma_0}{H} \tag{3.3.4}$$

where the tangent modulus H is defined as the slope of the stress-strain curve. Figure 3.6(b) shows an elastic linear strain hardening model. The tangent modulus H also is defined. However, the tangent modulus H involves some elastic contribution. The total strain ε for elastic linear strain hardening materials is

$$\varepsilon = \frac{\sigma_0}{E} + \frac{\sigma - \sigma_0}{H} \tag{3.3.5}$$

Linear strain hardening usually is considered for analytical modeling or for saving computational time because the stiffness matrix of the elastic-plastic finite element formulation during iterations remains unchanged for continued plastic loading.

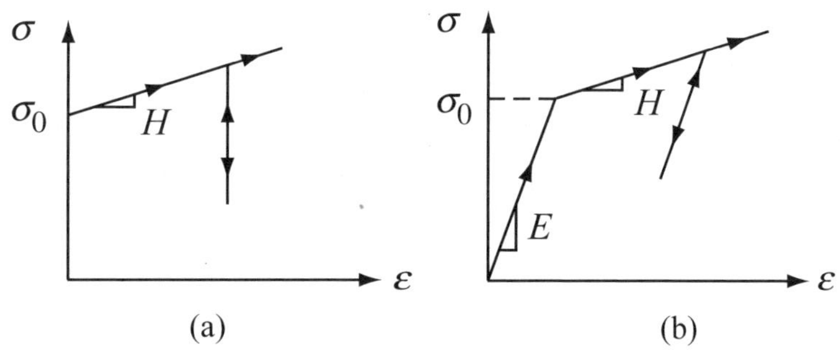

Figure 3.6 (a) Rigid linear strain hardening model, and (b) elastic linear strain hardening model.

When nonlinear strain hardening is needed, a power-law strain hardening model typically is used. Figure 3.7(a) shows a pure power-law strain hardening model. Figure 3.7(b) shows an elastic power-law strain hardening model. Figure 3.7(c) shows a Ramberg-Osgood model.

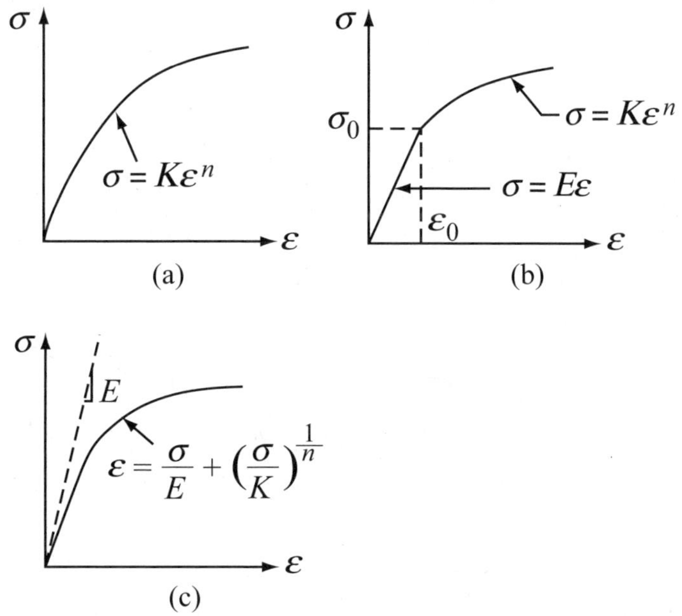

Figure 3.7 (a) Pure power-law strain hardening model, (b) elastic power-law strain hardening model, and (c) Ramberg-Osgood stress-strain model.

For pure power-law strain hardening materials, no elastic strain is considered

$$\sigma = K\varepsilon_p^n = K\varepsilon^n \tag{3.3.6}$$

or

$$\varepsilon = \varepsilon_p = \left(\frac{\sigma}{K}\right)^{\frac{1}{n}} \tag{3.3.7}$$

Here, n is the hardening exponent, and K is a material constant. For elastic power-law strain hardening materials,

$$\sigma = E\varepsilon \quad \text{for } \sigma < \sigma_0 \tag{3.3.8}$$

$$\sigma = K\varepsilon^n \quad \text{for } \sigma > \sigma_0 \tag{3.3.9}$$

or

$$\varepsilon = \frac{\sigma}{E} \quad \text{for } \varepsilon < \varepsilon_0 \tag{3.3.10}$$

$$\varepsilon = \varepsilon_e + \varepsilon_p = \left(\frac{\sigma}{K}\right)^{\frac{1}{n}} \quad \text{for } \varepsilon > \varepsilon_0 \tag{3.3.11}$$

Here,

$$\varepsilon_0 = \frac{\sigma_0}{E} \tag{3.3.12}$$

When $\sigma = \sigma_0$, then $\varepsilon = \varepsilon_0$. At this point of the stress-strain curve, both the linear and the power-law stress-strain relations must be satisfied. This leads to

$$K = \sigma_0 \left(\frac{E}{\sigma_0}\right)^n \tag{3.3.13}$$

Note that

$$\varepsilon_e = \frac{\sigma}{E} \tag{3.3.14}$$

From Eq. 3.3.11, the plastic strain ε_p can be obtained for a given σ as

$$\varepsilon_p = \left(\frac{\sigma}{K}\right)^{\frac{1}{n}} - \frac{\sigma}{E} \quad \text{for } \varepsilon > \varepsilon_0 \tag{3.3.15}$$

For Ramberg-Osgood materials, the total strain ε is the sum of the elastic strain ε_e and the plastic strain ε_p as

$$\varepsilon = \varepsilon_e + \varepsilon_p \tag{3.3.16}$$

where the elastic strain ε_e is

$$\varepsilon_e = \frac{\sigma}{E} \tag{3.3.17}$$

and the plastic strain ε_p is

$$\varepsilon_p = \left(\frac{\sigma}{K}\right)^{\frac{1}{n}} \tag{3.3.18}$$

This leads to

$$\varepsilon = \frac{\sigma}{E} + \left(\frac{\sigma}{K}\right)^{\frac{1}{n}} \tag{3.3.19}$$

Alternatively, we can write

$$\frac{\varepsilon}{\varepsilon_0} = \frac{\sigma}{\sigma_0} + \alpha\left(\frac{\sigma}{\sigma_0}\right)^{\overline{n}} \tag{3.3.20}$$

where ε_0 represents the reference strain, σ_0 represents the reference stress, α represents a material constant, and $\overline{n}$ represents the hardening exponent ($\overline{n} > 1$). Usually, we take

$$\varepsilon_0 = \frac{\sigma_0}{E} \tag{3.3.21}$$

The last model that we discuss here is the Voce law, as shown in Figure 3.8. This law usually is adopted for aluminum sheets. The stress σ is written with respect to the plastic strain ε_p as

$$\sigma = A - Be^{-C\varepsilon_p} \tag{3.3.22}$$

where A and B are material constants. As shown in Figure 3.8, the stress approaches to a saturation stress when the plastic strain becomes large. In this figure, $A - B$ represents the yield stress at $\varepsilon_p = 0$, and A represents the saturation stress when the plastic strain ε_p becomes infinity.

Finally, the tangent modulus E_t of the uniaxial stress-strain curve is a useful parameter in developing a rate-independent elastic-plastic incremental constitutive relation. The tangent modulus E_t is defined as

$$E_t = \frac{d\sigma}{d\varepsilon} \tag{3.3.23}$$

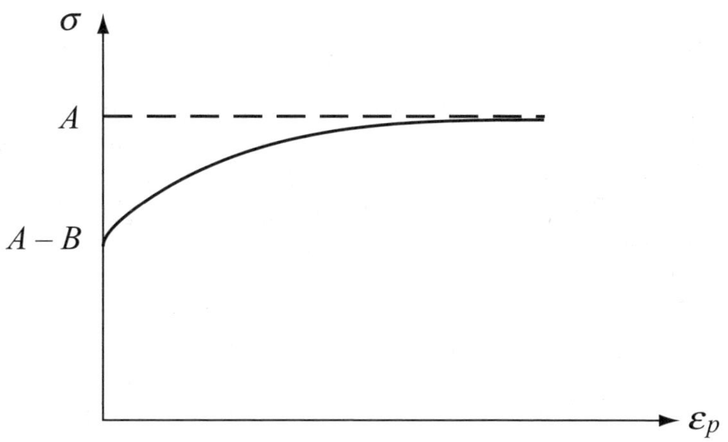

Figure 3.8 The Voce law with a saturation stress.

The tangent modulus E_p for the stress-plastic strain curve is defined as

$$E_p = \frac{d\sigma}{d\varepsilon_p} \tag{3.3.24}$$

or

$$\frac{1}{E_p} = \frac{d\varepsilon_p}{d\sigma} \tag{3.3.25}$$

Based on Eq. 3.3.23, E_p can be related to E_t by

$$\frac{1}{E_p} = \frac{1}{E_t} - \frac{1}{E} \tag{3.3.26}$$

3.4 Yield Functions Under Multiaxial Stresses

The uniaxial stress-strain relation from a tensile test gives the initial yield stress σ_0. The initial yield stress is an important material property that usually is listed in material handbooks. However, the material elements in the structural members under plastic deformation usually are not subjected to the uniaxial stress state. Instead, the material elements often are subjected to multiaxial loading conditions. Therefore, we need to determine the yield behavior under multiaxial loading conditions. First we determine the initial yield behavior for a material element under multiaixal stresses. The yield function f for a material element under multiaxial stresses can be expressed as

$$f(\boldsymbol{\sigma}) = 0 \tag{3.4.1}$$

where $\boldsymbol{\sigma}$ represents the stress tensor. In general, $\boldsymbol{\sigma}$ is the Cauchy stress tensor based on the deformed configuration because plastic flow usually depends critically on the current plastic state. For a three-dimensional material element, the Cauchy stress tensor $\boldsymbol{\sigma}$ has only six components because of symmetry. Therefore, the yield function determines the so-called yield surface in the six-dimensional hyperspace. There are certain requirements for the yield surface, based on the maximum plastic work inequality.

3.4.1 Maximum Plastic Work Inequality

Figure 3.9 schematically shows a yield surface Ω that is described by the yield function $f = 0$ in the stress space. Here, $\boldsymbol{\sigma}$ represents the stress on the yield surface, and $\boldsymbol{\sigma}^0$ represents a stress inside the yield surface. The maximum plastic work inequality [Drucker, 1951; Rice, 1970] indicates that the dyadic product of $\boldsymbol{\sigma} - \boldsymbol{\sigma}^0$ and $d\boldsymbol{\varepsilon}^p$ must be equal to or larger than 0

$$(\boldsymbol{\sigma} - \boldsymbol{\sigma}^0) : d\boldsymbol{\varepsilon}^p \geq 0 \tag{3.4.2}$$

where $d\boldsymbol{\varepsilon}^p$ is the plastic strain increment under the stress $\boldsymbol{\sigma}$.

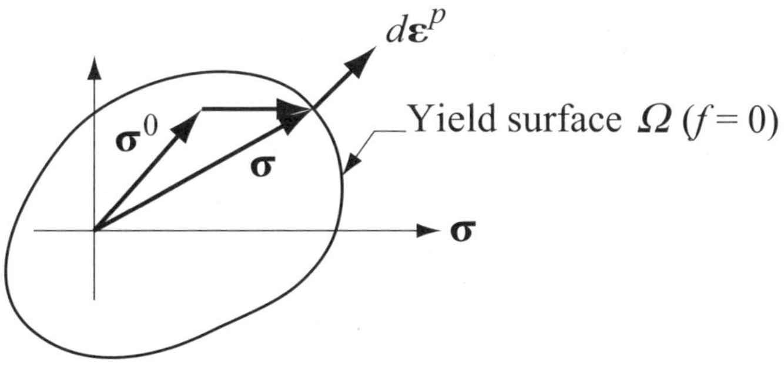

Figure 3.9 A yield surface Ω described by $f = 0$ in the stress space.

The two consequences of the maximum plastic work inequality are that the plastic flow must satisfy the normality flow rule for a smooth yield surface, and the yield surface must be convex. As shown in Figure 3.9, a smooth yield surface Ω is described by $f = 0$. For all possible cases of $\boldsymbol{\sigma}^0$, the maximum plastic work inequality in Eq. 3.4.2 must be satisfied. Therefore, the plastic strain increment $d\boldsymbol{\varepsilon}^p$ must be normal to the yield surface Ω. Consequently, the plastic strain increment $d\boldsymbol{\varepsilon}^p$ can be derived from the yield function f as

$$d\varepsilon_{ij}^p = d\lambda \frac{\partial f}{\partial \sigma_{ij}} \tag{3.4.3}$$

where $d\lambda$ represents a scalar factor of proportionality. The plastic flow described by Eq. 3.4.3 gives the so-called normality flow or the associated flow rule. When the yield function has a corner or vertex, the outward normal to the yield surface cannot be defined. For example, the yield surface has vertices when different crystalline slip systems are considered for single crystals. As shown in Figure 3.10, if we have a non-convex yield surface, then the maximum plastic work inequality in Eq. 3.4.2 can be violated. Therefore, the yield surface must be convex.

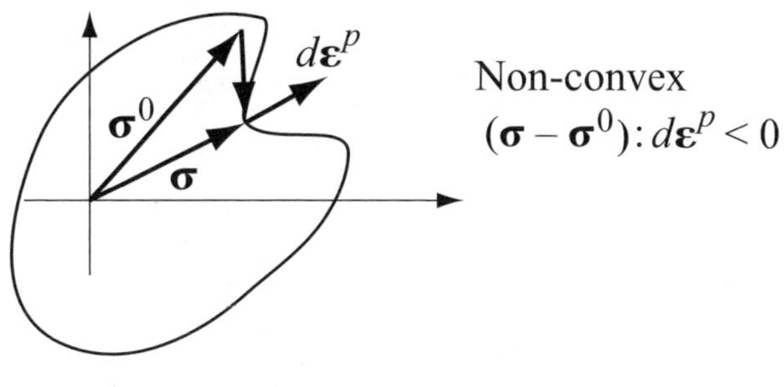

Figure 3.10 A non-convex yield surface.

3.4.2 Yield Functions for Isotropic Materials

For isotropic materials, we can always find three principal stresses σ_1, σ_2, and σ_3 for a given stress tensor $\boldsymbol{\sigma}$. Therefore, the yield function f can be expressed in terms of these three principal stresses as

$$f(\boldsymbol{\sigma}) = f(\sigma_1, \sigma_2, \sigma_3) = 0 \tag{3.4.4}$$

Because the principal stresses σ_1, σ_2, and σ_3 can be obtained from the three invariants I_1, I_2, and and I_3 of the stress tensor $\boldsymbol{\sigma}$, the yield function f can be expressed as

$$f(\boldsymbol{\sigma}) = f(I_1, I_2, I_3) = 0 \tag{3.4.5}$$

where the three invariants I_1, I_2, and and I_3 are the coefficients of the characteristic equation

$$-\sigma_p^3 + I_1\sigma_p^2 - I_2\sigma_p + I_3 = 0 \tag{3.4.6}$$

Here, the three invariants I_1, I_2, and and I_3 are

$$I_1 = \sigma_{kk} = \sigma_{11} + \sigma_{22} + \sigma_{33} \tag{3.4.7}$$

$$I_2 = \frac{1}{2}\left(\sigma_{ii}\sigma_{jj} - \sigma_{ij}\sigma_{ij}\right) \tag{3.4.8}$$

$$I_3 = \det|\boldsymbol{\sigma}| \tag{3.4.9}$$

The three invariants I_1, I_2, and and I_3 can be expressed in terms of the three principal stresses σ_1, σ_2, and σ_3 as

$$I_1 = \sigma_1 + \sigma_2 + \sigma_3 \tag{3.4.10}$$

$$I_2 = \sigma_1\sigma_2 + \sigma_2\sigma_3 + \sigma_3\sigma_1 \tag{3.4.11}$$

$$I_3 = \sigma_1\sigma_2\sigma_3 \tag{3.4.12}$$

In dense metals at low temperature, plastic deformation comes mainly from dislocation motion or plastic slip. The dislocation motion or plastic slip is not significantly affected by the hydrostatic tension or pressure. Therefore, within the context of classical plasticity theories, yielding is assumed not to be affected by any amount of hydrostatic stress or pressure. Therefore, the yield function can be expressed in terms of two invariants I_2 and I_3 as

$$f(I_2, I_3) = 0 \tag{3.4.13}$$

Now, we consider the deviatoric stress tensor s_{ij}, which contains no contribution from the hydrostatic tension or pressure

$$s_{ij} = \sigma_{ij} - \frac{1}{3}\sigma_{kk}\delta_{ij} \tag{3.4.14}$$

The stress invariants I_1', I_2', and I_3' for the deviatoric stress tensor s_{ij} are

$$I_1' = trace(\mathbf{s}) = s_{kk} = 0 \tag{3.4.15}$$

$$I_2' = \frac{1}{2}\left(s_{ii}s_{jj} - s_{ij}s_{ij}\right) = -\frac{1}{2}s_{ij}s_{ij} \tag{3.4.16}$$

$$I_3' = \det|\mathbf{s}| = \frac{1}{3}s_{ij}s_{jk}s_{kl} \tag{3.4.17}$$

Then, the yield function can be written in terms of the two invariants I_2' and I_3' as

$$f\left(I_2', I_3'\right) = 0 \tag{3.4.18}$$

3.4.2.1 von Mises Yield Condition

Now consider the stress state that first satisfies the initial yield condition where the plastic yielding starts. The von Mises yield condition is expressed in terms of the second invariant of the deviatoric stress tensor σ_{ij}'. The second invariant J_2 is now expressed as

$$J_2 = \frac{1}{2}s_{ij}s_{ij} \tag{3.4.19}$$

When the value of the J_2 reaches a critical value τ_0^2, the initial yield condition is satisfied.

$$J_2 = \frac{1}{2}s_{ij}s_{ij} = \tau_0^2 \tag{3.4.20}$$

where τ_0 can be identified as the shear yield stress. The von Mises yield condition can be written as

$$f = \frac{1}{2}s_{ij}s_{ij} - \tau_0^2 = 0 \tag{3.4.21}$$

The yield condition also can be written in the component form as

$$f = \frac{1}{6}\left[\left(\sigma_{11} - \sigma_{22}\right)^2 + \left(\sigma_{22} - \sigma_{33}\right)^2 + \left(\sigma_{33} - \sigma_{11}\right)^2 + 6\left(\sigma_{12}^2 + \sigma_{23}^2 + \sigma_{31}^2\right)\right] - \tau_0^2 = 0 \tag{3.4.22}$$

The yield condition can be expressed in terms of the principal stresses σ_1, σ_2, and σ_3 as

$$f = \frac{1}{6}\left[\left(\sigma_1 - \sigma_2\right)^2 + \left(\sigma_2 - \sigma_3\right)^2 + \left(\sigma_3 - \sigma_1\right)^2\right] - \tau_0^2 = 0 \tag{3.4.23}$$

Because uniaixial tensile tests usually are performed to obtain the stress-strain curve of the material, the tensile yield stress σ_0 can be related to the shear yield stress τ_0 through the yield condition in Eq. 3.4.23 as

$$\sigma_0 = \sqrt{3}\tau_0 \tag{3.4.24}$$

Therefore, the von Mises yield condition can be written in terms of the tensile yield stress σ_0 as

$$f = \frac{3}{2}s_{ij}s_{ij} - \sigma_0^2 = 0 \tag{3.4.25}$$

The yield condition can be written in the component form as

$$f = \frac{1}{2}\left[\left(\sigma_{11} - \sigma_{22}\right)^2 + \left(\sigma_{22} - \sigma_{33}\right)^2 + \left(\sigma_{33} - \sigma_{11}\right)^2 + 6\left(\sigma_{12}^2 + \sigma_{23}^2 + \sigma_{31}^2\right)\right] - \sigma_0^2 = 0 \tag{3.4.26}$$

The yield condition can be expressed in terms of the principal stresses σ_1, σ_2, and σ_3 as

$$f = \frac{1}{2}\left[\left(\sigma_1 - \sigma_2\right)^2 + \left(\sigma_2 - \sigma_3\right)^2 + \left(\sigma_3 - \sigma_1\right)^2\right] - \sigma_0^2 = 0 \tag{3.4.27}$$

Note that the yield surface based on the von Mises yield condition represents a circular cylinder with the axis in the direction of $\sigma_1 = \sigma_2 = \sigma_3$ in the three-dimensional principal stress space.

3.4.2.2 Tresca Yield Condition

The Tresca yield condition is a maximum shear stress criterion. The maximum shear stress τ_{max} for a given stress tensor can be obtained conveniently in terms of the maximum difference of the three principal stresses σ_1, σ_2, and σ_3 as

$$\tau_{max} = \frac{1}{2}\max\left(\left|\sigma_1 - \sigma_2\right|, \left|\sigma_2 - \sigma_3\right|, \left|\sigma_3 - \sigma_1\right|\right) = \tau_0 \quad (3.4.28)$$

When the maximum shear stress τ_{max} reaches the shear yield stress τ_0, the Tresca yield condition is satisfied. Therefore, the Tresca yield condition is written as

$$\max\left(\left|\sigma_1 - \sigma_2\right|, \left|\sigma_2 - \sigma_3\right|, \left|\sigma_3 - \sigma_1\right|\right) = 2\tau_0 \quad (3.4.29)$$

Because uniaixial tensile tests usually are performed to obtain the stress-strain curve of the material, the tensile yield stress σ_0 can be related to the shear yield stress τ_0 through the yield condition in Eq. 3.4.29 as

$$\sigma_0 = 2\tau_0 \quad (3.4.30)$$

Therefore, the Tresca yield condition can be written as

$$\max\left(\left|\sigma_1 - \sigma_2\right|, \left|\sigma_2 - \sigma_3\right|, \left|\sigma_3 - \sigma_1\right|\right) = \sigma_0 \quad (3.4.31)$$

Note that the Tresca yield condition is based on the difference of the principal stresses, which is not affected by any superposed hydrostatic stress. Therefore, the Tresca yield condition is not dependent on the hydrostatic stress as is the von Mises yield condition. Note that the yield surface based on the Tresca yield condition represents a hexagonal prism with the axis in the direction of $\sigma_1 = \sigma_2 = \sigma_3$ in the three-dimensional principal stress space.

3.4.2.3 Plane Stress Yield Conditions for Isotropic Materials

When sheet metal is under forming operations, due to the relatively small dimension in the thickness direction, the sheet metal usually is assumed under plane stress conditions. In Figure 3.11, a sheet element is considered under plane stress conditions. Usually, we use x_1 and x_2 to represent the in-plane coordinates, and x_3 to represent the out-of-plane coordinate. Under plane stress conditions, the out-of-plane normal stress σ_{33} and the out-of-plane shear stresses σ_{13} and σ_{23} are assumed to be zero. The von Mises yield condition in Eq. 3.4.26 is reduced to

$$f = \sigma_{11}^2 + \sigma_{22}^2 - \sigma_{11}\sigma_{22} + 3\sigma_{12}^2 - \sigma_0^2 = 0 \quad (3.4.32)$$

However, there are still three stresses in the yield condition. To visualize the yield condition in a two-dimensional plot, the yield condition is now expressed in terms of the two in-plane principal stresses σ_1 and σ_2 as

$$f = \sigma_1^2 + \sigma_2^2 - \sigma_1\sigma_2 - \sigma_0^2 = 0 \quad (3.4.33)$$

The yield condition can be plotted in terms of σ_1 and σ_2 as an ellipse in Figure 3.12. The ellipse represents the intercept of the yield surface of a circular cylinder shape with the plane described by $x_3 = 0$.

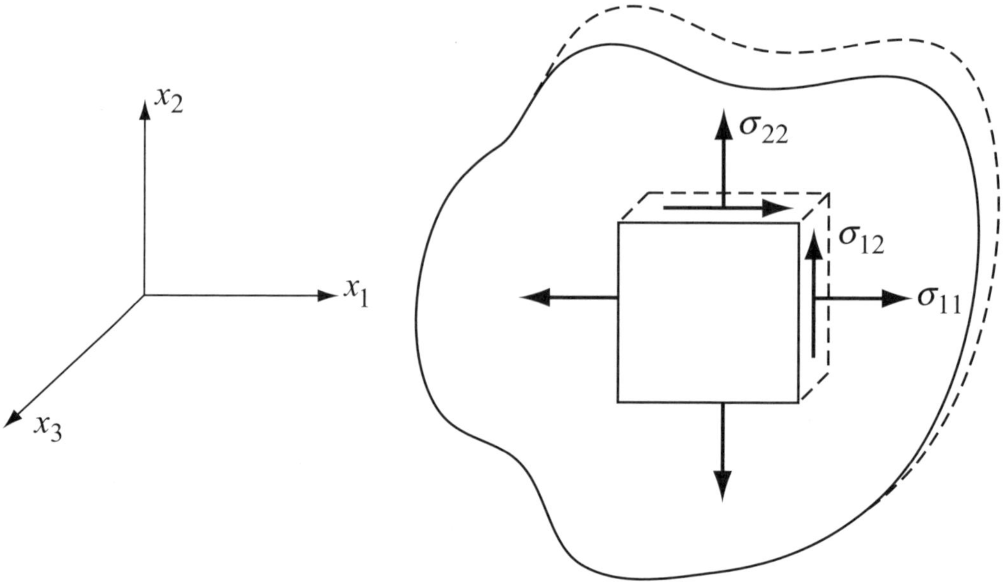

Figure 3.11 A sheet element under plane stress conditions.

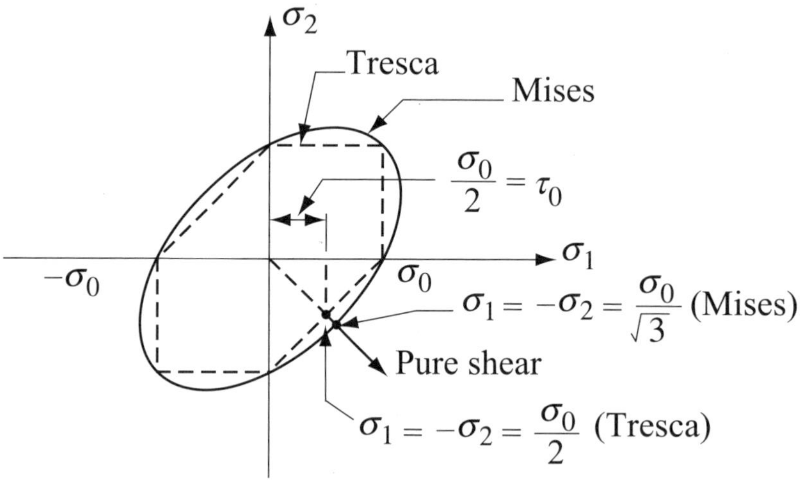

Figure 3.12 The von Mises and Tresca yield conditions.

Now we consider the Tresca yield condition in terms of the in-plane principal stresses σ_1 and σ_2 as

$$\max\left(\left|\sigma_1 - \sigma_2\right|, \left|\sigma_2\right|, \left|\sigma_1\right|\right) = \sigma_0 \tag{3.4.34}$$

For all possible combinations of the in-plane principal stresses σ_1 and σ_2, Eq. 3.4.34 becomes

$$\left|\sigma_1 - \sigma_2\right| = \sigma_0 \tag{3.4.35}$$

$$|\sigma_2| = \sigma_0 \tag{3.4.36}$$

$$|\sigma_1| = \sigma_0 \tag{3.4.37}$$

Equations 3.4.35, 3.4.36, and 3.4.37 represent six straight lines in the $\sigma_1 - \sigma_2$ plane. Because we consider that the material is loaded from the stress state $\sigma_1 = \sigma_2 = 0$, the Tresca yield condition is now represented by a hexagon as the yield contour, as shown in Figure 3.12. The hexagon represents the intercept of the yield surface of a hexagonal prism shape with the plane described by $x_3 = 0$.

Consider a material element that is under the monotonically increasing pure shear stress state. The stress state can be represented by the principal stresses σ_1 and σ_2 with

$$\sigma_1 = -\sigma_2 \tag{3.4.38}$$

As shown in Figure 3.12, when the stress state starts from $\sigma_1 = \sigma_2 = 0$ under the condition given by Eq. 3.4.38 with $\sigma_1 > 0$, the Tresca yield condition will be satisfied first. As shown in Figure 3.12, the location of the Tresca yield condition is

$$\sigma_1 = -\sigma_2 = \frac{\sigma_0}{2} \tag{3.4.39}$$

As the load continues to increase, the von Mises yield condition will then be satisfied. The location of the von Mises yield condition is

$$\sigma_1 = -\sigma_2 = \frac{\sigma_0}{\sqrt{3}} \tag{3.4.40}$$

As shown in Figure 3.12, the Tresca yield contour is smaller than the von Mises yield contour. Therefore, when the Tresca yield condition is used to determine the yield behavior, the prediction usually is conservative.

3.4.3 Yield Functions for Anisotropic Materials

Sheet metals usually display a certain extent of plastic anisotropy due to rolling processes. For simplicity, we usually model sheet metals as orthotropic materials. The additional plastic anisotropy developed due to large plastic strains during forming typically is too complex to be included in the practical application of sheet forming simulations. Many anisotropic yield functions are available; for example, see Hill [1948], Hershey [1954], Hosford [1972], Hill [1979], Logan and Hosford [1980], Karafillis and Boyce [1993], Barlat *et al.* [1997], Barlat *et al.* [2003] and Barlat [2005]. However, we concentrate on discussions of the general anisotropic yield functions that can be implemented into finite element codes for sheet metal forming simulations. First we discuss Hill's quadratic yield function, which is the most widely used yield function to model sheet metal forming. Then, we discuss the high-order yield function of Barlat *et al.* [2003], which is less complex than the one proposed by Barlat *et al.* [2005].

3.4.3.1 Hill Quadratic Yield Condition for Orthotropic Materials

To characterize the orthotropy of sheet metals, symmetry planes first must be defined. We consider an orthotropic sheet metal element with respect to the Cartesian coordinates x_1, x_2, and x_3, which coincide with the orthotropic symmetry axes X_1, X_2, and X_3, as shown in Figure 3.13.

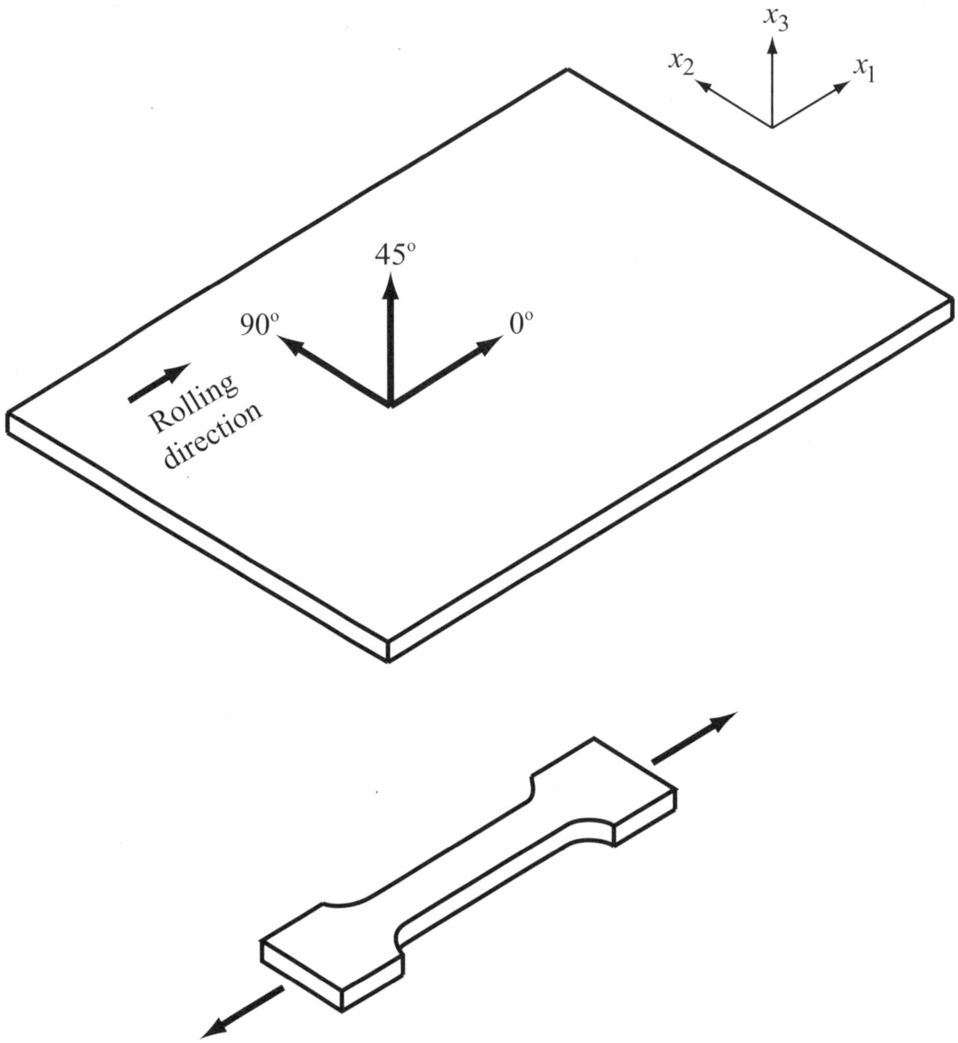

Figure 3.13 A sheet metal element as an orthotropic material with the Cartesian axes x_1, x_2, and x_3 coincident with the symmetry axes X_1, X_2, and X_3.

For orthotropic materials, Hill proposed a quadratic yield condition [Hill, 1948] as

$$\begin{aligned} f &= F(\sigma_{22}-\sigma_{33})^2 + G(\sigma_{33}-\sigma_{11})^2 + H(\sigma_{11}-\sigma_{22})^2 \\ &\quad + 2L\sigma_{23}^2 + 2M\sigma_{31}^2 + 2N\sigma_{12}^2 - \bar{\sigma}^2 = 0 \end{aligned} \tag{3.4.41}$$

where $\bar{\sigma}$ represents a reference yield stress. Here, F, G, H, L, M, and N are material constants that can be determined, in principle, by tensile and shear tests with respect to different orientations. Note that when the chosen Cartesian coordinates x_1, x_2, and x_3 do not coincide with the symmetry axes X_1, X_2, and X_3, the yield condition will be much more complex due to the transformation of stresses to the chosen coordinate system.

In the sheet metal forming industry, an anisotropy parameter R is defined as the ratio of the transverse plastic strain rate to the thickness plastic strain rate from uniaxial tensile tests of sheet specimens. Here, $\bar{\sigma}$ usually is taken as the yield stress in the rolling (X_1) direction. We can express the material constants F, G, H, L, M, and N in terms of the anisotropy parameters R_0, R_{45}, and R_{90} that represent the values of R when the tensile axis is at 0°, 45°, and 90° from the rolling (X_1 or x_1) direction, respectively, as shown in Figure 3.13. For simplicity, the material constants L, M, and N related to the shear responses of σ_{23}, σ_{31}, and σ_{12} are taken to be identical according to the computational results of the anisotropic plastic behavior of a body-centered cubic (bcc) sheet metal after plane strain compression using a polycrystal model as reported in Liao *et al.* [1998]. Another assumption also is taken to obtain the values of the material constants L and M in the yield condition. Appropriate material constants associated with σ_{23} and σ_{31} can be assigned when experimental data are available.

Based on the associated flow rule, the plastic strain rates $\dot{\varepsilon}^p_{11}$, $\dot{\varepsilon}^p_{22}$, $\dot{\varepsilon}^p_{33}$, $\dot{\varepsilon}^p_{23}$, $\dot{\varepsilon}^p_{31}$, and $\dot{\varepsilon}^p_{12}$ can be obtained as

$$\dot{\varepsilon}^p_{11} = \dot{\lambda}\frac{\partial f}{\partial \sigma_{11}} = \dot{\lambda}\left[-2G\left(\sigma_{33} - \sigma_{11}\right) + 2H\left(\sigma_{11} - \sigma_{22}\right)\right] \tag{3.4.42}$$

$$\dot{\varepsilon}^p_{22} = \dot{\lambda}\frac{\partial f}{\partial \sigma_{22}} = \dot{\lambda}\left[2F\left(\sigma_{22} - \sigma_{33}\right) - 2H\left(\sigma_{11} - \sigma_{22}\right)\right] \tag{3.4.43}$$

$$\dot{\varepsilon}^p_{33} = \dot{\lambda}\frac{\partial f}{\partial \sigma_{33}} = \dot{\lambda}\left[-2F\left(\sigma_{22} - \sigma_{33}\right) + 2G\left(\sigma_{33} - \sigma_{11}\right)\right] \tag{3.4.44}$$

$$2\dot{\varepsilon}^p_{23} = \dot{\lambda}\frac{\partial f}{\partial \sigma_{23}} = \dot{\lambda}\left[4L\sigma_{23}\right] \tag{3.4.45}$$

$$2\dot{\varepsilon}^p_{31} = \dot{\lambda}\frac{\partial f}{\partial \sigma_{31}} = \dot{\lambda}\left[4M\sigma_{31}\right] \tag{3.4.46}$$

$$2\dot{\varepsilon}^p_{12} = \dot{\lambda}\frac{\partial f}{\partial \sigma_{12}} = \dot{\lambda}\left[4N\sigma_{12}\right] \tag{3.4.47}$$

where $\dot{\lambda}$ is a scalar factor of proportionality.

First we consider a uniaxial loading in the X_1 direction, which is at 0° from the rolling direction. In this case, $\sigma_{11} \neq 0$ and $\sigma_{22} = \sigma_{33} = \sigma_{23} = \sigma_{31} = \sigma_{12} = 0$, R_0 can be obtained as

$$R_0 = \frac{\dot{\varepsilon}^p_{22}}{\dot{\varepsilon}^p_{33}} = \frac{H}{G} \tag{3.4.48}$$

Then we consider a uniaxial loading in the X_2 direction, which is at 90° from the rolling direction. In this case, $\sigma_{22} \neq 0$ and $\sigma_{11} = \sigma_{33} = \sigma_{23} = \sigma_{31} = \sigma_{12} = 0$, R_{90} can be obtained as

$$R_{90} = \frac{\dot{\varepsilon}_{11}^{p}}{\dot{\varepsilon}_{33}^{p}} = \frac{H}{F} \tag{3.4.49}$$

Consider that the coordinate system rotates 45° counterclockwise with respect to the X_3 axis to a new coordinate system of x_1', x_2', and x_3'. Now we consider a uniaxial loading in the x_1' direction. In this case, the stress transformation gives the stress components with respect to the X_1, X_2, and X_3 coordinates as $\sigma_{11} = \sigma_{22} = \sigma_{12} = \sigma/2$ and $\sigma_{33} = \sigma_{31} = \sigma_{23} = 0$. Here, σ represents the tensile stress. The plastic strain rates can be obtained from Eqs. 3.4.42, 3.4.43, and 3.4.47 as

$$\varepsilon_{11}^{p} = \dot{\lambda} G \sigma \tag{3.4.50}$$

$$\dot{\varepsilon}_{22}^{p} = \dot{\lambda} F \sigma \tag{3.4.51}$$

$$\dot{\varepsilon}_{12}^{p} = \dot{\lambda} N \sigma \tag{3.4.52}$$

Based on the strain transformation, the transverse plastic strain rate $\dot{\varepsilon}_{22}'^{p}$ with respect to the coordinate system of x_1', x_2', and x_3' is

$$\dot{\varepsilon}_{22}'^{p} = \dot{\lambda} \frac{1}{2} \left[G\sigma + F\sigma - 2N\sigma \right] \tag{3.4.53}$$

Because $\dot{\varepsilon}_{33}'^{p} = \dot{\varepsilon}_{33}^{p}$, we can write the through-thickness plastic strain rate $\dot{\varepsilon}_{33}'^{p}$ with respect to the coordinate system of x_1', x_2', and x_3' as

$$\dot{\varepsilon}_{33}'^{p} = \dot{\lambda} \left[-F\sigma - G\sigma \right] \tag{3.4.54}$$

Therefore, R_{45} can be written as

$$R_{45} = \frac{\dot{\varepsilon}_{22}'^{p}}{\dot{\varepsilon}_{33}'^{p}} = \frac{2N - (F + G)}{2(F + G)} \tag{3.4.55}$$

As mentioned, we take σ_0 in Eq. 3.4.41 as the yield stress in the X_1 direction. Then,

$$G + H = 1 \tag{3.4.56}$$

We can solve for F, G, H, and N using Eqs. 3.4.46, 3.4.47, 3.4.53, and 3.4.54 as

$$F = \frac{R_0}{R_{90}\left(1 + R_0\right)} \tag{3.4.57}$$

$$G = \frac{1}{1 + R_0} \tag{3.4.58}$$

$$H = \frac{R_0}{1 + R_0} \tag{3.4.59}$$

$$N = \frac{\left(2R_{45} + 1\right)\left(R_0 + R_{90}\right)}{2R_{90}\left(1 + R_0\right)} \tag{3.4.60}$$

Substituting Eqs. 3.4.55 through 3.4.58 into Eq. 3.4.41 and assuming $L = M = N$ gives the yield condition as

$$\begin{aligned} f = \frac{1}{1 + R_0}\Bigg[& \frac{R_0}{R_{90}}\left(\sigma_{22} - \sigma_{33}\right)^2 + \left(\sigma_{33} - \sigma_{11}\right)^2 + R_0\left(\sigma_{11} - \sigma_{22}\right)^2 \\ & + \frac{\left(2R_{45} + 1\right)\left(R_0 + R_{90}\right)}{R_{90}}\left(\sigma_{23}^2 + \sigma_{31}^2 + \sigma_{12}^2\right)\Bigg] - \sigma_0^2 = 0 \end{aligned} \tag{3.4.61}$$

Again we consider a uniaxial loading in the X_1 direction, which is at 0° from the rolling direction. In this case, $\sigma_{11} \neq 0$ and $\sigma_{22} = \sigma_{33} = \sigma_{23} = \sigma_{31} = \sigma_{12} = 0$. If we denote the yield stress in the X_1 direction as X, the yield function can be written as

$$\begin{aligned} f = \frac{1}{1 + R_0}\Bigg[& \frac{R_0}{R_{90}}\left(\sigma_{22} - \sigma_{33}\right)^2 + \left(\sigma_{33} - \sigma_{11}\right)^2 + R_0\left(\sigma_{11} - \sigma_{22}\right)^2 \\ & + \frac{\left(2R_{45} + 1\right)\left(R_0 + R_{90}\right)}{R_{90}}\left(\sigma_{23}^2 + \sigma_{31}^2 + \sigma_{12}^2\right)\Bigg] - X^2 = 0 \end{aligned} \tag{3.4.62}$$

For sheet metal forming applications, the yield condition under plane stress conditions usually is needed. Plane stress conditions for thin sheets require

$$\sigma_{33} = \sigma_{31} = \sigma_{32} = 0 \tag{3.4.63}$$

Under plane stress conditions, Eq. 3.4.62 reduces to

$$f = \frac{1}{1+R_0}\left[\frac{R_0}{R_{90}}\sigma_{22}^2 + \sigma_{11}^2 + R_0\left(\sigma_{11} - \sigma_{22}\right)^2 + \frac{\left(2R_{45}+1\right)\left(R_0 + R_{90}\right)}{R_{90}}\sigma_{12}^2\right] - X^2 = 0 \tag{3.4.64}$$

When the planar isotropy is considered, $R_0 = R_{45} = R_{90} = R$. Here, R can be considered as the normal anisotropy parameter. Here, X represents the uniaxial tensile yield stress for all in-plane directions. The yield criterion becomes

$$f = \sigma_{11}^2 + \sigma_{22}^2 - \frac{2R}{1+R}\sigma_{11}\sigma_{22} + \frac{2(1+2R)}{1+R}\sigma_{12}^2 - X^2 = 0 \tag{3.4.65}$$

In general, sheet metal usually has planar anisotropy, such that $R_0 \neq R_{45} \neq R_{90}$. To take advantage of the planar isotropy or the transverse anisotropy, the weighted average $\overline{R}$ of R_0, R_{45}, and R_{90} usually is used for the value of R for the normal anisotropy parameter as

$$\overline{R} = \frac{R_0 + 2R_{45} + R_{90}}{4} \tag{3.4.66}$$

When $R = 1$, the yield condition reduces to the von Mises yield condition under plane stress conditions as

$$f = \sigma_{11}^2 + \sigma_{22}^2 - \sigma_{11}\sigma_{22} + 3\sigma_{12}^2 - X^2 = 0 \tag{3.4.67}$$

For materials with normal anisotropy, consider the special cases where the X_1 and X_2 coordinates coincide with the principal stress directions. The yield condition becomes

$$f = \sigma_{11}^2 + \sigma_{22}^2 - \frac{2R}{1+R}\sigma_{11}\sigma_{22} - X^2 = 0 \tag{3.4.68}$$

Rearranging the equation gives the yield condition as

$$f = \sigma_{11}^2 + \sigma_{22}^2 + R\left(\sigma_{11} - \sigma_{22}\right)^2 - (1+R)X^2 = 0 \tag{3.4.69}$$

Logan and Hosford [1980] took a higher exponent m larger than 2 in the yield condition based on their upper bound yield surfaces. The yield condition in Eq. 3.4.69 becomes

$$f = \left|\sigma_{11}\right|^m + \left|\sigma_{22}\right|^m + R\left|\sigma_{11} - \sigma_{22}\right|^m - (1+R)X^m = 0 \tag{3.4.70}$$

Logan and Hosford [1980] suggested $m = 6$ for body-centered cubic (bcc) polycrystals and $m = 8$ for face-centered cubic (fcc) polycrystals.

3.4.3.2 A General Plane Stress Anisotropic Yield Condition

Many attempts have been made to characterize the anisotropic plastic behavior of sheet metals. A review of these attempts can be found in Barlat *et al.* [2003]. A general anisotropic yield criterion was proposed by Barlat *et al.* [1997] where a higher exponent for stress parameters can be taken for aluminum alloys. However, the yield condition appears to have some numerical difficulties when implemented in three-dimensional finite element simulations of sheet forming processes. Also, the convexity of the yield condition was not proved. Recently, Barlat *et al.* [2003] proposed a general anisotropic yield condition based on the combination of two linear transformations of two isotropic yield functions. The convexity of the yield condition can be proved. Here, we summarize the plane stress version of the yield condition, the Yld2000-2d condition, proposed by Barlat *et al.* [2003]. The plane stress yield condition can be fitted by eight measurable anisotropic material properties.

Under plane stress conditions, only three in-plane stresses σ_{11}, σ_{22}, and σ_{12} are needed to be considered. The Yld2000-2d anisotropic yield criterion starts with the deviatoric stresses. The deviatoric stresses s_{11}, s_{22}, and s_{12} can be obtained from the in-plane stresses σ_{11}, σ_{22}, and σ_{12} as

$$\begin{bmatrix} s_{11} \\ s_{22} \\ s_{12} \end{bmatrix} = \begin{bmatrix} 2/3 & -1/3 & 0 \\ -1/3 & 2/3 & 0 \\ 0 & 0 & 1 \end{bmatrix} \begin{bmatrix} \sigma_{11} \\ \sigma_{22} \\ \sigma_{12} \end{bmatrix} \tag{3.4.71}$$

The yield function Φ can be obtained from

$$\Phi = \Phi' + \Phi'' = 2\bar{\sigma}^a \tag{3.4.72}$$

where a represents the exponent that is a material constant, and $\bar{\sigma}$ represents a reference yield stress. Here,

$$\Phi' = |X_1' - X_2'|^a \tag{3.4.73}$$

$$\Phi'' = |2X_2'' + X_1''|^a + |2X_1'' + X_2''|^a \tag{3.4.74}$$

Here, X_i' and X_i'' $(i = 1, 2)$, are the principal values of two linear transformations of the three plane stress deviatoric stresses s_{11}, s_{22}, and s_{12}. The two linear transformations are

$$\begin{bmatrix} X_{11}' \\ X_{22}' \\ X_{12}' \end{bmatrix} = \begin{bmatrix} C_{11}' & C_{12}' & 0 \\ C_{21}' & C_{22}' & 0 \\ 0 & 0 & C_{66}' \end{bmatrix} \begin{bmatrix} s_{11} \\ s_{22} \\ s_{12} \end{bmatrix} \tag{3.4.75}$$

$$\begin{bmatrix} X''_{11} \\ X''_{22} \\ X''_{12} \end{bmatrix} = \begin{bmatrix} C''_{11} & C''_{12} & 0 \\ C''_{21} & C''_{22} & 0 \\ 0 & 0 & C''_{66} \end{bmatrix} \begin{bmatrix} s_{11} \\ s_{22} \\ s_{12} \end{bmatrix} \tag{3.4.76}$$

where C'_{ij} and C''_{ij} are material constants. Here, $\bar{\sigma}$ is taken as the yield stress in the rolling direction. Because Φ' depends on $X'_1 - X'_2$, only three C'_{ij} are independent. $C'_{12} = C'_{21} = 0$ can be imposed. Therefore, eight independent material constants are needed in the Yld2000-2d yield condition. Three yield stresses σ_0, σ_{45}, and σ_{90}, and three anisotropy parameters R_0, R_{45}, and R_{90} obtained from tensile tests at 0°, 45°, and 90° with respect to the rolling direction are used. The balanced biaxial yield stress σ_b under balanced biaxial stress tests also is used. The strain ratio $R_b = \dot{\varepsilon}_{yy}/\dot{\varepsilon}_{xx}$ under balanced biaxial stress tests also is used as an input parameter. We can calculate the material constants C'_{11}, C'_{22}, C'_{66}, C''_{11}, C''_{12}, C''_{21}, C''_{22}, and C''_{66} from σ_0, σ_{45}, σ_{90}, σ_b, R_0, R_{45}, R_{90}, and R_b. Figure 3.14 shows a comparison of the yield contours based on the Yld2000-2d yield condition and Hill's quadratic yield condition for aluminum alloy 6016-T4 sheets. The experimental results also are shown (Barlat *et al.* [2003]). In this figure, the Yld2000-2d yield condition can be used to model the yield behavior well. The yield contour of the Yld2000-2d yield condition gives a much high curvature near the balanced biaxial yield stress state. As shown in Figure 3.14, Hill's quadratic yield condition cannot be used to model the yield behavior near the balanced biaxial stress state. A more general yield function, denoted Yld2004, recently was proposed by Barlat *et al.* [2005].

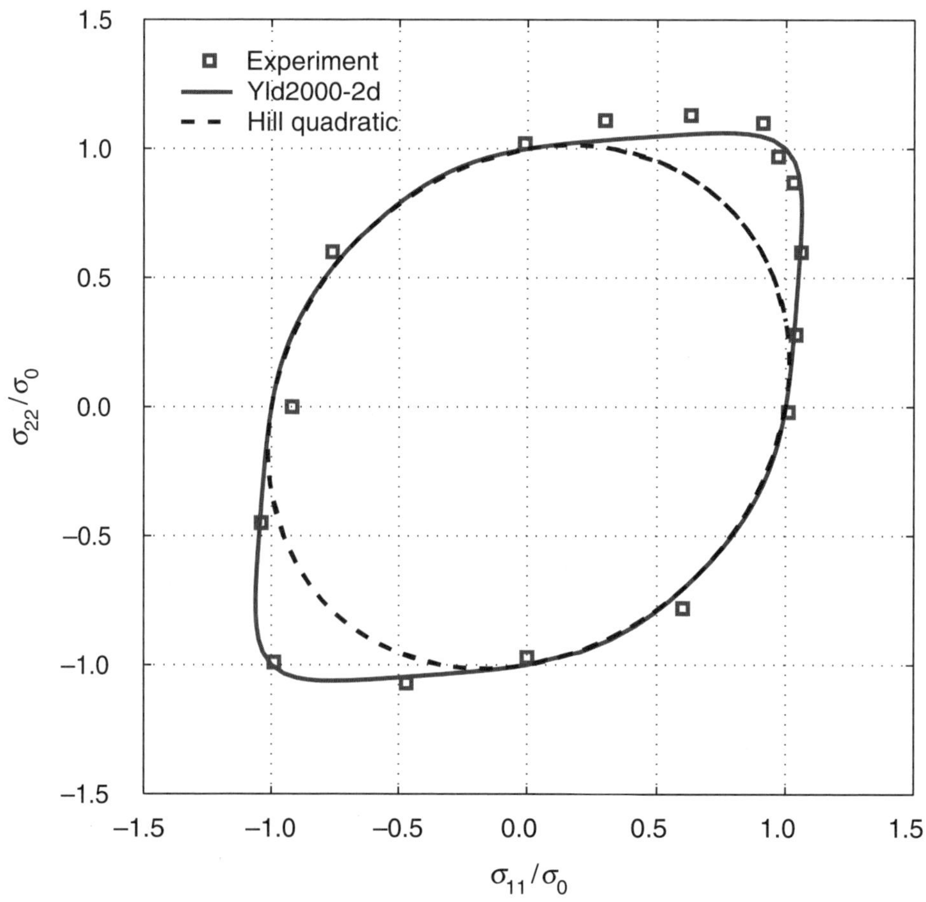

Figure 3.14 A comparison of the yield contours based on the Yld2000-2d yield condition and Hill's quadratic yield condition for aluminum 6016-T4 sheets.

3.5 Evolution of Yield Surface

Different hardening rules were proposed in the past to specify the evolution of the yield surface during plastic deformation. In general, these hardening rules allow the yield surface to expand (isotropic hardening), contract (isotropic softening), translate (kinematic hardening), and change shape in the stress space. The movement of the yield surface can be, for example, in the direction of the plastic strain rate [Prager, 1955] or in the direction of the vector connecting the center of the yield surface to the current loading point [Ziegler, 1959]. Under monotonic increasing nearly proportional loading conditions, the isotropic hardening rule appears to give a good approximation of the material plastic hardening behavior. On the other hand, the kinematic hardening rule was proposed to model Bauschinger's effect (where the yield stress is reduced under reversed plastic loading conditions), which cannot be modeled by the isotropic hardening rule. However, substantial deviations of actual plastic behavior from those predicted by both isotropic and kinematic hardening rules often are observed under complex plastic deformation histories. Different plasticity theories were proposed to model the cyclic hardening behavior; for example, see Armstrong and Frederick [1966], Mroz [1967], Dafalias and Popov [1976], and Chaboche [1977]. In this book, we will concentrate on the hardening modeling of Chu [1984], which is a variation of the hardening model proposed by Mroz [1967].

Based on the cyclic plastic behavior of metals observed in experiments, Mroz [1967] introduced the concept of a field of work hardening moduli corresponding to an infinite number of initially concentric hypersurfaces in the deviatoric stress space. In his model, the translation of each surface does not intersect each other but these surfaces are consecutively in contact with each other. Because each of these surfaces is given an equivalent work hardening modulus, a unique stress-strain relationship can be determined. Due to the computational complexity, Mroz's model was modified by Krieg [1975] and Dafalias and Popov [1975], where all surfaces in Mroz's model were replaced by an inner surface and an outer surface with a continuum of intermediate surfaces that are analytically determined. In this section, we will introduce the general concepts of isotropic hardening, kinematic hardening, and the Mroz hardening rules. Chu [1984] proposed an anisotropic hardening rule for incompressible von Mises materials, based on the concept of Mroz [1967]. Recently, Choi [2007] proposed an anisotropic hardening rule for both pressure-insensitive and pressure-sensitive materials with associated and non-associated flow rules based on the concept of Mroz [1967]. These rules are quite convenient for implementation into finite element codes. Chu's anisotropic hardening rule will be detailed in the next section.

First we considered the so-called nearly proportional increasing stressing or straining conditions where the ratios of the stresses or strains remain nearly fixed during deformation. Under such loading conditions, an isotropic hardening theory typically is used. The evolution of the yield surface is illustrated schematically in Figure 3.15. As shown in the figure, when the stresses increase, the stress state first will meet the initial yield surface. As the stresses continue to increase, plastic strain hardening takes place. The yield surface expands isotropically in the stress space. When unloading takes place, the plastic reloading will not occur until the stress state meets the current expanded yield surface.

When loading, unloading, and plastic reloading are encountered under cyclic loading conditions, a kinematic hardening theory may be used. The evolution of the yield surface is illustrated schematically in Figure 3.16. In this case, the yield surface does not change its size but only translates in the stress space, as shown in Figure 3.16. When unloading takes place, the plastic reloading will not occur until the stress state meets

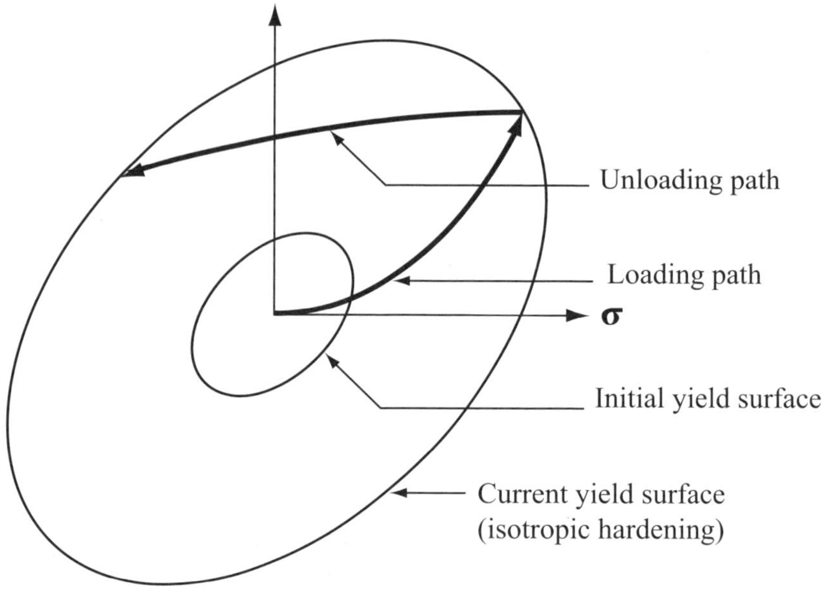

Figure 3.15 Evolution of the yield surface according to the isotropic hardening theory.

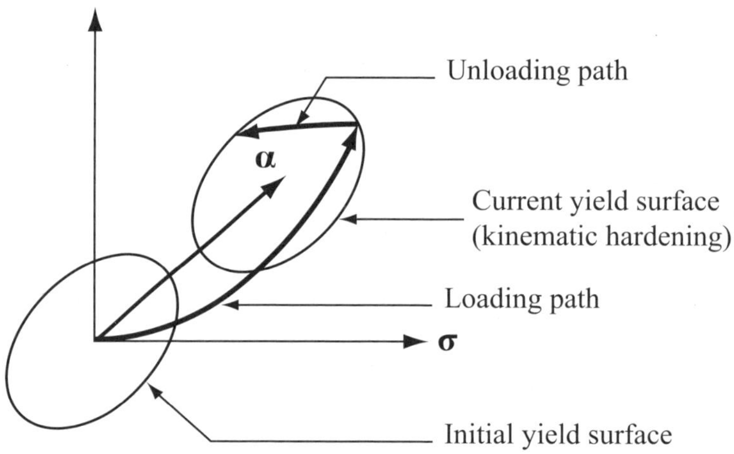

Figure 3.16 Evolution of the yield surface according to the kinematic hardening theory.

the current yield surface. In this figure, a back stress tensor $\boldsymbol{\alpha}$ is defined as the center of the current yield surface to represent the movement of the yield surface.

A combined kinematic and isotropic theory can be used where the yield surface size is interpolated between the size of the initial yield surface and the size of the corresponding isotropic hardening yield surface. The center of the yield surface moves in the stress space. The combined kinematic and isotropic hardening theory is illustrated schematically in Figure 3.17.

During sheet metal forming operations, springback of sheet metals is important for the final workpiece dimension. Sheet metals may experience plastic loading, elastic unloading, and plastic reloading, especially when sheet metals pass the draw-beads. Later in this chapter, an anisotropic hardening theory based on the work of Mroz [1967] and Chu [1984] is explained in detail for the finite element formulation. The basic concepts of Mroz [1967] are explained here. Based on the observations of the uniaxial cyclic

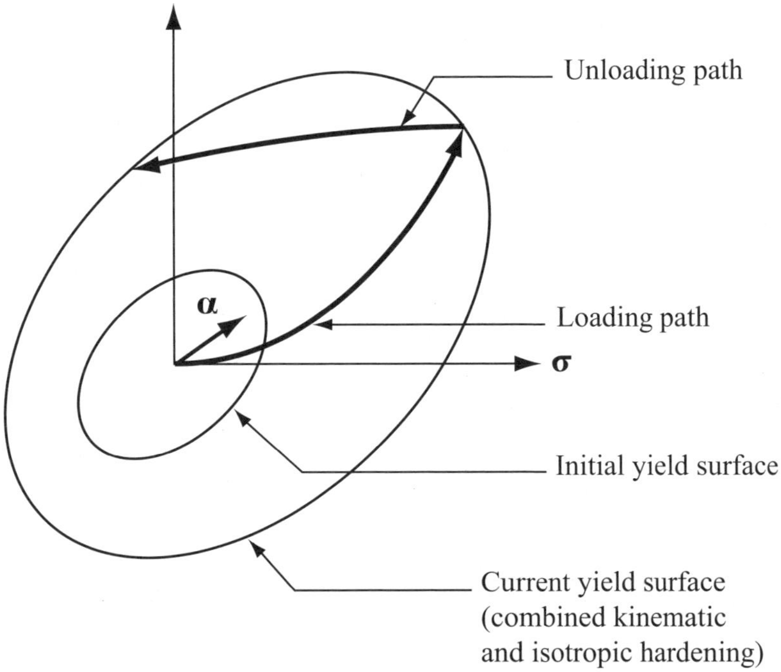

Figure 3.17 Evolution of the yield surface according to a combined isotropic and kinematic hardening theory.

plastic behavior of metals, Mroz [1967] introduced the concept of a field of work hardening moduli that are defined from an infinite number of initial yield surfaces centered at the zero stress state in the stress space. The general rule that governs the evolution of yield surfaces is that the yield surfaces must move as rigid bodies with the current stress point and that the surfaces cannot pass through one another. Therefore, the surfaces become mutually tangent to each other at the current stress point. When elastic unloading/plastic reloading occurs, the current stress point first will become plastic when the stress point encounters the smallest yield surface. The active yield surface has the initial yield stress to represent its size. This surface will be translated with the current stress point until the next larger surface is reached. These surfaces that are tangent at the current stress point then will be translated together until the next larger surface is reached. This process repeats until the next unloading/reloading occurs.

Each of these surfaces has an equivalent work hardening modulus. Because a surface is permitted only rigid body translation, its size or the flow stress may be used as a parameter to determine the modulus. When multiple yield surfaces are tangent to one another at the current stress state, the work hardening modulus for continued plastic loading is determined by the largest yield surface in contact. The smaller yield surfaces can become active again whenever unloading or reloading takes place. As shown in Figure 3.18, as the material element is under loading conditions, the yield surface expands isotropically, which follows the isotropic hardening theory. However, as unloading takes place, the active yield surface is tangent to the current stress point but with the size of the initial yield surface. Therefore, as unloading takes place, elastic unloading occurs, as shown in Figure 3.19. The largest yield surface now becomes a memory yield surface. The plastic reloading will take place when the stress state meets the current yield surface with the initial yield surface size. As loading continues, the active yield surface expands but is still tangent to the point where the unloading takes place, as shown in Figure 3.20. However, during the deformation history, memory yield surfaces are memorized and are used to determine the subsequent plastic response.

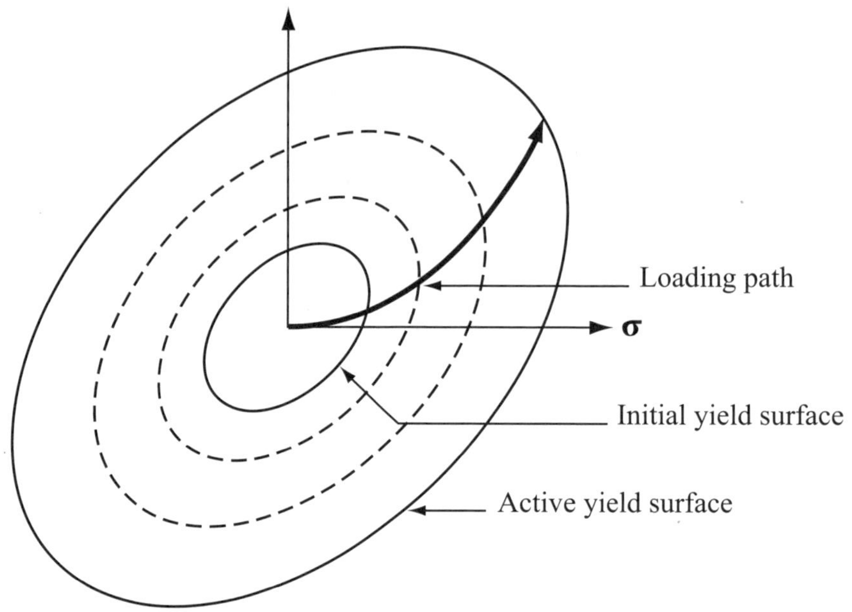

Figure 3.18 Under Mroz's rule, the yield surface expands isotropically and follows the isotropic hardening theory under loading conditions.

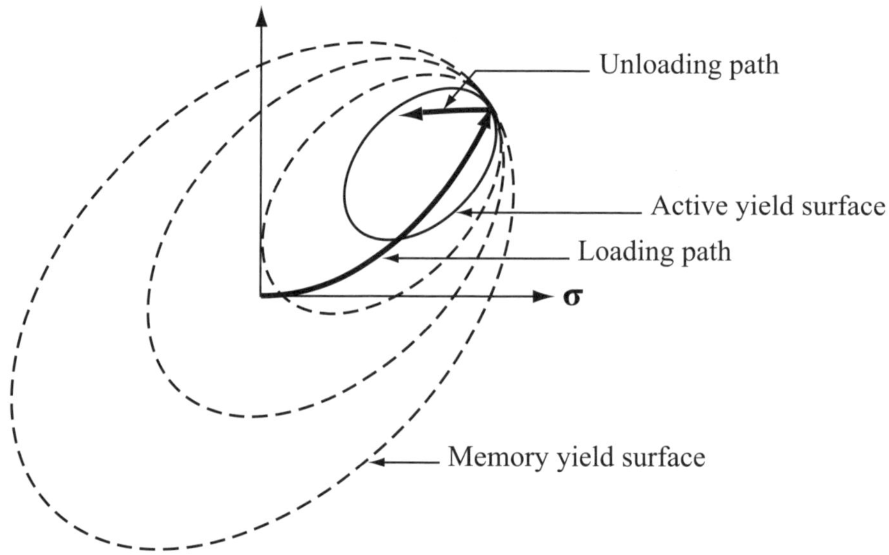

Figure 3.19 Under Mroz's rule, the active yield surface is tangent to the current stress point but with the size of the initial yield surface as unloading takes place. The largest yield surface now becomes a memory yield surface.

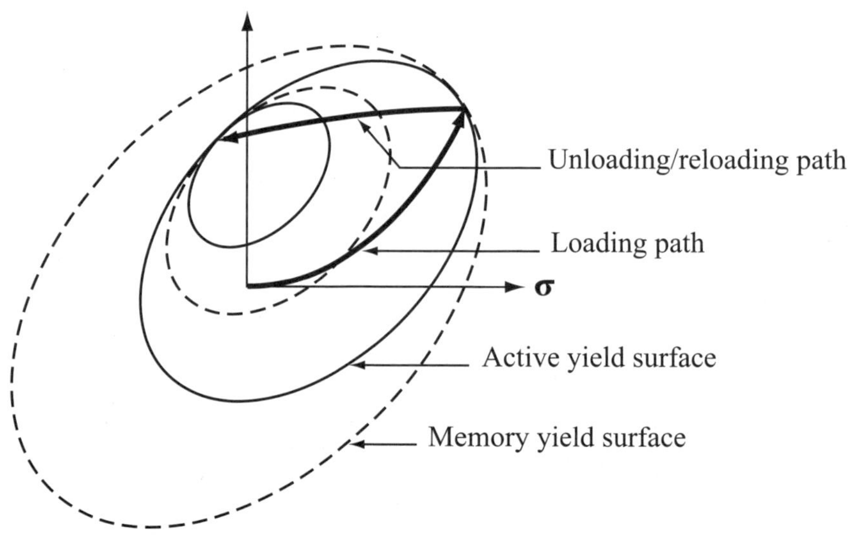

Figure 3.20 Under Mroz's rule, as unloading continues, the active yield surface expands but is still tangent to the point where unloading takes place.

3.6 Isotropic Hardening Based on the von Mises Yield Condition

Extensive plastic deformation occurs during sheet metal forming processes. However, the elastic deformation of sheets must be considered because buckling occurs at low strain levels. The elastic deformation of sheets also must be considered because an accurate prediction of the springback of stamped parts in the subsequent analysis after forming is important to control the part dimensions. Hooke's law for isotropic materials in the rate form generally is used to describe the elastic part of the elastic-plastic constitutive law. We will follow the same strategy here. For the plastic part of the elastic-plastic constitutive law, the von Mises yield condition is the most widely used to model the plastic behavior of metals. We will begin with the von Mises yield condition in developing elastic-plastic constitutive laws in this book. However, sheet metals typically are produced from rolling processes, and they generally are anisotropic before the forming process. Consequently, we must consider the plastic anisotropy in elastic-plastic constitutive equations. We will concentrate on the constitutive laws based on Hill's quadratic anisotropic yield condition [Hill, 1948]. Because we need to analyze springback possibly involving reversed plastic flow that can occur in draw operations, anisotropic hardening rules instead of the conventional isotropic hardening rule should be used. Therefore, we discuss the anisotropic hardening rule based on the work of Tang [1990] for reversed plastic flow and springback analysis.

For metals, the plastic deformation depends on the current plastic state and the prior plastic deformation. When the geometry of the sheet metal is referred to the current geometry and/or the small strain theory is used, the additive theory of Lee [1969] is a good approximation to express the total strain rate as the sum of the elastic and plastic strain rates. We start here to present the constitutive equations within the context of the small strain theory. The small strain version of the constitutive laws can readily be extended for shells with consideration of large deformation.

Within the context of the small strain theory, the small strain rates $\dot{\varepsilon}_{ij}$ can be decomposed into the elastic part $\dot{\varepsilon}_{ij}^{e}$ and the plastic part $\dot{\varepsilon}_{ij}^{p}$ as

$$\dot{\varepsilon}_{ij} = \dot{\varepsilon}_{ij}^{e} + \dot{\varepsilon}_{ij}^{p} \tag{3.6.1}$$

The elastic strain rates $\dot{\varepsilon}_{ij}^{e}$ can be related to the Cauchy stress rates $\dot{\sigma}_{ij}$ by

$$\dot{\varepsilon}_{ij}^{e} = C_{ijkl}^{e}\dot{\sigma}_{kl} \tag{3.6.2}$$

where C_{ijkl}^{e} represents the elastic compliance tensor. For isotropic elastic materials, C_{ijkl}^{e} can be written as

$$C_{ijkl}^{e} = \frac{1}{E}\left(\frac{1+\nu}{2}\left(\delta_{ik}\delta_{jl} + \delta_{il}\delta_{jk}\right) - \nu\delta_{ij}\delta_{kl}\right) \tag{3.6.3}$$

For isotropic hardening based on the von Mises yield condition, the yield function f can be expressed as

$$f = \frac{3}{2}s_{ij}s_{ij} - \bar{\sigma}^2 = 0 \tag{3.6.4}$$

where $\bar{\sigma}$ represents the effective stress, and s_{ij} represent the deviatoric stresses. Note that the second invariant of the deviatoric stress is defined as

$$J_2 \equiv \frac{1}{2}s_{ij}s_{ij} \tag{3.6.5}$$

Therefore, the von Mises yield condition can be expressed in terms of J_2 as

$$f = 3J_2 - \bar{\sigma}^2 = 0 \tag{3.6.6}$$

The effective stress $\bar{\sigma}$ can be written as

$$\bar{\sigma} = \sqrt{\frac{3}{2}s_{ij}s_{ij}} \tag{3.6.7}$$

or

$$\bar{\sigma} = \sqrt{3J_2} \tag{3.6.8}$$

According to the associated or normality flow rule, the plastic strain increments $\dot{\varepsilon}_{ij}^p$ can be expressed as

$$\dot{\varepsilon}_{ij}^p = \dot{\lambda}\frac{\partial f}{\partial \sigma_{ij}} \tag{3.6.9}$$

where $\dot{\lambda}$ represents a scalar factor of proportionality. The equivalence of the plastic work rate $\dot{W}^p$ also is needed to derive the constitutive law

$$\dot{W}^p = \sigma_{ij}\dot{\varepsilon}_{ij}^p = \bar{\sigma}\dot{\bar{\varepsilon}}^p \tag{3.6.10}$$

Here, $\dot{W}^p$ represents the plastic work rate, and $\dot{\bar{\varepsilon}}^p$ represents the effective plastic strain rate. Based on the equivalence of the plastic work rate, the flow rule, and the definition of the effective stress $\bar{\sigma}$, the effective plastic strain rate $\dot{\bar{\varepsilon}}^p$ can be derived as

$$\dot{\bar{\varepsilon}}^p = \sqrt{\frac{2}{3}\dot{\varepsilon}_{ij}^p\dot{\varepsilon}_{ij}^p} \tag{3.6.11}$$

Within the context of the small strain approach or the updated Lagrangian approach, the effective plastic strain $\bar{\varepsilon}^p$ can be integrated from the effective plastic strain rate $\dot{\bar{\varepsilon}}^p$ over the deformation history as

$$\bar{\varepsilon}^p = \int \dot{\bar{\varepsilon}}^p dt \tag{3.6.12}$$

Based on the definition of the effective plastic strain rate $\dot{\bar{\varepsilon}}^p$ and the effective stress $\bar{\sigma}$, the scalar factor of proportionality $\dot{\lambda}$ can be derived as

$$\dot{\lambda} = \frac{\dot{\bar{\varepsilon}}^p}{2\bar{\sigma}} \tag{3.6.13}$$

Note that $\dot{\lambda}$ is a positive scalar. The flow rule can be written as

$$\dot{\varepsilon}_{ij}^p = p_{ij}\dot{\bar{\varepsilon}}^p \tag{3.6.14}$$

where

$$p_{ij} = \frac{3}{2}\frac{s_{ij}}{\bar{\sigma}} \tag{3.6.15}$$

Here, p_{ij} represents a normalized outward normal tensor to the yield surface in the stress space.

To derive the constitutive law, the consistency condition for continued plastic loading is needed. For plastic loading, the yield condition must be satisfied. Therefore,

$$\dot{f} = \frac{\partial f}{\partial \sigma_{ij}}\dot{\sigma}_{ij} + \frac{\partial f}{\partial \bar{\sigma}}\dot{\bar{\sigma}} = 0 \tag{3.6.16}$$

Substituting the yield function into Eq. 3.6.16, the consistency condition becomes

$$\dot{f} = 3s_{ij}\dot{\sigma}_{ij} - 2\bar{\sigma}\dot{\bar{\sigma}} = 0 \tag{3.6.17}$$

The consistency condition can be written as

$$\dot{\bar{\sigma}} = p_{ij}\dot{\sigma}_{ij} \tag{3.6.18}$$

The effective stress $\bar{\sigma}$ and the effective plastic strain $\bar{\varepsilon}^p$ are assumed to be related to the flow stress σ and the plastic strain ε^p of the uniaxial tensile stress-strain curve as

$$\bar{\sigma} = \sigma \tag{3.6.19}$$

$$\bar{\varepsilon}^p = \varepsilon^p \tag{3.6.20}$$

Because the flow stress rate $\dot{\sigma}$ can be related to the plastic strain rate $\dot{\varepsilon}^p$ as

$$\dot{\varepsilon}^p = \dot{\sigma}\left(\frac{1}{E_t} - \frac{1}{E}\right) \tag{3.6.21}$$

where E_t represents the tangent modulus as defined in Eq. 3.3.23 for a given σ ($= \bar{\sigma}$), and E represents the Young's modulus. Therefore, we take the relation of the effective stress rate $\dot{\bar{\sigma}}$ and the effective plastic strain rate $\dot{\bar{\varepsilon}}^p$ as

$$\dot{\bar{\varepsilon}}^p = \dot{\bar{\sigma}}\left(\frac{1}{E_t} - \frac{1}{E}\right) \tag{3.6.22}$$

Substituting Eq. 3.6.22 into Eq. 3.6.14 with the use of Eq. 3.6.18 gives

$$\dot{\varepsilon}_{ij}^p = \left(\frac{1}{E_t} - \frac{1}{E}\right) p_{ij} p_{kl} \dot{\sigma}_{kl} \tag{3.6.23}$$

Under plastic loading conditions, $\dot{\sigma}_{ij}\dot{\varepsilon}_{ij}^p \geq 0$. Combining Eqs. 3.6.1, 3.6.2, and 3.6.23 gives

$$\dot{\varepsilon}_{ij} = C_{ijkl}\dot{\sigma}_{kl} \tag{3.6.24}$$

where the compliance tensor C_{ijkl} is defined as

$$\begin{aligned} C_{ijkl} = {} & \frac{1}{E}\left(\frac{(1+\nu)}{2}\left(\delta_{ik}\delta_{jl} + \delta_{il}\delta_{jk}\right) - \nu\delta_{ij}\delta_{kl}\right) \\ & + \left(\frac{1}{E_t} - \frac{1}{E}\right) p_{ij} p_{kl} \end{aligned} \tag{3.6.25}$$

Inverting Eq. 3.6.24 gives

$$\dot{\sigma}_{ij} = D_{ijkl}\dot{\varepsilon}_{kl} \tag{3.6.26}$$

where the tangent stiffness tensor D_{ijkl} is

$$D_{ijkl} = \frac{E}{1+\nu}\left(\frac{1}{2}\left(\delta_{ik}\delta_{jl} + \delta_{il}\delta_{jk}\right) + \frac{\nu}{1-2\nu}\delta_{ij}\delta_{kl} - g^{-1}s_{ij}s_{kl}\right) \tag{3.6.27}$$

Here,

$$g^{-1} = \frac{1}{\left(1 + \frac{2}{3}\frac{1+\nu}{E/E_t - 1}\right) s_{ij}s_{ij}} \tag{3.6.28}$$

The tangent modulus E_t can be derived from the uniaxial stress-strain relation. For example, the uniaxial stress-strain relation in the plastic range can be written in a power-law form as

$$\sigma = K\varepsilon^n \tag{3.6.29}$$

We assume the effective stress-strain relation as

$$\bar{\sigma} = K\bar{\varepsilon}^n \tag{3.6.30}$$

Differentiating both sides of Eq. 3.6.30 with respect to time yields

$$\dot{\sigma} = K\varepsilon^{n-1}\dot{\varepsilon} \tag{3.6.31}$$

The tangent modulus E_t now can be expressed as

$$E_t = \frac{\dot{\sigma}}{\dot{\varepsilon}} = K^{\frac{1}{n}}\sigma^{\frac{n-1}{n}} \tag{3.6.32}$$

Therefore, the tangent modulus E_t can be expressed in terms of a given $\bar{\sigma}$ for a deformation history

$$E_t = \frac{\dot{\sigma}}{\dot{\varepsilon}} = K^{\frac{1}{n}}\bar{\sigma}^{\frac{n-1}{n}} \tag{3.6.33}$$

Under the elastic unloading (when the direction of the stress rate is into the yield surface) or neutral loading (when the direction of the stress rate is parallel to the yield surface), there is no plastic strain rate. Therefore, the last term due to the plastic strain rate in Eq. 3.6.27 is dropped, and D_{ijkl} is reduced to the elastic material tangent stiffness tensor as

$$D_{ijkl} = D^e_{ijkl} = \frac{E}{1+\nu}\left(\frac{1}{2}\left(\delta_{ik}\delta_{jl} + \delta_{il}\delta_{jk}\right) + \frac{\nu}{1-2\nu}\delta_{ij}\delta_{kl}\right) \tag{3.6.34}$$

When the plastic strains become large in simulations of sheet metal forming processes, we must use an objective rate of a stress tensor. The Jaumann rate of the Cauchy stress tensor, $\breve{\sigma}_{ij}$, can be used. Therefore, Eq. 3.6.26 is changed to

$$\breve{\sigma}_{ij} = D_{ijkl}d_{ij} \tag{3.6.35}$$

where d_{ij} represents the rate of the deformation tensor, which is defined as

$$d_{ij} = \frac{1}{2}\left(\dot{u}_{i,j} + \dot{u}_{j,i}\right) \tag{3.6.36}$$

where $\dot{u}_{i,j}$ is the displacement rate gradient tensor. The Jaumann rate of the Cauchy stress tensor [Eringen, 1967] is

$$\breve{\sigma}_{ij} = \dot{\sigma}_{ij} + \sigma_{ik}\omega_{kj} - \sigma_{kj}\omega_{ik} \tag{3.6.37}$$

where the spin rate tensor ω_{ij} is defined as

$$\omega_{ij} = \frac{1}{2}\left(\dot{u}_{i,j} - \dot{u}_{j,i}\right) \tag{3.6.38}$$

When the kinematic hardening rule is used, $\overline{\sigma}$ is a constant and is equal to the initial yield stress σ_0, and a different hardening parameter can be used to benchmark with the uniaxial tensile stress-strain curve.

3.7 Anisotropic Hardening Based on the von Mises Yield Condition

A few cycles of plastic loading, elastic unloading, and plastic reloading occur in the forming simulation of a typical automotive panel from the binder wrap to the springback after the flanging operation. For example, in a draw operation in which a portion of a sheet involves stretching, bending over a die corner with a small radius, and unbending along a draw wall or straightening an initial wrinkling, there is a cycle of stresses from plastic loading, elastic unloading, and plastic reloading. Therefore, the isotropic hardening rule in plasticity theory is no longer adequate for obtaining good numerical results in the analysis of a draw operation. For example, Figure 3.21(a) shows a schematic for a uniaxial cyclic stress-strain relation under a specific cyclic strain history. In the figure, X_0 represents the initial yield stress. In this figure, the Bauschinger effects are specified in the curve. Figure 3.21(b) shows a uniaxial stress-strain curve under the given cyclic strain history when the isotropic hardening rule is used to model the stress response. The stresses based on the isotropic hardening rule in Figure 3.21(b) are quite different from those shown in Figure 3.21(a), which generally agrees with the experimental observations.

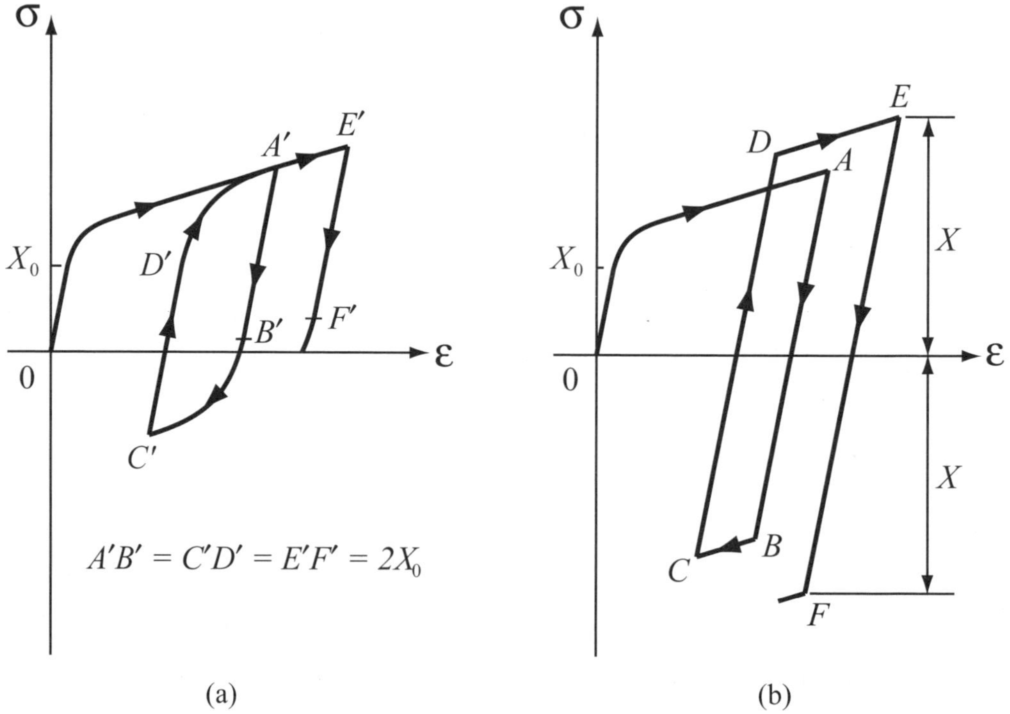

Figure 3.21 (a) A schematic of a uniaxial stress-strain curve, and (b) a uniaxial stress-strain curve based on the isotropic hardening model.

The simplest anisotropic hardening rule is the kinematic rule by Prager [1955] and Ziegler [1959]. The kinematic hardening rule has been used to simulate the Bauschinger effect, which could not be predicted by the isotropic hardening rule. However, for complex loading histories, actual material behavior can deviate substantially from that predicted by the kinematic hardening rule. In addition, there is no definite method for determining the tangent modulus for a nonlinear hardening material. The hardening rule proposed by Mroz [1967] based on the observation of the cyclic stress-strain relation is selected here because the tangent modulus for a nonlinear hardening material by Mroz's hardening rule is easy to determine computationally. Mroz's rule can be used to model the uniaxial cyclic stress-strain relation shown in Figure 3.21(a). To avoid computational complexity, Krieg [1975], Dafalias and Popov [1975], Chu [1984], and Choi [2007] proposed yield surface evolution rules within the context of Mroz's hardening rule.

The modification of Mroz's hardening rule by Chu [1984] is described briefly here. Unlike the isotropic and kinematic hardening rules, in which a single yield surface is assumed to either expand or translate, respectively, as a result of plastic deformation, Mroz [1967] introduced the concept of a field of work hardening moduli that are defined by the configuration of an infinite number of initially concentric yield surfaces in the deviatoric stress space, as shown in Figure 3.22. The general rules governing the configuration change are that yield surfaces must move as rigid bodies with the loading point when it is in contact with them, and that the surfaces cannot pass through one another. Therefore, the surfaces will become mutually tangent at the loading point. As it passes from the elastic into the plastic region, the loading point first will encounter the smallest

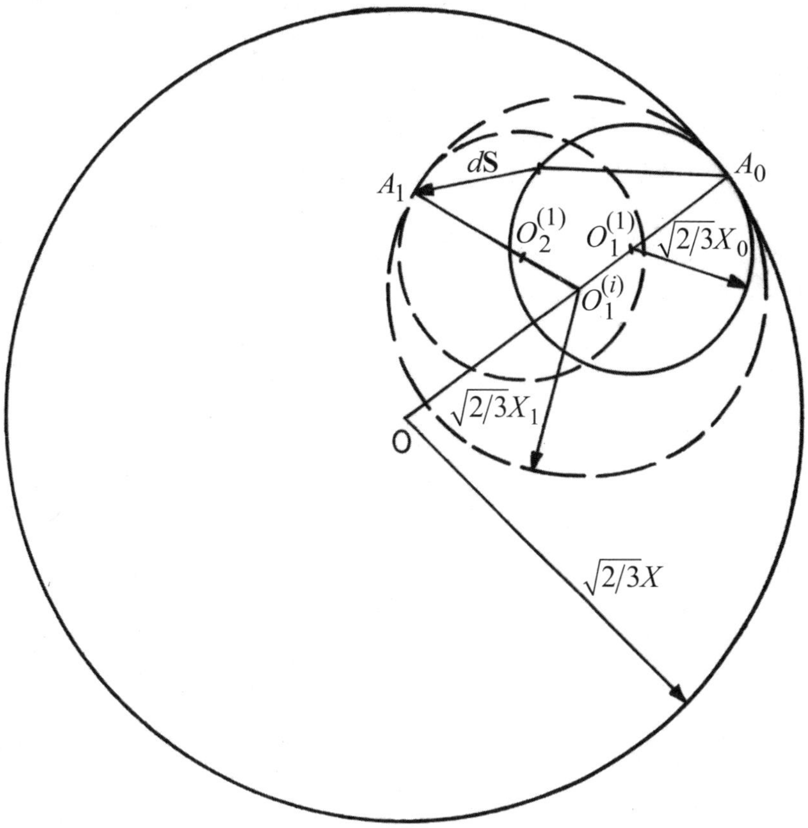

Figure 3.22 Yield surfaces in the deviatoric stress space during loading and unloading.

yield surface, which has a radius $\sqrt{2/3}X_0$ (X_0 is the initial yield stress). This surface will be pushed forward until the next larger surface is reached, and these two surfaces then will move forward together, and so forth. Each of these yield surfaces has a work hardening modulus depending on its radius $\sqrt{2/3}X$. Because a surface is permitted only rigid body motion, its size may be used as a parameter to determine the modulus. When the material is deep into the plastic range and multiple yield surfaces are tangent to one another at the loading point, the instantaneous modulus for monotonically increasing loading is the one associated with the largest yield surface in contact. This is the active yield surface. The smaller yield surfaces can become active again whenever unloading or reloading takes place.

The equations for change of yield surface size and center movement are derived as follows. The von Mises yield criterion is used. According to Mroz's rule, the yield function is written as

$$f = \frac{3}{2}\left(s_{ij} - \alpha_{ij}\right)\left(s_{ij} - \alpha_{ij}\right) - X^2 = 0 \tag{3.7.1}$$

where s_{ij} is the deviatoric stress tensor, α_{ij} is the position tensor of the center of the active yield surface in the deviatoric stress space, and X is the flow stress determined by the effective plastic strain. Note that the radius of the active yield surface is $\sqrt{2/3}X$, as shown in Figure 3.22. The differential form of the yield function is

$$\frac{3}{2}\left(s_{ij} - \alpha_{ij}\right)\left(ds_{ij} - d\alpha_{ij}\right) - XdX = 0 \tag{3.7.2}$$

where ds_{ij} represents the differential of the deviatoric stress tensor, and $d\alpha_{ij}$ is the differential of the center position tensor. The center of a yield surface is assumed to move along a unit tensor β_{ij} with a magnitude of $\sqrt{2/3}dX$ as

$$d\alpha_{ij} = \sqrt{2/3}dX\,\beta_{ij} \tag{3.7.3}$$

The selection of β_{ij} will be discussed later. Substituting this equation into Eq. 3.7.2 yields

$$dX = \frac{3}{2}\left(s_{ij} - \alpha_{ij}\right)ds_{ij}/\bar{X} \tag{3.7.4}$$

where

$$\bar{X} = X + \sqrt{3/2}\left(s_{ij} - \alpha_{ij}\right)\beta_{ij} \tag{3.7.5}$$

Inserting dX from Eq. 3.7.4 into Eq. 3.7.3 gives

$$d\alpha_{ij} = \frac{\sqrt{3/2}\left(s_{mn} - \alpha_{mn}\right)ds_{mn}}{\bar{X}}\beta_{ij} \tag{3.7.6}$$

If the associated flow rule is assumed, the elastic-plastic constitutive equation can be derived in a procedure similar to that for the isotropic hardening rule. Thus, the plastic strain rate $\dot{\varepsilon}_{ij}^{p}$ based on the active yield surface is

$$\dot{\varepsilon}_{ij}^{p} = p_{ij}\dot{\bar{\varepsilon}}^{p} \tag{3.7.7}$$

where

$$p_{ij} = \frac{3}{2}\frac{s_{ij} - \alpha_{ij}}{X} \tag{3.7.8}$$

An example for change of the active yield surfaces in a process for the initial loading, elastic unloading, plastic reloading, elastic unloading again, and then plastic reloading in the deviatoric stress space is illustrated in Figure 3.22 as follows.

1. Initial and Monotonically Increasing Loading

   The center of the initial yield surface is at the origin, and its radius is $\sqrt{2/3}X_0$. We monotonically increase the load to point A_0, where the radius of the yield surface is $\sqrt{2/3}X$. The unit tensor β_{ij} in Eq. 3.7.3 is along OA_0. The center of the smallest yield surface with radius $\sqrt{2/3}X_0$ moves to $O_1^{(1)}$. Note that the center of the active yield surface is still at the origin.

2. Elastic Unloading and Plastic Reloading

   We unload inside the smallest yield surface with center at $O_1^{(1)}$ and reload to A_1 with the differential of the deviatoric stress tensor ds_{ij}. Using this differential and the unit tensor β_{ij}, we can compute the center $O_1^{(i)}$ of the larger yield surface and its radius $\sqrt{2/3}X_1$ from Eqs. 3.7.3 through 3.7.5. The updated unit tensor β_{ij} is along $A_1O_1^{(i)}$, and the center of the updated smallest yield surface is along this line and thus can be located at $O_2^{(1)}$, as shown in Figure 3.22.

3. Elastic Unloading Again and Then Plastic Reloading

   If we unload again inside the newest smallest yield surface and reload into the plastic range again, the center of the active yield surface is along the line $O_2^{(1)}O_1^{(i)}$ This center cannot go beyond $O_1^{(i)}$ as shown in Figure 3.22. If it does (called "break"), the new center will lie on the line A_0O. If monotonically increasing loading occurs, the center of the active yield surface cannot go beyond O. Otherwise, the yield surface with the radius $\sqrt{2/3}X$ will be active and will move to tangent a yield surface with a larger radius than $\sqrt{2/3}X$, and its center is still at O.

3.8 Isotropic Hardening Based on the von Mises Yield Condition with Rate Sensitivity

Because the tool speed in a stretch/draw forming is as low as 200 mm/sec, the effects of the strain rate hardening on the yield stress and hardening characteristics usually are neglected in the initial stage of sheet forming simulations. However, we will discuss a numerical method that can be used conveniently to account for the rate sensitivity but is based on the rate-insensitive finite element formulations here. For most metals, Young's modulus and Poisson's ratio are assumed to be insensitive to the strain rate. Therefore, Eq. 3.6.2 is still valid for elastic deformation at different strain rates. Based on the experimental results of the uniaxial stress-strain relations, the flow stress usually is dependent on the amount of plastic strain and the plastic strain rate. Both power-law strain hardening and power-law strain rate hardening typically are assumed to fit the experimental data. The strain rate hardening rule usually can be written as

$$\bar{\sigma} = K(\dot{\bar{\varepsilon}}_{vp}/\gamma)^m \bar{\varepsilon}_{vp}^n \tag{3.8.1}$$

where $\bar{\sigma}$ is the effective stress, $\dot{\bar{\varepsilon}}_{vp}$ is the effective visco-plastic strain rate, γ is a reference effective strain rate, $\bar{\varepsilon}_{vp}$ is the effective visco-plastic strain, and K, m, and n are the material parameters determined by uniaxial tensile tests at various strain rates and strain values. Another commonly used example for rate-sensitive materials is the overstressed model of Wang and Wenner [1978]

$$\bar{\sigma} = K\left[\left(\bar{\varepsilon}_{vp} + \varepsilon_0\right)^n + \eta \ln\left(1 + \dot{\bar{\varepsilon}}_{vp}/\gamma\right)\right] \tag{3.8.2}$$

where K, ε_0, η, and n are material parameters to be determined by uniaxial tensile tests. We assume the flow rule based on the von Mises yield condition (now a plastic potential) is valid, and the effective visco-plastic strain still can be computed from Eq. 3.6.11. We then can add the rates of elastic and visco-plastic strains, and the total strain rate becomes

$$\dot{\varepsilon}_{ij} = C_{ijkl}^e \dot{\sigma}_{kl} + p_{ij} \dot{\bar{\varepsilon}}_{vp} \tag{3.8.3}$$

where the elastic compliance C_{ijkl}^e is expressed in Eq. 3.6.2.

Rearranging Eq. 3.8.3 yields

$$\dot{\sigma}_{ij} = D_{ijkl}^e\left[\dot{\varepsilon}_{kl} - p_{kl}\dot{\bar{\varepsilon}}_{vp}\right] \tag{3.8.4}$$

where D_{ijkl}^e is the elastic tangent stiffness tensor as expressed in Eq. 3.6.34 as

$$D_{ijkl}^e = \frac{E}{1+\nu}\left(\frac{1}{2}\left(\delta_{ik}\delta_{jl} + \delta_{il}\delta_{jk}\right) + \frac{\nu}{1-2\nu}\delta_{ij}\delta_{kl}\right) \tag{3.8.5}$$

When Eq. 3.8.4 is used in the implicit method (to be discussed in Chapter 7) to solve the equations of motion or equilibrium, the term of the visco-plastic strain depends on a solution that is unknown. We may use the values from the previous iteration to compute the visco-plastic strain and then move it to the right sides of the system of simultaneous equations. An iterative scheme can be set up. Because the elastic tangent stiffness tensor D_{ijkl}^e is extremely stiff, a very small time step must be used for rapid convergence of the iterations. This iterative method is not practical because of its slow convergence for a practical time step.

Here, we summarize a method by Peirce, Shih, and Needleman [1984], which is used to estimate the change of the effective visco-plastic strain $\bar{\varepsilon}_{vp}$ during the current time increment. This gives the material tangent stiffness, depending on the magnitude of the time increment Δt and on the material properties. To do so for most metals, the tangent stiffness for the elastic visco-plastic material is considerably reduced, such that the iteration becomes practical for an implicit method to solve the equations of motion or equilibrium.

Now, we present the method by Peirce, Shih, and Needleman [1984]. First, Taylor's expansion is used to estimate $\dot{\bar{\varepsilon}}_{vp}$ at time $t + \Delta t$ in terms of its value and the derivatives evaluated at time t. Using the approximation $\Delta x \approx \dot{x}\Delta t$, we express Eq. 3.8.4 in the incremental (finite difference) form as

$$\Delta\sigma_{ij} = D^{e}_{ijkl}\left[\Delta\varepsilon_{kl} - p_{kl}\Delta\bar{\varepsilon}_{vp}\right] \tag{3.8.6}$$

We now define

$$\Delta\bar{\varepsilon}_{vp} \equiv \bar{\varepsilon}_{vp}(t + \Delta t) - \bar{\varepsilon}_{vp}(t) \tag{3.8.7}$$

and interpolate

$$\Delta\bar{\varepsilon}_{vp} = \Delta t\left[(1-\theta)\dot{\bar{\varepsilon}}_{vp}(t) + \theta\dot{\bar{\varepsilon}}_{vp}(t + \Delta t)\right], \quad 0 < \theta < 1 \tag{3.8.8}$$

Substituting Eq. 3.8.4 into Eq. 3.6.18 gives

$$\dot{\bar{\sigma}} = p_{ij}D^{e}_{ijkl}\left(\dot{\varepsilon}_{kl} - p_{kl}\dot{\bar{\varepsilon}}_{vp}\right) \tag{3.8.9}$$

or in the finite difference form as

$$\Delta\bar{\sigma} = p_{ij}D^{e}_{ijkl}\left(\Delta\varepsilon_{kl} - p_{kl}\Delta\bar{\varepsilon}_{vp}\right) \tag{3.8.10}$$

Based on the Taylor series expansion, we have

$$\dot{\bar{\varepsilon}}_{vp}(t + \Delta t) \approx \dot{\bar{\varepsilon}}_{vp}(t) + \frac{\partial\dot{\bar{\varepsilon}}_{vp}}{\partial\bar{\sigma}}\Delta\bar{\sigma} + \frac{\partial\dot{\bar{\varepsilon}}_{vp}}{\partial\bar{\varepsilon}_{vp}}\Delta\bar{\varepsilon}_{vp} \tag{3.8.11}$$

where the partial derivatives $\partial\dot{\bar{\varepsilon}}_{vp}/\partial\bar{\sigma}$ and $\partial\dot{\bar{\varepsilon}}_{vp}/\partial\bar{\varepsilon}_{vp}$ are evaluated at time t from the uniaxial stress-strain relationship of the general form

$$\bar{\sigma} = \bar{\sigma}\left(\bar{\varepsilon}_{vp}, \dot{\bar{\varepsilon}}_{vp}\right) \tag{3.8.12}$$

Equations 3.8.1 and 3.8.2 are the special cases of Eq. 3.8.12. Note that the effective stress $\bar{\sigma}$ and the effective visco-plastic strain $\bar{\varepsilon}_{vp}$ here can be identified to be the uniaxial stress and strain from a tensile test, respectively.

Combining Eqs. 3.8.11, 3.8.10, and 3.8.8, we can solve for

$$\Delta\bar{\varepsilon}_{vp} = \Delta t\left[\frac{\dot{\bar{\varepsilon}}_{vp}(t)}{1+\xi} + \frac{\xi}{\psi(1+\xi)} p_{ij} D^{e}_{ijkl} \frac{\Delta\varepsilon_{kl}}{\Delta t}\right] \tag{3.8.13}$$

where

$$\psi = p_{ij} D^{e}_{ijkl} p_{kl} - \frac{\partial\dot{\bar{\varepsilon}}_{vp}}{\partial\bar{\varepsilon}_{vp}}\left(\frac{\partial\dot{\bar{\varepsilon}}_{vp}}{\partial\bar{\sigma}}\right)^{-1} \tag{3.8.14}$$

and

$$\xi = (\theta\Delta t)\psi\frac{\partial\dot{\bar{\varepsilon}}_{vp}}{\partial\bar{\sigma}} \tag{3.8.15}$$

Substituting $\Delta\bar{\varepsilon}_{vp}$ from Eq. 3.8.13 into Eq. 3.8.6, we have the incremental form of the constitutive equation

$$\Delta\sigma_{ij} = D^{evp}_{ijkl}\Delta\varepsilon_{kl} - \frac{\dot{\bar{\varepsilon}}_{vp}(t)\Delta t}{(1+\xi)} D^{e}_{ijkl} p_{kl} \tag{3.8.16}$$

where

$$D^{evp}_{ijkl} = D^{e}_{ijkl} - \frac{\xi}{\psi(1+\xi)} D^{e}_{ijmn} p_{mn} p_{rs} D^{e}_{rskl} \tag{3.8.17}$$

and D^{e}_{ijkl} is the elastic material tangent stiffness tensor in Eq. 3.8.4. Dividing both sides of Eq. 3.8.16 by Δt and setting Δt to approach zero, we obtain the rate form

$$\dot{\sigma}_{ij} = D^{evp}_{ijkl}\dot{\varepsilon}_{kl} - \frac{\dot{\bar{\varepsilon}}_{vp}(t)}{(1+\xi)} D^{e}_{ijkl} p_{kl} \tag{3.8.18}$$

Similar to the rate insensitive case, we may extend Eq. 3.8.18 to a large strain formulation by changing $\dot{\sigma}_{ij}$ to $\breve{\sigma}_{ij}$ and $\dot{\varepsilon}_{ij}$ to d_{ij}, respectively. For the explicit method (also to be discussed in Chapter 7), because the Euler's forward integration method is used to solve the equations of motion, a very small time step must be taken to satisfy the Courant-Fredrich-Lewy condition of numerical stability. Therefore, there is no problem in obtaining an accurate stress increment by numerical integration of Eq. 3.8.4 with such a small step.

3.9 Isotropic and Anisotropic Hardening Based on the Hill Quadratic Anisotropic Yield Condition

As already discussed, the Hill quadratic anisotropic yield condition with normal anisotropy under plane stress conditions can be written as

$$f = \sigma_{11}^2 + \sigma_{22}^2 - \frac{2R}{1+R}\sigma_{11}\sigma_{22} + \frac{2(1+2R)}{1+R}\sigma_{12}^2 - X^2 = 0 \tag{3.9.1}$$

Although the advantage of using the deviatoric stresses for isotropic materials is lost for anisotropic materials, there is still some convenience in applying them in computing the effective stress and the effective plastic strain rate. For example, based on the flow rule, generalized deviatoric stress components can be defined as

$$s_{11} = \frac{1+R}{2(1+2R)}\frac{\partial f}{\partial \sigma_{11}} = \sigma_{11} - \frac{R}{1+2R}(\sigma_{11} + \sigma_{22}) \tag{3.9.2}$$

$$s_{22} = \frac{1+R}{2(1+2R)}\frac{\partial f}{\partial \sigma_{22}} = \sigma_{22} - \frac{R}{1+2R}(\sigma_{11} + \sigma_{22}) \tag{3.9.3}$$

$$s_{33} = -(s_{11} + s_{22}) = -s_{\alpha\alpha} \tag{3.9.4}$$

$$s_{12} = \frac{1+R}{1+2R}\frac{\partial f}{\partial \sigma_{12}} = \sigma_{12} \tag{3.9.5}$$

Note that $\sigma_{33} = 0$, but $s_{33} \neq 0$. These stresses in Eqs. 3.9.2 through 3.9.5 are identical to those in Eq. 3.4.14 for $R = 1$.

Similar to that of isotropic materials, an effective stress $\bar{\sigma}$ can be defined in terms of a modified J_2 as

$$\bar{\sigma} = \sqrt{(1+2R)J_2} \tag{3.9.6}$$

where

$$J_2 = \frac{1}{1+R}\left[s_{\alpha\beta}s_{\alpha\beta} + Rs_{\alpha\alpha}s_{\beta\beta}\right] \tag{3.9.7}$$

and the yield function is modified to

$$f = (1+2R)J_2 - X^2 = 0 \tag{3.9.8}$$

The flow rule now can be written as

$$\dot{\varepsilon}_{11}^{p} = \dot{\lambda}\frac{\partial f}{\partial \sigma_{11}} = 2\dot{\lambda}\frac{1+2R}{1+R}s_{11} \tag{3.9.9}$$

$$\dot{\varepsilon}_{22}^{p} = \dot{\lambda}\frac{\partial f}{\partial \sigma_{22}} = 2\dot{\lambda}\frac{1+2R}{1+R}s_{22} \tag{3.9.10}$$

$$2\dot{\varepsilon}_{12}^{p} = \dot{\lambda}\frac{\partial f}{\partial \sigma_{12}} = 4\dot{\lambda}\frac{1+2R}{1+R}s_{12} \tag{3.9.11}$$

Combining Eqs. 3.9.6, 3.9.7, and 3.9.9 through 3.9.11 gives the scalar factor of proportionality $\dot{\lambda}$ as

$$\dot{\lambda} = \frac{1}{2\bar{\sigma}}\frac{1+R}{(1+2R)^{1/2}}\left[\left(\dot{\varepsilon}_{11}^{p}\right)^2 + \left(\dot{\varepsilon}_{22}^{p}\right)^2 + \frac{2R}{1+R}\dot{\varepsilon}_{11}^{p}\dot{\varepsilon}_{22}^{p} + \frac{2}{1+R}\left(\dot{\varepsilon}_{12}^{p}\right)^2\right]^{1/2} \tag{3.9.12}$$

Based on the equivalence of the plastic work rate, the work-conjugate effective plastic strain rate $\dot{\bar{\varepsilon}}^{p}$ can be derived as

$$\dot{\bar{\varepsilon}}^{p} = \frac{1+R}{(1+2R)^{1/2}}\left[\left(\dot{\varepsilon}_{11}^{p}\right)^2 + \left(\dot{\varepsilon}_{22}^{p}\right)^2 + \frac{2R}{1+R}\dot{\varepsilon}_{11}^{p}\dot{\varepsilon}_{22}^{p} + \frac{2}{1+R}\left(\dot{\varepsilon}_{12}^{p}\right)^2\right]^{1/2} \tag{3.9.13}$$

Also note that the effective plastic strain rate for $R = 1$ in Eq. 3.9.13 is identical to that in Eq. 3.6.11 (setting $\dot{\varepsilon}_{33}^{p} = -\left(\dot{\varepsilon}_{11}^{p} + \dot{\varepsilon}_{22}^{p}\right)$ and $\dot{\varepsilon}_{\alpha 3} = 0$ under plane stress conditions) as a special case. Note that Eq. 3.6.12 is still valid to compute the total effective plastic strain.

Using Eqs. 3.9.12 and 3.9.13, the proportional parameter is expressed as

$$\dot{\lambda} = \frac{\dot{\bar{\varepsilon}}^{p}}{2X} \tag{3.9.14}$$

In Chapter 5, we present a shell theory with consideration of the two transverse shear components. Therefore, we include the transverse shear components in the finite element formulation of plastic deformation. Another assumption also is taken to obtain the values of the material constants M and N in the yield condition in Eq. 3.4.41, based on the existing test results from sheet specimens. Appropriate material constants associated with σ_{23} and σ_{31} can be assigned when experimental data are available. The yield function is modified as

$$f = \sigma_{11}^2 + \sigma_{22}^2 - \frac{2R}{1+R}\sigma_{11}\sigma_{22} + \frac{2(1+2R)}{1+R}\sigma_{12}^2 + \frac{6}{1+R}\left(\sigma_{23}^2 + \sigma_{31}^2\right) - X^2 = 0 \tag{3.9.15}$$

For the yield condition in Eq. 3.9.15, we take the values of M and N in the yield condition in Eq. 3.4.41 and assign the values so that the shear stresses σ_{23} and σ_{31} are related to the uniaxial yield stress σ_{33} in the X_3 direction by a usual factor of $\sqrt{3}$, based on the usual isotropic von Mises yield condition. Based on the flow rule, the rates of plastic shear strains corresponding to σ_{23} and σ_{31} are

$$\dot{\varepsilon}_{23}^p = 2\dot{\lambda}\frac{3}{1+R}\sigma_{23} \tag{3.9.16}$$

$$\dot{\varepsilon}_{31}^p = 2\dot{\lambda}\frac{3}{1+R}\sigma_{31} \tag{3.9.17}$$

The proportional parameter in Eq. 3.9.14 is still good for this case. Note that

$$s_{23} = \sigma_{23} \tag{3.9.18}$$

$$s_{31} = \sigma_{31} \tag{3.9.19}$$

Therefore,

$$\dot{\varepsilon}_{23}^p = 2\dot{\lambda}\frac{3}{1+R}s_{23} \tag{3.9.20}$$

$$\dot{\varepsilon}_{31}^p = 2\dot{\lambda}\frac{3}{1+R}s_{31} \tag{3.9.21}$$

Finally, the effective stress is

$$\bar{\sigma} = \left[\sigma_{11}^2 + \sigma_{22}^2 - \frac{2R}{1+R}\sigma_{11}\sigma_{22} + \frac{2(1+2R)}{1+R}\sigma_{12}^2 + \frac{6}{1+R}\left(\sigma_{23}^2 + \sigma_{31}^2\right)\right]^{1/2} \tag{3.9.22}$$

The effective plastic strain rate is defined as

$$\dot{\bar{\varepsilon}}^P = \frac{1+R}{(1+2R)^{1/2}}\left[\left(\dot{\varepsilon}_{11}^p\right)^2 + \left(\dot{\varepsilon}_{22}^p\right)^2 + \frac{2R}{1+R}\dot{\varepsilon}_{11}^p\dot{\varepsilon}_{22}^p + \frac{2}{1+R}\left(\dot{\varepsilon}_{12}^p\right)^2 + \frac{2}{3}\frac{1+2R}{1+R}\left(\left(\dot{\varepsilon}_{23}^p\right)^2 + \left(\dot{\varepsilon}_{31}^p\right)^2\right)\right]^{1/2} \tag{3.9.23}$$

Because we want to include the normal anisotropy of a sheet that is under plane stress conditions, there is no advantage to defining the yield surface in the deviatoric stress space. We now define the yield surface in the Cauchy stress space. Based on Chu's hardening rule [Chu, 1984] and taking care of the transversely anisotropic property, Hill's quadratic yield condition for sheet metals with normal anisotropy is modified as

$$f\left(r_{\alpha\beta}\right) = r_{11}^2 + r_{22}^2 - \frac{2R}{1+R} r_{11} r_{22} + \frac{2\left(1+2R\right)}{1+R} r_{12}^2 - X^2 = 0 \tag{3.9.24}$$

where

$$r_{\alpha\beta} = \sigma_{\alpha\beta} - \alpha_{\alpha\beta} \tag{3.9.25}$$

Here, $\alpha_{\alpha\beta}$ is the position tensor for the center of the active yield surface in the Cauchy stress space, and $\alpha_{\alpha\beta}$ represents the back stress. We can replace $\sigma_{\alpha\beta}$ by $r_{\alpha\beta}$ in all equations for both the isotropic hardening and anisotropic rules presented in the previous section.

3.10 Plastic Localization and Forming Limit Diagram

Figure 3.23 shows a material element having an imperfection band under plane stress conditions. In this figure, **N** is the normal vector to the band in the undeformed state. The band angle θ, which is used to represent the orientation of the imperfection band, is the angle between the direction of **N** and a loading direction of interest, which is taken as the X_2 direction, as shown in Figure 3.23. Homogeneous deformations inside

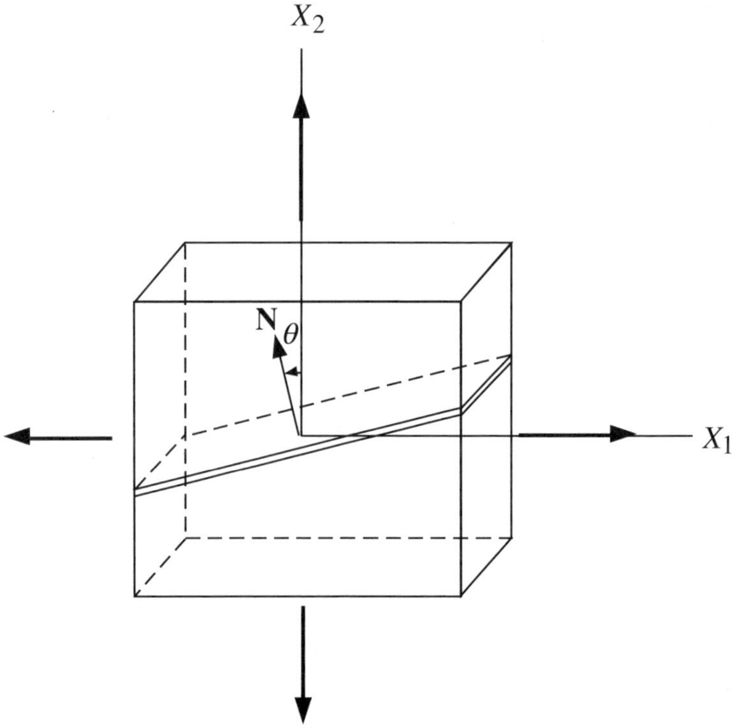

Figure 3.23 A material element having an imperfection band under plane stress conditions.

and outside the band are assumed to occur throughout a deformation history. Here, a superscript or subscript "b" represents a quantity inside the band, whereas a superscript or subscript "o" represents a quantity outside the band.

As the deformation proceeds, compatibility requires [Rice, 1976]

$$\dot{\mathbf{F}}^{b} = \dot{\mathbf{F}}^{o} + \mathbf{CN} \tag{3.10.1}$$

where $\dot{\mathbf{F}}$ represents the rate of the deformation gradient tensor, and $\mathbf{C}$ represents a vector denoting the discontinuity across the band. Also, equilibrium requires that the normal tractions are continuous over the band interface

$$\mathbf{N} \cdot \dot{\mathbf{t}}^{b} = \mathbf{N} \cdot \dot{\mathbf{t}}^{o} \tag{3.10.2}$$

where $\dot{\mathbf{t}}$ represents the rate of the first Piola-Kirchhoff stress tensor $\mathbf{t}$. Combining the rate form of the compatibility equations in Eq. 3.10.1, the rate form of the equilibrium equations in Eq. 3.10.2, and the constitutive relation between $\dot{\mathbf{t}}$ and $\dot{\mathbf{F}}$ gives a set of equations for the components of $\mathbf{C}$.

From the bifurcation viewpoint, when a possible solution exists for the components of $\mathbf{C}$, plastic localization is said to be possible. From the initial imperfection viewpoint, for a given a prescribed deformation history outside the band $\dot{\mathbf{F}}^{o}$, the initial conditions, and some form of imperfection, the set of equations for the components of $\mathbf{C}$ can be solved incrementally to determine the deformation history inside the band. The condition of failure or plastic localization is reached when the ratio of the normal component of the deformation gradient rate in the direction of $\mathbf{N}$ inside the band to that outside the band becomes infinity. Note that a reduced thickness in the imperfection band was assumed in the approach of Marciniak and Kuczynski [1967] to examine the necking instability of sheet metals. Many other types of imperfections, such as reduced yield stress or hardening exponent, also can be assumed to analyze the necking instability under biaxial straining conditions. For example, see Needleman and Triantafyllidis [1978]. Marciniak and Kuczynski's initial imperfection approach has been widely adopted to simulate the forming limit diagram that is used to provide a guideline on the forming limit of sheet metals under stretching conditions.

The forming limit diagram is widely used in the sheet forming industry to specify the forming limit of sheet metals [Keeler and Backofen, 1964; Goodwin, 1968]. A typical strain-based forming limit diagram from limiting dome height (LDH) tests for aluminum 5754 sheets is shown in Figure 3.24. The strain-based forming limit diagram is expressed in terms of the two principal engineering strains E_{11} and E_{22} in the X_1 and X_2 directions. On the right side or the stretching side of the forming limit diagram, the experimental results indicate that the necking band is perpendicular to the major principal strain direction X_2. During a proportional deformation history, the principal stretching direction does not change, and the necking band angle remains unchanged. On the left side or the draw side of the forming limit diagram, the experimental results indicate that the necking band is inclined to the major principal strain direction X_2. During a proportional deformation history, the necking band angle changes. The rotation of the band and an objective stress rate such as a Jaumann stress rate must be considered to simulate correctly the necking band formation. Recently, a stress-based forming limit criterion was proposed by Stoughton [2000].

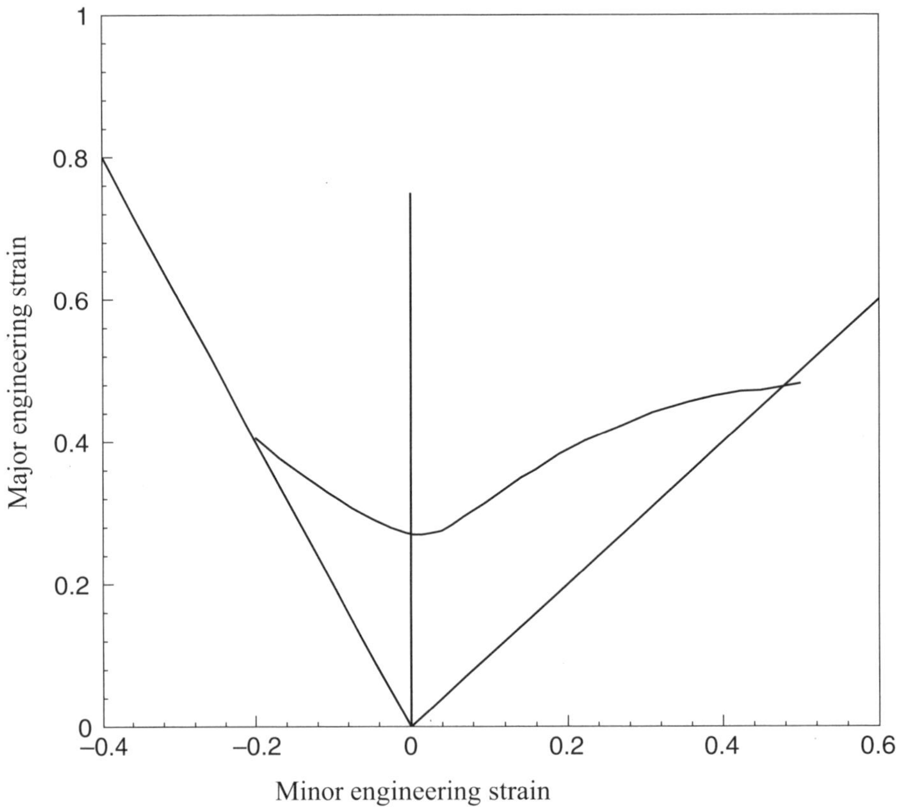

Figure 3.24 A typical strain-based forming limit diagram from limiting dome height (LDH) tests for aluminum 5754 sheets.

During the forming operation, a critical material element may experience significant nonproportional loading conditions. In this case, the effects of the loading paths on the forming limit must be considered. For example, Graf and Hosford [1993 and 1994] have examined the effects of strain paths on the forming limit. Huang *et al.* [2000] examined the nonproportional loading paths of a critical sheet element in the forming operation of a fender and modeled the failure of the critical sheet element using an initial imperfection approach.

For low-carbon steel sheets or more ductile aluminum 5754 sheets, the sheet metal fails mostly by necking. The Marciniak and Kuczynski approach based on a plane stress formulation for necking instability is quite suitable. However, for some aluminum sheets such as aluminum 6111 sheets where the ductility is quite limited, the sheet fails by shear localization inclined to the sheet surface. In this case, the failure mode is three-dimensional in nature. Both the necking and shear localization must be considered to model the failure process correctly. A combined necking and shear localization analysis was proposed to model the failure process (Chien *et al.* [2004]).

3.11 Modeling of Failure Processes

An accurate description of the plastic behavior and ductile failure processes of sheet metals under biaxial loading conditions is necessary for accurate prediction of the stretching failure in sheet metal forming processes. Ductile failure processes in metals usually involve nucleation, growth, and coalescence of microvoids. Figure 3.25 schematically

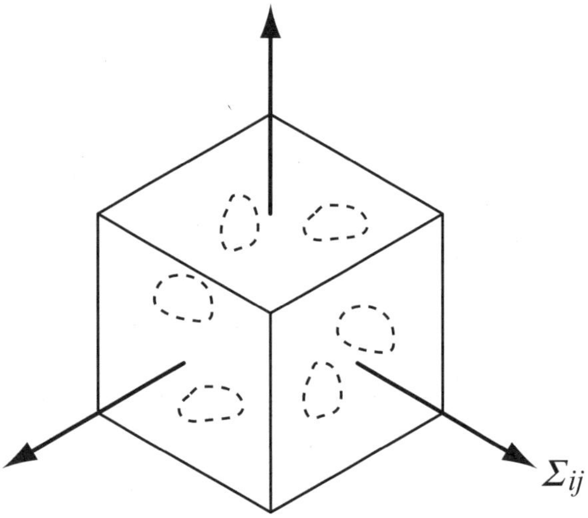

Figure 3.25 A material element containing a few voids.

shows a material element containing a few voids. To model the plastic flow and failure of these ductile materials with voids, Gurson [1977] conducted an upper bound analysis of simplified models containing voids and proposed an approximate yield condition for porous materials where the matrices obey the von Mises yield condition. Gurson's yield condition is expressed as

$$\Phi = \Phi(\Sigma, \bar{\sigma}, f) = \left(\frac{\Sigma_e}{\bar{\sigma}}\right)^2 + 2f\cosh\left(\frac{3}{2}\frac{\Sigma_m}{\bar{\sigma}}\right) - 1 - f^2 = 0 \tag{3.11.1}$$

where Φ represents the yield function, Σ represents the macroscopic average stress tensor, f represents the void volume fraction, $\bar{\sigma}$ represents the average flow stress of the matrix, Σ_m $(= \Sigma_{kk}/3)$ represents the macroscopic mean stress, and Σ_e represents the macroscopic effective stress based on the von Mises yield condition as

$$\Sigma_e = \left(\frac{3}{2}\Sigma_{ij}\Sigma_{ij}\right)^{\frac{1}{2}} \tag{3.11.2}$$

Figure 3.26 shows schematically the Gurson yield condition plotted in the $\Sigma_m - \Sigma_e$ plane for different values of the void volume fraction f. When f is equal to 0, the yield condition reduces to the von Mises yield condition. As shown in Figure 3.26, the yield contour for $f = 0$ is represented by a horizontal line. However, when f is not equal to 0, the yield condition is not only a function of the macroscopic effective stress Σ_e but also a function of the macroscopic mean stress Σ_m and the void volume fraction f. The yield contour for $f \neq 0$ now intercepts with the Σ_m axis. As f increases, the yield contour becomes smaller, and the load-carrying capacity of the material decreases.

Tvergaard [1981 and 1982] introduced three additional fitting parameters in the Gurson yield condition by comparing the results of shear band instability in square arrays of cylindrical holes and axisymmetric spherical voids based on his finite element models with those based on the Gurson yield condition. The yield condition is now written as

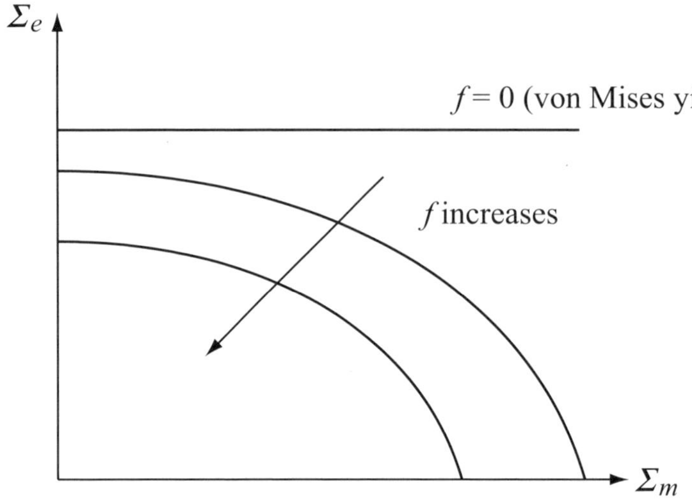

Figure 3.26 The Gurson yield condition for different values of the void volume fraction f.

$$\Phi = \Phi(\Sigma, \bar{\sigma}, f) = \left(\frac{\Sigma_e}{\bar{\sigma}}\right)^2 + 2q_1 f \cosh\left(q_2 \frac{3}{2}\frac{\Sigma_m}{\bar{\sigma}}\right) - 1 - q_3 f^2 = 0 \qquad (3.11.3)$$

where q_1, q_2, and q_3 are the fitting parameters. For sheet metals, when the plastic flow of the matrix is characterized by Hill's anisotropic yield function, a Gurson-type yield function for porous materials with consideration of planar plastic anisotropy is proposed by Wang *et al.* [2004] as

$$\begin{aligned}\Phi &= \Phi(\Sigma, \bar{\sigma}, f) \\ &= \left(\frac{\Sigma_e}{\bar{\sigma}}\right)^2 + 2q_1 f \cosh\left(q_2 \sqrt{\frac{1+2\bar{R}}{6(1+\bar{R})}}\frac{3\Sigma_m}{\bar{\sigma}}\right) - 1 - q_3 f^2 = 0\end{aligned} \qquad (3.11.4)$$

where Σ_e represents the macroscopic effective stress, $\bar{\sigma}$ is the matrix flow stress under in-plane uniaxial loading in the rolling direction, f is the void volume fraction, $\bar{R}$ is an average anisotropic parameter, Σ_m is the macroscopic mean stress, and q_1, q_2, and q_3 are the fitting parameters that are determined by a finite element analysis of a unit cell with a spherical void. The macroscopic effective stress Σ_e is defined as

$$\begin{aligned}\Sigma_e = \Bigg\{\frac{1}{1+R_0}\Bigg[&\frac{R_0}{R_{90}}(\Sigma_{22} - \Sigma_{33})^2 + (\Sigma_{33} - \Sigma_{11})^2 + R_0(\Sigma_{11} - \Sigma_{22})^2 \\ &+ \frac{(2R_{45}+1)(R_0 + R_{90})}{R_{90}}\left(\Sigma_{23}^2 + \Sigma_{31}^2 + \Sigma_{12}^2\right)\Bigg]\Bigg\}^{1/2}\end{aligned} \qquad (3.11.5)$$

The average anisotropic parameter $\bar{R}$, which typically is used to characterize the average planar anisotropy, is defined as

$$\bar{R} = \frac{R_0 + 2R_{45} + R_{90}}{4} \tag{3.11.6}$$

When planar isotropy is considered, $R_0 = R_{45} = R_{90} = R$. Then the anisotropic Gurson yield function becomes the one proposed by Liao *et al.* [1997]

$$\Phi = \left(\frac{\Sigma_e}{\bar{\sigma}}\right)^2 + 2q_1 f \cosh\left(q_2 \sqrt{\frac{1+2R}{6(1+R)}} \frac{3\Sigma_m}{\bar{\sigma}}\right) - 1 - q_3 f^2 = 0 \tag{3.11.7}$$

Here, the macroscopic effective stress Σ_e in Eq. 3.11.7 becomes

$$\begin{aligned}\Sigma_e = \Bigg\{\frac{1}{1+R}\Big[&(\Sigma_{22} - \Sigma_{33})^2 + (\Sigma_{33} - \Sigma_{11})^2 + R(\Sigma_{11} - \Sigma_{22})^2 \\ &+ 2(2R+1)\left(\Sigma_{23}^2 + \Sigma_{31}^2 + \Sigma_{12}^2\right)\Big]\Bigg\}^{1/2}\end{aligned} \tag{3.11.8}$$

Based on the work of Liao *et al.* [1997], the basic form of the yield condition in Eq. 3.11.7 also works for planar isotropic matrix materials characterized by a non-quadratic yield condition [Hill, 1979]. For a more complex non-quadratic yield condition such as Yld2000-2d proposed by Barlat *et al.* [2003], the basic form of the yield condition should be similar to that found in Eq. 3.11.4.

For isotropic von Mises materials, $R = 1$. Then the original Gurson yield function with the three fitting parameters is recovered as

$$\Phi = \left(\frac{\Sigma_e}{\bar{\sigma}}\right)^2 + 2q_1 f \cosh\left(q_2 \frac{3\Sigma_m}{2\bar{\sigma}}\right) - 1 - q_3 f^2 = 0 \tag{3.11.9}$$

Here, the macroscopic effective stress Σ_e in Eq. 3.11.9 becomes

$$\begin{aligned}\Sigma_e = \Bigg\{\frac{1}{2}\Big[&(\Sigma_{22} - \Sigma_{33})^2 + (\Sigma_{33} - \Sigma_{11})^2 + (\Sigma_{11} - \Sigma_{22})^2 \\ &+ 6\left(\Sigma_{23}^2 + \Sigma_{31}^2 + \Sigma_{12}^2\right)\Big]\Bigg\}^{1/2}\end{aligned} \tag{3.11.10}$$

The Gurson yield condition with consideration of plastic anisotropy can be used to model the necking instability and/or the shear localization, as outlined in the last section. More information can be found in Huang *et al.* [2000] and Chien *et al.* [2004].

3.12 References

Armstrong, P.J. and Frederick, C.O., "A Mathematical Representation of the Multiaxial Bauschinger Effects," Central Electricity Generating Board Report RD/B/N 731, 1966.

Barlat, F., Aretz, H., Yoon, J.W., Karabin, M.E., Brem, J.C., and Dick, R.E., "Linear Transformation-Based Anisotropic Yield Function, *International Journal of Plasticity*, Vol. 21, 2005, p. 1009.

Barlat, F., Brem, J.C., Yoon, J.W., Chung, K., Dick, R.E., Lege, D.J., Pourboghrat, F., Choi, S.-H., and Chu, E., "Plane Stress Yield Function for Aluminum Alloy Sheets—Part 1: Theory," *International Journal of Plasticity*, Vol. 19, 2003, p. 1297.

Barlat, F., Maeda, Y., Chung, K., Yanagawa, M., Brem, J.C., Hayashida, Y., Lege, D.J., Matsui, K., Murtha, S.J., Hattori, S., Becker, R.C., and Makosey, S., "Yield Function Development for Aluminum Alloy Sheets," *Journal of the Mechanics and Physics of Solids*, Vol. 45, 1997, p. 1727.

Chaboche, J.L., "Viscoplastic Constitutive Equations for the Description of Cyclic and Anisotropic Behavior of Metals," *Bull. de l'Acad. Polonaise des Sciences, Sevie Sc. et Techn*. Vol. 25, 1977, p. 33.

Chien, W.-Y., Pan, J., and Tang, S.C., "A Combined Necking and Shear Localization Analysis for Aluminum Sheets Under Biaxial Stretching Conditions," *International Journal of Plasticity*, Vol. 20, 2004, p. 1953.

Choi, K.S., "Residual Stress near Fillet and Roller Fatigue in Crankshaft Fillet Rolling Process," Ph.D. thesis, University of Michigan, Ann Arbor, MI, 2007.

Chu, C.C., "A Three Dimensional Model of Anisotropic Hardening in Metals and Its Application to the Analysis of Sheet Metal Formability," *Journal of the Mechanics and Physics of Solids*, Vol. 32, 1984, p. 197.

Dafalias, Y.F. and Popov, E.P., "A Model of Nonlinear Hardening Materials for Complex Loading," *Acta Mechanica*, Vol. 21, 1975, p. 173.

Dafalias, Y.F. and Popov, E.P., "Plastic Internal Variables Formalism of Cyclic Plasticity," *Journal of Applied Mechanics*, Vol. 98, 1976, p. 645.

Drucker, D.C., "A More Fundamental Approach to Plastic Stress-Strain Relations," Proceedings of the 1st U.S. National Congress of Applied Mechanics, American Society of Mechanical Engineers, 1951, p. 487.

Eringen, A.C., *Mechanics of Continua*, John Wiley & Sons, New York, 1967.

Goodwin, G.M., "Application of Strain Analysis to Sheet Metal Forming Problems in the Press Shop," SAE Paper No. 680093, Society of Automotive Engineers, Warrendale, PA, 1968.

Graf, A. and Hosford, W.F., "Calculation of Forming Limit Diagrams for Changing Strain Paths," *Metallurgical Transactions A*, Vol. 24, 1993, p. 2497.

Graf, A. and Hosford, W.F., "The Influence of Strain-Path Changes on Forming Limit Diagrams of Al 6111 T4," *International Journal of Mechanical Sciences*, Vol. 36, 1994, p. 897.

Gurson, A.L., "Continuum Theory of Ductile Rupture by Void Growth: Part I—Yield Criteria and Flow Rules for Porous Ductile Media," *Journal of Engineering Materials and Technology*, Vol. 99, 1977, p. 2.

Hershey, A.V., "The Plasticity of an Isotropic Aggregate of Anisotropic Face-Centered Cubic Crystals," *Journal of Applied Mechanics*, Vol. 76, 1954, p. 241.

Hill, R., *The Mathematical Theory of Plasticity*, Clarendon Press, Oxford, U.K., 1950.

Hill, R., "Theoretical Plasticity of Textured Aggregates," *Mathematical Proceedings of the Cambridge Philosophical Society*, Vol. 85, 1979, p. 179.

Hill, R., "A Theory of the Yielding and Plastic Flow of Anisotropic Metals," *Proceedings of the Royal Society of London,* Series A, Vol. 193, 1948, p. 281.

Hosford, W.F., "A Generalized Isotropic Yield Criterion," *Journal of Applied Mechanics*, Vol. 69, 1972, p. 607.

Huang, H.-M., Pan, J., and Tang, S.C., "Failure Prediction in Anisotropic Sheet Metals Under Forming Operations with Consideration of Rotating Principal Stretch Directions," *International Journal of Plasticity*, Vol. 16, 2000, p. 611.

Karafillis, A.P. and Boyce, M.C., "A General Anisotropic Yield Criterion Using Bounds and a Transformation Weighting Tensor," *Journal of the Mechanics and Physics of Solids*, Vol. 41, 1993, p. 1859.

Keeler, S.P. and Backofen, W.A., "Plastic Instability and Fracture in Sheets Stretched over Rigid Punches, *ASM Transactions Quarterly*, Vol. 56, 1964, p. 25.

Krieg, R.D., "A Practical Two Surface Plasticity Theory," *Journal of Applied Mechanics*, Vol. 42, 1975, p. 641.

Lee, E.H., "Elastic-Plastic Deformation at Finite Strains," *Journal of Applied Mechanics*, Vol. 36, 1969, p. 1.

Liao, K.-C., Friedman, P.A., Pan, J., and Tang, S.C., "Texture Development and Plastic Anisotropy of B.C.C. Strain Hardening Sheet Metals," *International Journal of Solids and Structures*, Vol. 35, 1998, p. 5205.

Liao, K.-C., Pan, J., and Tang, S.C., "Approximate Yield Criteria for Anisotropic Porous Ductile Sheet Metals," *Mechanics of Materials*, Vol. 26, 1997, p. 213.

Logan, R.W. and Hosford, W.F., "Upper-Bound Anisotropic Yield Locus Calculations Assuming <111> Pencil Glide," *International Journal of Mechanical Sciences*, Vol. 22, 1980, p. 419.

Marciniak, Z. and Kuczynski, K., "Limit Strains in the Processes of Stretch-Forming Sheet Metal," *International Journal of Mechanical Sciences*, Vol. 9, 1967, p. 609.

Mroz, Z., "On the Description of Anisotropic Work-Hardening," *Journal of the Mechanics and Physics of Solids*, Vol. 15, 1967, p. 163.

Needleman, A. and Triantafyllidis, N., "Void Growth and Local Necking in Biaxial Stretched Sheets," *Journal of Engineering Materials and Technology*, Vol. 100, 1978, p. 164.

Peirce, D., Shih, C.F., and Needleman, A., "A Tangent Modulus Method for Rate Dependent Solids," *Computers & Structures*, Vol. 18, 1984, p. 875.

Prager, W., "The Theory of Plasticity: A Survey of Recent Achievements," Proceedings of the Institution of Mechanical Engineers, Vol. 169, 1955, p. 41.

Rice, J.R., "The Localization of Plastic Deformation," Proceedings of the 14th International Congress on Theoretical and Applied Mechanics, Vol. I, Koiter, W.T. (ed.), Delft, North-Holland, 1976, p. 207.

Rice, J.R., "On the Structure of Stress-Strain Relations for Time-Dependent Plastic Deformation in Metals," *Journal of Applied Mechanics*, Vol. 37, 1970, p. 728.

Stoughton, T.B., "A General Forming Limit Criterion for Sheet Metal Forming," *International Journal of Mechanical Sciences*, Vol. 42, 2000, p. 1.

Tang, S.C., "An Anisotropic Hardening Rule for the Analysis of Sheet Metal Forming Operations," Advanced Technology of Plasticity 1990, Proceedings of the Third International Conference on the Technology of Plasticity (3ICTP), Kyoto, Japan, Vol. 3, 1990, p. 1149.

Tvergaard, V., "Influence of Voids on Shear Band Instabilities Under Plane Strain Conditions," *International Journal of Fracture*, Vol. 17, 1981, p. 389.

Tvergaard, V., "On Localization in Ductile Materials Containing Spherical Voids," *International Journal of Fracture*, Vol. 18, 1982, p. 237.

Wang, D.-A., Pan, J., and Liu, S.-D., "An Anisotropic Gurson Yield Criterion for Porous Ductile Sheet Metals with Planar Anisotropy," *International Journal of Damage Mechanics*, Vol. 13, 2004, p. 7.

Wang, N.-M. and Wenner, M.L., "Elastic-Viscoplastic Analyses of Simple Stretch Forming Problems," in *Mechanics of Sheet Metal Forming*, Symposium on Mechanics of Sheet Metal Forming, Warren, MI, 1977, Plenum Press, New York, 1978, p. 367.

Ziegler, H., "A Modification of Prager's Hardening Rule," *Quarterly of Applied Mathematics*, Vol. 17, 1959, p. 55.

CHAPTER

4

Mathematical Models for Sheet Metal Forming Processes

4.1 Governing Equations for Simulation of Sheet Metal Forming Processes

To analyze various forming operations in Chapter 1, we need to establish the governing equations of motion in a three-dimensional (3-D) space. We first review the equations of motion for 3-D continua and then discretize the equations so that we can use numerical methods to solve these equations. For analyses of sheet metal forming processes, neither displacements nor external forces are specified in most cases. We must solve a surface contact problem with friction between the tool and the sheet surfaces to determine the contact forces. In this chapter, we will discuss the formulation for contact problems.

4.2 Equations of Motion for Continua

We adopt a three-dimensional rectangular Cartesian coordinate system in space as the reference coordinate system to derive the equations of motion. Because of extensive plastic deformation in forming processes, which is path dependent and nonconservative, we use the principle of virtual work as the basis to establish the equations of motion as

$$\int_v \rho \ddot{u}_i \, \delta u_i \, dv + \int_v \kappa \dot{u}_i \, \delta u_i \, dv + \int_v \sigma_{ij} \delta \varepsilon_{ij} \, dv = \int_v f_b^i \, \delta u_i \, dv + \int_s f_s^i \, \delta u_i \, ds + \int_c f_c^i \, \delta u_i \, ds \tag{4.2.1}$$

In Eq. 4.2.1, v represents the current deformed volume of a sheet, s is the current deformed surface of a sheet, c is the region for surface contact, ρ is the mass density of the sheet, and κ is the damping parameter of the sheet. Note that we assume that

kdv and ρdv are independent of deformation. The definitions of the remaining field variables in Eq. 4.2.1 in the component form are as follows:

σ_{ij} = Cauchy stress tensor

f_b^i = body force vector (including weight)

f_s^i = surface force vector (including moment components)

f_c^i = contact force vector

δu_i = virtual displacement vector

$\delta \varepsilon_{ij}$ = virtual strain tensor due to virtual displacement vector δu_i

$\ddot{u}_i$ = acceleration vector

$\dot{u}_i$ = velocity vector

Note that the subscripts or indices for these variables range from 1 to 3. Because Eq. 4.2.1 is a nonlinear functional equation, we discretize it using the finite element method to adopt numerical methods to solve the unknown displacements u_i on a digital computer.

4.3 Equations of Motion in Discrete Form

We use the finite element method to discretize the spatial variables in the equations of motion from the virtual work equation in Eq. 4.2.1 and the finite difference method (to be discussed in Chapter 7) to approximate the velocity and acceleration vector fields in terms of the discrete spatial variables and time increments. In Chapter 6, we will discuss the reason for the use of the finite element method in simulations of sheet metal forming processes instead of the finite difference method to discretize the spatial variables.

Because we emphasize shell elements in our sheet metal forming analyses, the equations of motion are formulated for finite element models with shell elements. Therefore, the displacement, velocity, or acceleration vectors containing rotation in addition to translation are adopted in the equations of motion or equilibrium. These vectors are called the generalized displacement, velocity, or acceleration vectors. Consequently, the generalized surface force vector f_s^i includes the moment components as well.

In the finite element method, we interpolate the displacement field u_i of the sheet from the generalized displacement vector $\mathbf{U}$ at a set of discrete points known as nodes by the use of interpolation functions (or shape functions). We express the relationship in the vector and matrix form as

$$\mathbf{u} = \mathbf{\Phi U} \qquad (4.3.1a)$$

where $\mathbf{\Phi}$ is the interpolation matrix, and the displacement field u_i are the components of the displacement vector $\mathbf{u}$. The virtual displacement vector $\delta\mathbf{u}$ is related to the virtual generalized displacement vector $\delta\mathbf{U}$ as

$$\delta\mathbf{u} = \mathbf{\Phi}\delta\mathbf{U} \qquad (4.3.1b)$$

where $\mathbf{\Phi}$ is a function of the coordinates of the position vector for the material point of interest. Note that $\mathbf{U}$ contains discrete variables of translation and variables other than those of translation at all nodes. This is the reason that we call $\mathbf{U}$ the generalized displacement vector. We apply Eq. 4.3.1 to discretize the spatial variables in the virtual work equation in Eq. 4.2.1.

4.3.1 Internal Nodal Force Vector

The term $\int_v \sigma_{ij}\delta\varepsilon_{ij}\,dv$ in Eq. 4.2.1 is the virtual work done by the stresses. To discretize this term, the form of the formulation depends on the kind of finite elements used. We will postpone the discretization until Chapter 6, where two different thin shell elements are discussed. However, at the present time for deriving the equations of motion, we may express the term in the following discrete form as

$$\int_v \sigma_{ij}\delta\varepsilon_{ij}\,dv = (\mathbf{Fi})^T\,\delta\mathbf{U} \tag{4.3.2}$$

where $\mathbf{Fi}$ is the generalized internal nodal force vector including the moment components, and $\delta\mathbf{U}$ is the generalized virtual displacement vector including variables of translation and variables other than those of translation at all nodes. The superscript "T" represents the transpose of $\mathbf{Fi}$ in the vector form. For simplicity, we drop the word "generalized" throughout the remainder of this book. From this time, the displacement vector for all nodes is understood for the generalized displacement vector.

4.3.2 External Nodal Force Vector

The term $\int_v f_b^i\,\delta u_i\,dv + \int_s f_s^i\,\delta u_i\,ds$ in Eq. 4.2.1 is the virtual work done by the body and surface forces. We express the virtual work in the discrete form using the interpolation function in Eq. 4.3.1b as

$$\left(\int_v \mathbf{\Phi}^T\mathbf{f}_b dv + \int_s \mathbf{\Phi}^T\mathbf{f}_s ds\right)\delta\mathbf{U} = (\mathbf{Fe})^T\,\delta\mathbf{U} \tag{4.3.3}$$

where $\mathbf{f}_b$ and $\mathbf{f}_s$ represent the body and surface force vectors, respectively, and $\mathbf{Fe}$ is the external nodal force vector, which may include the moment components. Note that f_b^i and f_s^i in Eq. 4.2.1 represent the components of $\mathbf{f}_b$ and $\mathbf{f}_s$, respectively. From Eq. 4.3.3, we can obtain the external nodal force vector as

$$\mathbf{Fe} = \int_v \mathbf{\Phi}^T\mathbf{f}_b dv + \int_s \mathbf{\Phi}^T\mathbf{f}_s ds \tag{4.3.4}$$

4.3.3 Contact Nodal Force Vector

The term $\int_v f_c^i\delta u_i\,dv$ in Eq. 4.2.1 is the virtual work done by the contact forces. We express the virtual work in the discrete form using the interpolation function in Eq. 4.3.1b as

$$\left(\int_c \mathbf{\Phi}^T \mathbf{f}_c ds\right)\delta\mathbf{U} = (\mathbf{Fc})^T \delta\mathbf{U} \tag{4.3.5}$$

where $\mathbf{f}_c$ represents the contact force vector, and **Fc** is the contact nodal force vector, which in this book does not include the moment components. Note that f_c^i in Eq. 4.2.1 represents the components of $\mathbf{f}_c$. Therefore, we modify the interpolation matrix $\mathbf{\Phi}$ with the components to interpolate the moments being set to zero. From Eq. 4.3.5, we can obtain the contact nodal force vector as

$$\mathbf{Fc} = \int_s \mathbf{\Phi}^T \mathbf{f}_c ds \tag{4.3.6}$$

We use the Coulomb friction law to compute the contact forces. The formulation for computing contact forces based on the Coulomb friction law will be discussed in detail in Section 4.5 of this chapter.

4.3.4 Mass and Damping Matrices

The first two terms on the left side of Eq. 4.2.1 are due to the inertia and the damping under dynamic loading conditions. We want to discretize these two dynamic force fields in terms of the variables at the finite element nodes. We interpolate the velocity and acceleration fields from the velocities and the accelerations, respectively, at the nodes by using the same interpolation function as that for the displacement field in Eq. 4.3.1a as

$$\dot{\mathbf{u}} = \mathbf{\Phi}\dot{\mathbf{U}} \tag{4.3.7a}$$

$$\ddot{\mathbf{u}} = \mathbf{\Phi}\ddot{\mathbf{U}} \tag{4.3.7b}$$

where $\dot{\mathbf{u}}$ and $\ddot{\mathbf{u}}$ are the velocity and acceleration vector fields, respectively, and $\dot{\mathbf{U}}$ and $\ddot{\mathbf{U}}$ are the velocity and acceleration vectors, respectively, of the nodes in a finite element model.

The virtual work by the inertia forces is discretized using the interpolation function in Eqs. 4.3.7b and 4.3.1b as

$$\int_v \rho \ddot{u}_i \, \delta u_i \, dv = \left(\mathbf{M}\ddot{\mathbf{U}}\right)^T \delta\mathbf{U} \tag{4.3.8}$$

where the mass matrix **M** is defined as

$$\mathbf{M} = \int_v \rho \mathbf{\Phi}^T \mathbf{\Phi} dv \tag{4.3.9}$$

Note that the mass matrix **M** is symmetric. The virtual work by the damping forces is discretized using the interpolation function in Eqs. 4.3.7a and 4.3.1b as

$$\int_v \kappa \dot{u}_i \, \delta u_i \, dv = \left(\mathbf{C}\dot{\mathbf{U}}\right)^T \delta\mathbf{U} \tag{4.3.10}$$

where the damping matrix **C** is defined as

$$C = \int_v \kappa \mathbf{\Phi}^T \mathbf{\Phi} dv \qquad (4.3.11)$$

Note that the damping matrix **C** also is symmetric.

4.3.5 Equations of Motion in Matrix Form

Now we can combine the virtual work done by the internal, external, contact, inertia, and damping forces in Eqs. 4.3.2, 4.3.3, 4.3.5, 4.3.8, and 4.3.10 and invoke the condition that $\delta\mathbf{U}$ is arbitrary to obtain the discrete equations of motion from Eq. 4.2.1 as

$$\mathbf{M\ddot{U}} + \mathbf{C\dot{U}} + \mathbf{Fi} = \mathbf{Fe} + \mathbf{Fc} \qquad (4.3.12)$$

Once the discrete equations of motion in Eq. 4.3.12 are set up, we can use numerical methods to integrate the equations as functions of the time t step by step. With the given initial conditions of the displacement vector $\mathbf{U}_0$ and the velocity vector $\dot{\mathbf{U}}_0$, the acceleration vector $\ddot{\mathbf{U}}$ can be computed from Eq. 4.3.12 with the specified external nodal force vector and the tool position at time 0. Using a time step size Δt, we may compute the response at any time $n\Delta t$ $(n = 1, 2, 3 ...)$ until a desired solution is reached. During the computation, we may change the step size Δt for accuracy and efficiency.

The speed of the tool hitting a sheet is approximately 200 mm/sec for typical automotive sheet stamping processes. Therefore, the inertia and the damping forces of the sheet can be neglected under such a low tool speed in the equations of motion. In this case, the equations of motion are reduced to the equations of equilibrium as

$$\mathbf{Fi} = \mathbf{Fe} + \mathbf{Fc} \qquad (4.3.13)$$

Because this set of equations is highly nonlinear, incremental numerical solutions with several hundred steps usually are required for accuracy. The solutions are called the quasi-static solutions. Because no dynamic effects are considered, undesirable oscillations usually can be avoided for the quasi-static solutions.

4.4 Tool Surface Models

Because thin sheets are more flexible than massive forming tools, the forming tools usually are assumed to be rigid. Therefore, forming tools are modeled as rigid surfaces for simplifying the contact conditions with sheets. On some areas, the contact pressure can be extremely high. It is feasible to model the forming tool as a deformable body only for the area of high contact pressure. A Winkler-type model instead of a three-dimensional elastic-plastic continuum model for the tool is suggested for a simpler computation. In this type of model, forces acting at a point of a deformable body generate deformation only at that point.

During a sheet contacting a rigid tool surface, only the sheet deforms. Therefore, we may establish a model of the tool surface with a mesh of rigid flat elements, triangular and/or quadrilateral, for simplicity, to cover the entire tool surface. Although these elements are C^0 continuous, they are accurate enough if we generate fine meshes in the areas of tight radii on the tool surface. In general, a flat element across an area of a tool surface with a slope change of less than 16° is sufficiently accurate. As will be shown in Chapter 7 in the section on solving contact problems, flat elements can avoid computational complexity.

Because tool surfaces usually are geometrically complex for automotive sheet metal parts, generation of a mesh to cover an entire tool surface requires a huge amount of work, even by an experienced engineer. Fortunately, interactive computer software is available to accomplish this tedious task.

4.5 Surface Contact with Friction

When a sheet contacts a tool surface, a contact force vector has a normal component from the reaction of the tool support and a tangential component from the friction of the interface between the sheet and the tool surface. In this book, only translational degrees of freedom are considered in the formulation of contact problems.

Figure 4.1 shows the positions of a sheet and a tool (punch) and a global Cartesian coordinate system. Referring to this figure, we note that $\mathbf{x}$ is the position vector of a generic point on the sheet, $\mathbf{x}_t$ is the position vector of a generic point on the tool surface, and $\mathbf{n}$ is the normal at the generic point on the tool surface. In reference to the global Cartesian coordinate system, the initial tool surface is expressed as

$$x_3 = S(x_\alpha) \tag{4.5.1}$$

where the subscript α has a range of 1 to 2. We assume the tool travels downward with the speed $\dot{U}_t$, without any rotation and without a change in its shape. Here, $\dot{U}_t$ is a scalar and is always positive. Therefore, the position of the traveling tool surface at time t is

$$x_3 = S(x_\alpha) - z(t) \tag{4.5.2a}$$

where the variable $z(t)$ is the tool travel distance. Hence, $z = \int_0^t \dot{U}_t dt$. For a stationary tool surface ($\dot{U}_t = 0$ or $z(t) = 0$), the position of the tool surface is independent of time t and is expressed as

$$x_3 = S(x_\alpha) \tag{4.5.2b}$$

Because the tool is assumed to be rigid and x_α are independent of time t, the material derivative of the tool position from Eq. 4.5.2a is

$$\frac{D\mathbf{x}_t}{Dt} = -\dot{U}_t \mathbf{e}_3 \tag{4.5.3}$$

In the analysis of sheet forming, the contact forces f_c^i in Eq. 4.2.1 are not specified explicitly in the contact area between the sheet and the tool surface. The contact forces

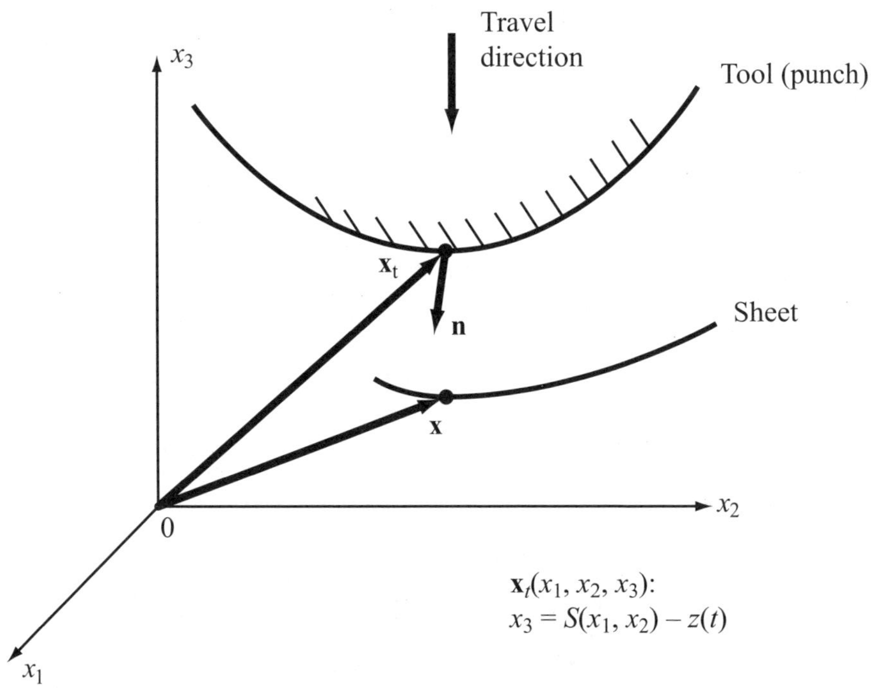

Figure 4.1 Tool and sheet positions.

from the contact pressure p and the frictional force vector $\mathbf{t}$ are not known. Additional conditions due to contact are required to solve for these unknowns. One of them is the friction law. Although much research has been done in modeling frictional forces other than the classical Coulomb friction law, the research results still are not universally accepted. In this book, we use the classical Coulomb friction law with a constant coefficient of friction. Based on past research, the Coulomb friction law appears to be good enough for the purpose of sheet forming simulations for design.

In the vector notation, the Coulomb friction law for sliding contact is

$$\mathbf{t} = -\mu p \frac{\dot{\mathbf{u}}_r}{\left\|\dot{\mathbf{u}}_r\right\|}, \quad \left\|\dot{\mathbf{u}}_r\right\| \neq 0 \tag{4.5.4}$$

where $\mathbf{t}$ is the frictional force vector, μ is the coefficient of friction (may be different for a die and a punch surface), p is the contact pressure (always positive), and $\dot{\mathbf{u}}_r$ is the velocity of a contact point on a sheet relative to the tool travel. Here, the notation $\| \ \|$ represents the magnitude of the vector quantity enclosed. When the relative velocity equals zero, the contact is sticking and

$$\left\|\mathbf{t}\right\| - \mu p < 0 \tag{4.5.5}$$

The other contact condition that must be considered is the geometric constraint where a contact point on the sheet cannot penetrate into the tool that in this book is assumed to be rigid. For simplicity, we assume that the shell does not have any thickness for consideration of contact. Therefore, the shell surface to contact the tool surface may be the deformed reference surface of the shell instead of the top or bottom surfaces of the shell. This simplification is valid only for thin sheets.

As an approximation, we use the finite element method instead of Eq. 4.5.4 to compute the contact forces. Following Wang and Budiansky [1978], we discretize the contact forces and express the contact nodal force vector $\mathbf{Fc}^l$ at node l in a finite element model of the sheet contacting with the tool surface.

We use the finite element method to compute the contact nodal force vector (no moment involved in contact) at each contact node. $\mathbf{Fc}$ represents the contact nodal force vector at all nodes contacting with the tool surfaces. The contact nodal force vector $\mathbf{Fc}^l$ at contact node l is expressed by

$$\mathbf{Fc}^l = P^l\mathbf{n} + \mathbf{T}^l \tag{4.5.6}$$

where P^l is the nodal force due to the contact pressure, and $\mathbf{n}$ is the normal to the tool surface at contact point l. The frictional nodal force vector $\mathbf{T}^l$ is

$$\mathbf{T}^l = -\mu P^l \frac{\dot{\mathbf{U}}_r^l}{\left\|\dot{\mathbf{U}}_r^l\right\|}, \qquad \left\|\dot{\mathbf{U}}_r^l\right\| \neq 0 \tag{4.5.7}$$

where $\dot{\mathbf{U}}_r^l$ is the relative velocity vector at contact node l on the reference surface of a deformed shell. For $\left\|\dot{\mathbf{U}}_r^l\right\| \neq 0$, the frictional force vector can be determined by this equation, and this is the case of sliding contact. For $\left\|\dot{\mathbf{U}}_r^l\right\| = 0$, the frictional force vector cannot be determined by this equation, and this is the case of sticking contact. In this case,

$$\left\|\mathbf{T}^l\right\| - \mu P^l < 0 \tag{4.5.8}$$

and the frictional force vector can be determined by computing the components of the reaction in the tangential directions.

Because a sheet cannot penetrate into the tool, the constraint condition is that the gap function at node l is

$$g_l(\mathbf{x}) = (\mathbf{x} - \mathbf{x}_t) \cdot \mathbf{n} \geq 0 \tag{4.5.9}$$

where $\mathbf{x}$ is the position vector of a possible contact node on the reference surface of a deformed shell, $\mathbf{x}_t$ is the position vector of the point on the tool surface such that $\left\|\mathbf{x} - \mathbf{x}_t\right\|$ is minimum, and $\mathbf{n}$ is the normal of the tool surface at the point $\mathbf{x}_t$. All are evaluated at time t. Note that for a rigid tool surface as considered in this book, $\mathbf{x}_t$ is not dependent on any deformation. There are various formulas to compute the contact nodal force vector $\mathbf{Fc}^l = P^l\mathbf{n} + \mathbf{T}^l$ at a contact node l. These formulas are described in Sections 4.5.1 through 4.5.3.

4.5.1 Formulation for the Direct Method

First we describe the direct method. In this method, we introduce the contact nodal force vector in Eq. 4.5.6 and the constraint condition in Eq. 4.5.9 to solve three more unknowns, which are P^l and the magnitude and direction of $\mathbf{T}^l$ on the tangent plane of the contact surface at the contact node l, in addition to the equations of motion. Therefore, the formulation outlined earlier is sufficient to generate solutions.

4.5.2 Formulation for the Lagrangian Multiplier Method

Now we discuss the Lagrangian multiplier method. Let the contact force vector **fc** at a generic contact point be

$$\mathbf{fc} = p\mathbf{n} + t_f\mathbf{s} \tag{4.5.10}$$

where **s** is a unit vector on the tangent surface of a tool passing point $\mathbf{x}_t$, and t_f is the frictional force on the tangent plane. For plane problems of line contact, **n** and **s** form a right-hand basis following Bathe [1996] and Bathe and Bouzinov [1997]. For three-dimensional simulations of forming processes, under surface contact conditions, we still can define **n** and **s** forming a right-hand basis on a plane determined by two unit vectors **n** and $\dot{\mathbf{u}}_r/\|\dot{\mathbf{u}}_r\|$, as shown in Figure 4.2. For sliding contact where $\|\dot{\mathbf{u}}_r\| \neq 0$, the Coulomb friction law in Eq. 4.5.4 gives

$$t_f\mathbf{s} = \mathbf{t} = -\mu p\frac{\dot{\mathbf{u}}_r}{\|\dot{\mathbf{u}}_r\|} \tag{4.5.11}$$

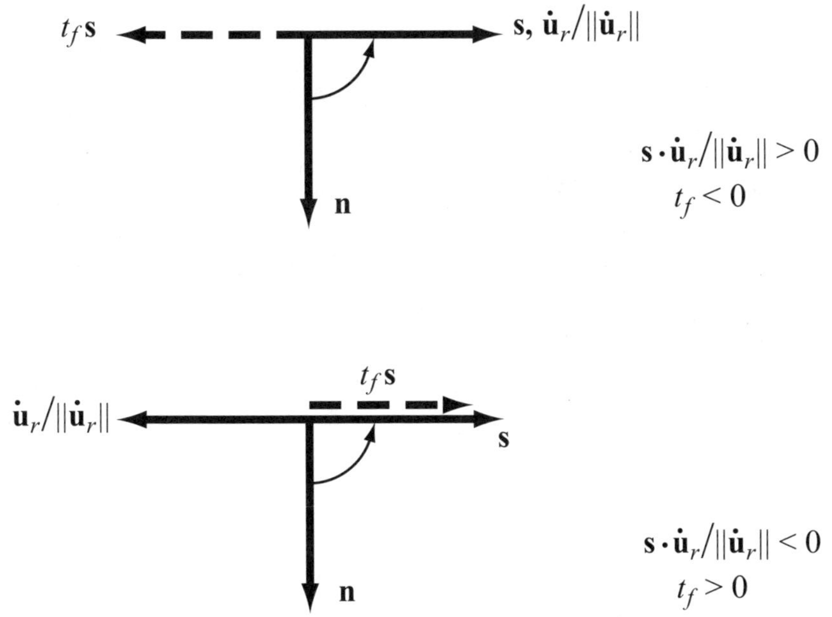

Figure 4.2 Definition of the vector **s** on the plane determined by the vectors **n** and $\dot{\mathbf{u}}_r$ for the Lagrangian multiplier method.

The tangent vector **s** cannot be determined for three-dimensional problems with sticking contact because the relative velocity $\|\dot{\mathbf{u}}_r\| = 0$. As mentioned previously, it can be determined by computing the components of the reaction in the tangential directions. Note that the lower-case $\dot{\mathbf{u}}_r$ is the relative velocity vector.

Following Bathe [1996], we introduce a Lagrangian multiplier to formulate the contact problem using the Lagrangian multiplier method. We set the Lagrangian multiplier λ such that

$$\begin{aligned} \lambda g &= 0 \\ \lambda &\geq 0 \\ g &\geq 0 \end{aligned} \tag{4.5.12}$$

where g is the gap function in Eq. 4.5.9, and the Lagrangian multiplier λ can be interpreted as the contact pressure p. The heavy solid lines with a slope discontinuity in Figure 4.3 show a schematic of λ as a function of g.

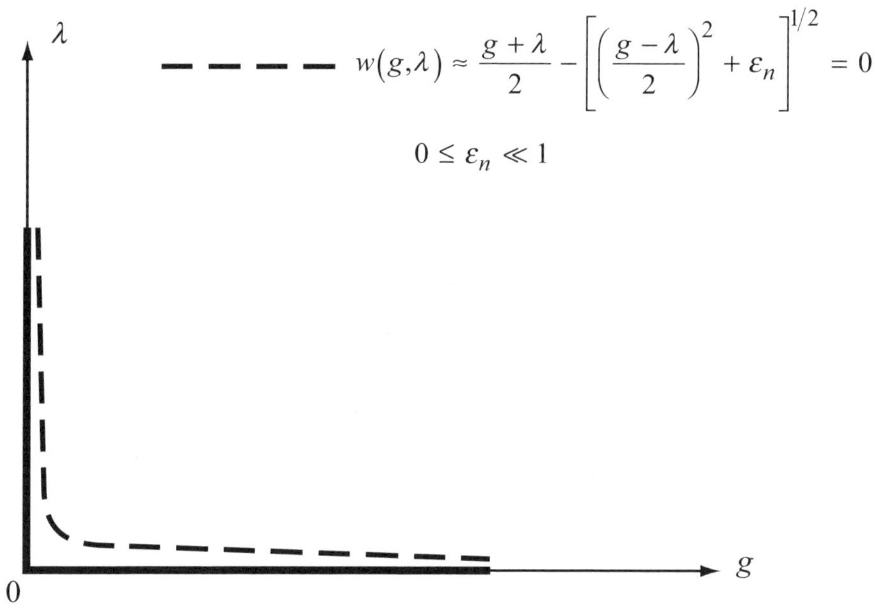

Figure 4.3 The Lagrangian multiplier λ versus the gap function g.

Now we change Bathe's sign and introduce another nondimensional variable τ as

$$\tau = -\frac{t_f}{\mu\lambda} \tag{4.5.13}$$

where $\mu\lambda > 0$ at a contact point. For sliding contact between the shell at a generic point $\mathbf{x}$ and the tool surface, $\|\dot{\mathbf{u}}_r\| \neq 0$. Therefore, Eq. 4.5.10 can be written in terms of λ and τ as

$$\mathbf{fc} = \lambda(\mathbf{n} - \mu\tau\mathbf{s}) \tag{4.5.14}$$

Combining Eqs. 4.5.11 and 4.5.13 gives

$$\tau \mathbf{s} = \frac{\dot{\mathbf{u}}_r}{\|\dot{\mathbf{u}}_r\|} \quad (4.5.15)$$

and $\mathbf{s}$ is parallel to $\dot{\mathbf{u}}_r$ but not necessarily pointing in the same direction, as shown in Figure 4.2. Bathe [1996] also introduced the relative speed of sliding contact as

$$\dot{u} = \dot{\mathbf{u}}_r \cdot \mathbf{s} \quad (4.5.16)$$

where $\dot{u}$ is a scalar but has a sign (positive or negative). Therefore, the Coulomb friction law can now be expressed for sticking contact as

$$\dot{u} = 0, \qquad |\tau| < 1 \quad (4.5.17a)$$

Although the frictional force vector is on the tangent plane, its direction cannot be determined (the vector $\mathbf{s}$ is not defined because $\|\dot{\mathbf{u}}_r\| = 0$).

For sliding contact, Eqs. 4.5.13, 4.5.15, and 4.5.16 give

$$\begin{aligned} &\dot{u} \neq 0 \\ &|\tau| = 1 \\ &sign(\tau) = sign(\dot{u}) \end{aligned} \quad (4.5.17b)$$

Therefore, Eq. 4.5.15 can be written as

$$\mathbf{s} = sign(\tau)\frac{\dot{\mathbf{u}}_r}{\|\dot{\mathbf{u}}_r\|} \quad (4.5.17c)$$

Combining Eqs. 4.5.16, 4.5.17b, and 4.5.17c yields

$$\dot{u} = sign(\tau)\|\dot{\mathbf{u}}_r\| = sign(\dot{u})\|\dot{\mathbf{u}}_r\| \quad (4.5.18)$$

The heavy solid lines with slope discontinuities in Figure 4.4 show a schematic of τ as a function of $\dot{u}$. In summary, we establish two contact conditions in terms of the Lagrangian multiplier λ and the frictional force variable τ as

$$w(g, \lambda) \equiv \lambda g = 0 \quad (4.5.19a)$$

and

$$v(\dot{u}, \tau) \equiv \tau - sign(\dot{u})1 = 0 \quad (4.5.19b)$$

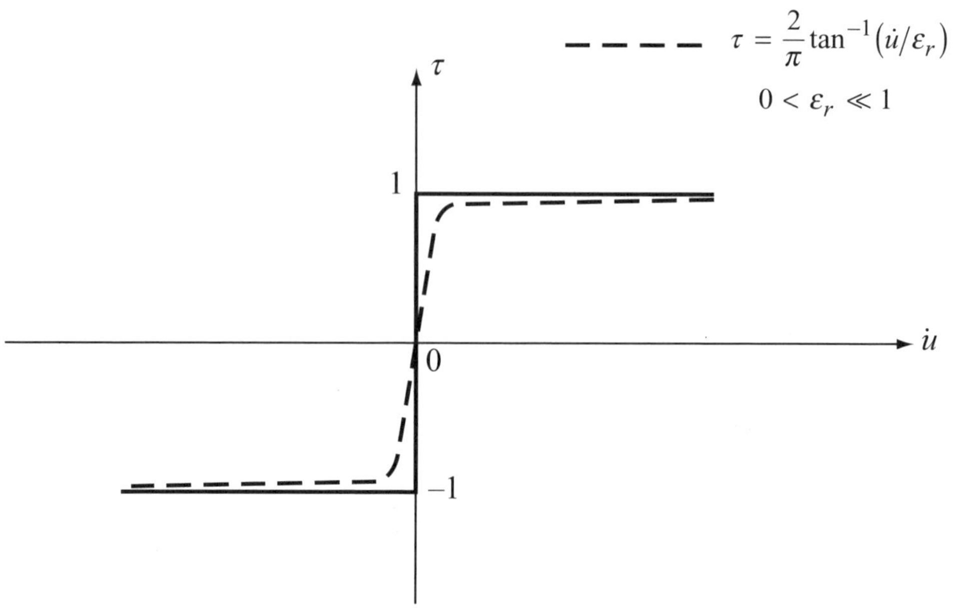

Figure 4.4 The frictional force variable τ versus the relative speed $\dot{u}$.

For the implicit method, which will be discussed in Chapter 7 of this book, the global tangent stiffness matrix is required. To establish the global tangent stiffness matrix, we must compute the partial derivatives of the contact conditions with respect to the nodal displacement component, the Lagrangian multiplier λ, and the frictional force variable τ. Note that in Eqs. 4.5.19a and 4.5.19b, $w(g, \lambda) \equiv \lambda g = 0$ is not differentiable at the origin, as shown by the heavy solid lines with a slope discontinuity in Figure 4.3, and $v(\dot{u}, \tau) \equiv \tau - sign(\dot{u})1 = 0$ is not differentiable at $\dot{u} = 0$, as shown by the heavy solid lines with slope discontinuities in Figure 4.4. In addition, the direction of the frictional force vector cannot be defined for sticking contact for three-dimensional simulations, and it is to be determined by the reaction on a tool surface. We must separate the sticking contact from the sliding contact. Therefore, the advantage to using the Lagrangian multiplier method is lost.

To make these contact conditions differentiable, Bathe and Bouzinov [1997] introduced some approximate functions (smooth and differentiable everywhere), such as

$$w(g, \lambda) \approx \frac{g + \lambda}{2} - \sqrt{\left(\frac{g - \lambda}{2}\right)^2 + \varepsilon_n} = 0, \quad 0 \le \varepsilon_n \ll 1 \tag{4.5.20a}$$

$$v(\dot{u}, \tau) \approx \tau - \frac{2}{\pi}\tan^{-1}\left(\frac{\dot{u}}{\varepsilon_r}\right) = 0, \quad 0 < \varepsilon_r \ll 1 \tag{4.5.20b}$$

or

$$v(\dot{u}, \tau) \approx \tau - \tan h\left(\frac{\dot{u}}{\varepsilon_r}\right) = 0, \quad 0 < \varepsilon_r \ll 1 \tag{4.5.20c}$$

The dashed lines in Figures 4.3 and 4.4 show λ versus g and τ versus $\dot{u}$ from the approximate equations in Eqs. 4.5.20a and 4.5.20b, respectively. These smooth functions do not satisfy Eqs. 4.5.19a and 4.5.19b exactly but with a small error (an order of the magnitude of ε_n or ε_r). Note that there are dimensional inconsistencies in Eq. 4.5.20. The variables λ, g, τ, and $\dot{u}$ should be nondimensionalized in actual computations. If we use the approximate equations for $v(\dot{u},\tau)$ in Eq. 4.5.20b, we also may avoid treating sticking contact separately because real sticking contact for Eq. 4.5.20b happens only at the origin, as shown in Figure 4.4. In that case, both $\dot{u}$ and τ are zero. Therefore, the approximate functions can be differentiated, and the advantage of the Lagrangian multiplier method can be kept for the use of the approximate equations.

Multiplying the above functions by the variations $\delta\lambda$ and $\delta\tau$, respectively, we establish the variational equation for the constraint conditions from the contact conditions as

$$\int_c \left[w(g,\lambda)\delta\lambda + v(\dot{u},\tau)\delta\tau \right] ds = 0 \tag{4.5.21}$$

Discretizing the virtual work term with the contact force in Eq. 4.5.10 using Eq. 4.5.13 and the variation equation for the constraint conditions in Eq. 4.5.21, we obtain the contact nodal forces for contact node l at time t as

$$\mathbf{Fc}^l = \lambda_l \left(\mathbf{n} - \mu \tau_l \mathrm{s} \right) \tag{4.5.22}$$

and the constraint conditions for contact node l at time t as

$$\mathbf{Fct}^l = \left[w(g_l,\lambda_l) \; v(\dot{u}_l,\tau_l) \right]^T = \mathbf{0} \tag{4.5.23}$$

The advantage of this method is that no approximation is used, except the two small numbers (ε_n and ε_r) being introduced to smooth the constraint functions. Therefore, the accuracy of this method is equivalent to that of using the direct method to solve a contact problem. However, new variables λ_l and τ_l for each contact point of a discrete system are introduced. In addition, the diagonal components of the tangent stiffness matrix for the implicit method of solution contain zeros, as shown in Chapter 7. Because the frictional force generates a nonconservative system, the global tangent stiffness matrix is no longer symmetric.

4.5.3 Formulation for the Penalty Method

From Eq. 4.5.6, the contact nodal force vector at contact node l in the finite element model contains two parts: the normal pressure and the frictional force, as

$$\mathbf{Fc}^l = P^l \mathbf{n} + \mathbf{T}^l \tag{4.5.24}$$

Following Simo and Laursen [1992], we define the Kuln-Tucker condition for the Coulomb friction law as

$$\Phi \equiv \left\| \mathbf{T}^l \right\| - \mu P^l \le 0 \tag{4.5.25a}$$

$$\dot{\mathbf{U}}_r^l = \xi \frac{\partial \Phi}{\partial \mathbf{T}^l}, \qquad \xi \geq 0 \tag{4.5.25b}$$

$$\xi \Phi = 0 \tag{4.5.25c}$$

where $\dot{\mathbf{U}}_r^l$ is the relative velocity of contact node l to the tool surface, and ξ is a scalar factor of proportionality.

Now we introduce the penalty parameters α_N and α_T, which are large positive numbers, as

$$P^l = \alpha_N \langle -g_l(\mathbf{x}) \rangle \tag{4.5.26a}$$

where $g_l(\mathbf{x})$ is the gap function as defined in Eq. 4.5.9, and the Macauley bracket is defined as

$$\langle -y \rangle = \frac{1}{2}\left[-y + |y|\right] \tag{4.5.26b}$$

Equation 4.5.25b is modified to

$$\dot{\mathbf{U}}_r^l - \xi \frac{\partial \Phi}{\partial \mathbf{T}^l} = \frac{1}{\alpha_T} \dot{\mathbf{T}}^l, \qquad \xi \geq 0 \tag{4.5.27}$$

We use the backward Euler integration scheme to integrate these equations and a return mapping strategy by Simo and Hughes [1991] to adjust $\Phi \leq 0$. At time $t = n\Delta t$, all variables including contact nodal forces are known. Now we would like to compute the variables at $t = (n+1)\Delta t$. In the iteration for the current time step, we set

$$P_{n+1}^{(k)} \approx \alpha_N \left\langle -g_l\left(\mathbf{x}_{n+1}^{(k-1)}\right)\right\rangle, \qquad \mathbf{x}_{n+1}^{(0)} = \mathbf{x}_n \tag{4.5.28}$$

at the kth iteration in the iterative scheme to solve the contact problem. Using the state of sticking contact, we establish a trial nodal frictional force vector $\overline{\mathbf{T}}_{n+1}^l$ as

$$\begin{aligned} \overline{\mathbf{T}}_{n+1}^l &= \mathbf{T}_n^l + \alpha_T \Delta \mathbf{U}_r^l \\ \Delta \mathbf{U}_r^l &= \mathbf{U}_{r_{n+1}}^{l(k-1)} - \mathbf{U}_{r_n}^l \end{aligned} \tag{4.5.29}$$

where the subscript r denotes the relative value, and $k-1$ denotes the $(k-1)$th iteration. The trial $\overline{\Phi}_{n+1}$ is

$$\overline{\Phi}_{n+1} \equiv \left\| \overline{\mathbf{T}}_{n+1}^l \right\| - \mu P_{n+1}^l \tag{4.5.30}$$

Using the return mapping strategy by Simo and Hughes [1991], we set

$$\mathbf{T}^l_{n+1} = \bar{\mathbf{T}}^l_{n+1} - \Delta\xi \frac{\bar{\mathbf{T}}^l_{n+1}}{\left\|\bar{\mathbf{T}}^l_{n+1}\right\|} \tag{4.5.31a}$$

where

$$\Delta\xi = \begin{Bmatrix} 0 & \text{if } \bar{\Phi}_{n+1} \le 0 \\ \dfrac{\bar{\Phi}_{n+1}}{\alpha_T} & \text{if } \bar{\Phi}_{n+1} > 0 \end{Bmatrix} \tag{4.5.31b}$$

With a straightforward check, we can show that the above formulation satisfies the Coulomb friction law.

The advantage of using this method is that no additional variables such as λ_l and τ_l for each contact node in the Lagrangian multiplier method are introduced. However, the accuracy of this formulation depends on the chosen value of the penalty parameters α_N and α_T. If these constants are too large, there could be numerical problems in the solution procedure. If these constants are too small, the accuracy of a solution might be sacrificed.

4.6 Draw-Bead Model

There are some grooves on the binder surface of a die. These grooves represent draw-beads, as mentioned in Chapter 1, to control the metal flow into the die cavity. Figure 4.5(a) shows the bead centerline on a binder surface of a die that is used to form an automotive fender. Figure 4.5(b) shows a schematic of a cross-sectional view of a punch, binder, sheet, and die for a draw operation after the binder closes and before the punch moves into the die cavity to form the sheet to the final shape. Figure 4.5(c) shows the cross-sectional view near the draw-bead. This cross section is perpendicular to the draw-bead centerline. In this figure, we denote two fixed points a and b as the boundary of the draw-bead groove on the binder surface of a die, while the material point A is on the blank outline.

Researchers have been trying to use thin shell finite elements to model a sheet passing through draw-bead grooves between the binder surface of a die and the binder ring surface in the simulation of a draw forming process. It is not practical. First, the geometry of a draw-bead is rather complex, with very tight corner radii. A significantly larger number of finite elements than those for the other areas must be used for the portion of the sheet passing through the draw-beads to obtain reliable results. Second, adaptive meshing (see Chapter 7) should be used to save computational time. When the sheet moves close to the draw-beads, we can refine the finite element model locally. When the sheet moves away from the draw-beads, we can restore the original coarse finite element model. This process would add more computational effort. Third, because corner radii of a draw-bead groove may be as tight as 1 mm, shell elements are not valid for modeling this kind of bending radii for sheet metals with the thickness of the same order as the bending radii. Rigorously speaking, three-dimensional finite elements should be used. For these reasons, the use of finite element methods to model the portions of a sheet passing through draw-beads is not recommended.

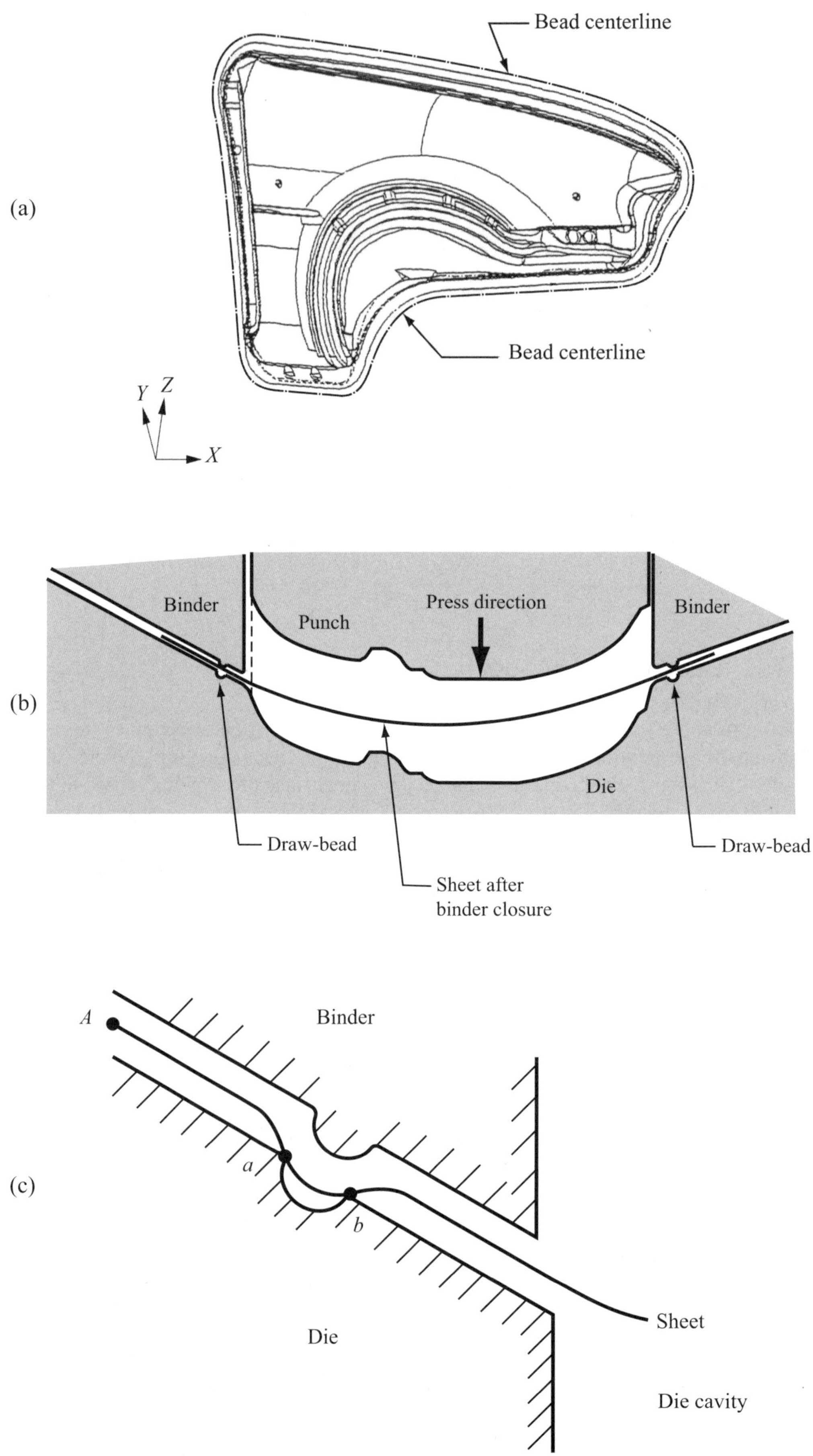

Figure 4.5 Draw-bead in a conventional draw die: (a) draw-bead centerline; (b) cross-sectional view of a punch, binder, sheet, and die; and (c) cross-sectional view near the draw-bead.

To overcome this difficulty for a practical solution, we ignore the geometrical complexity of draw-bead grooves and establish relatively simple mathematical models for draw-bead restraint forces. Draw-bead restraint forces are generated by the friction from the binder pressure and the bending and unbending of the sheet passing through a draw-bead. We may use an elastic-perfectly plastic spring to simulate a draw-bead restraint force on a section normal to the bead centerline. In the section, only local elongation of the spring affects the draw-bead restraint force. Deformation of neighboring sections does not influence the restraint force on this section.

Figure 4.6(a) shows a simple mathematical model for the draw-bead force for a section. In this model, the sheet under the binder is allowed to move only along the binder surface of a die. During a drawing process, a material point A on the blank outline moves toward the die cavity, as shown by the arrow in Figure 4.6(a). The original position x_A of the

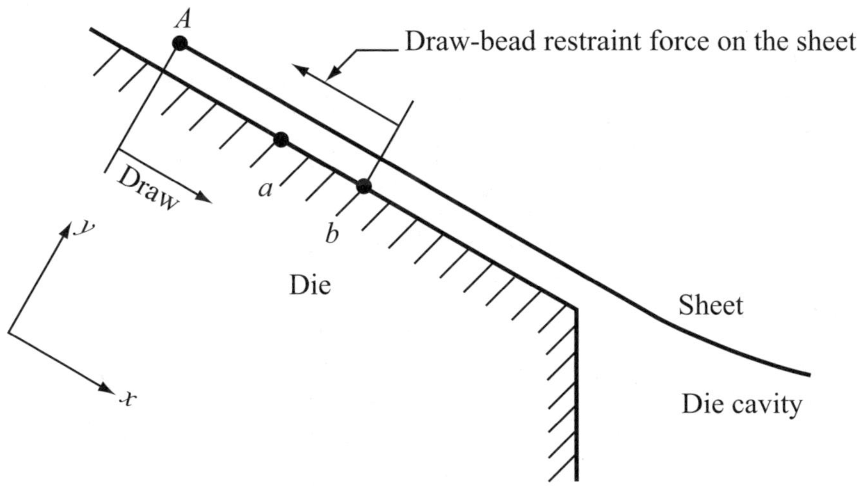

(a)

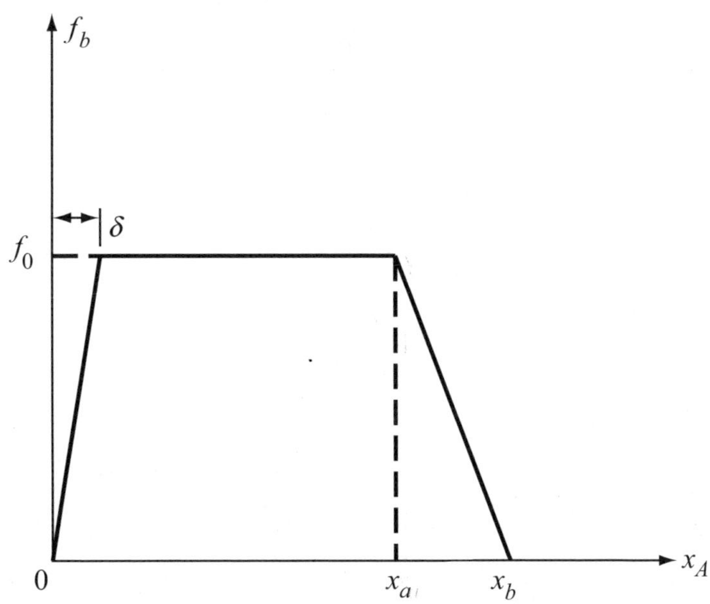

(b)

Figure 4.6 Draw-bead model and restraint force: (a) mathematical model for a draw-bead, and (b) draw-bead restraint force acting at a point on a sheet coinciding with point b on the binder surface.

material point A before the drawing is zero. Figure 4.6(b) shows that the draw-bead restraint force acts at a point on the sheet coinciding with point b with the coordinate x_b versus x_A, which represents the position of the traveling material point A. Here, a and b are the boundary points at the section across the draw-bead groove on the binder surface of a die. The draw-bead restraint force reaches the full amount f_0 after point A travels a small distance δ. The draw-bead restraint force at a point on the sheet coinciding with point b is at the full amount f_0 before the material point A is drawn to point a, as shown in Figure 4.7(a). As the material point A is continuously drawn and reaches a position between point a with the coordinate x_a and point b with the coordinate x_b, as shown in Figure 4.7(b), the draw-bead restraint force is reduced, as shown in Figure 4.6(b). After the material point A reaches and passes point b, as shown in Figure 4.7(c), the sheet at this section essentially passes through the draw-bead, and the draw-bead restraint force becomes zero.

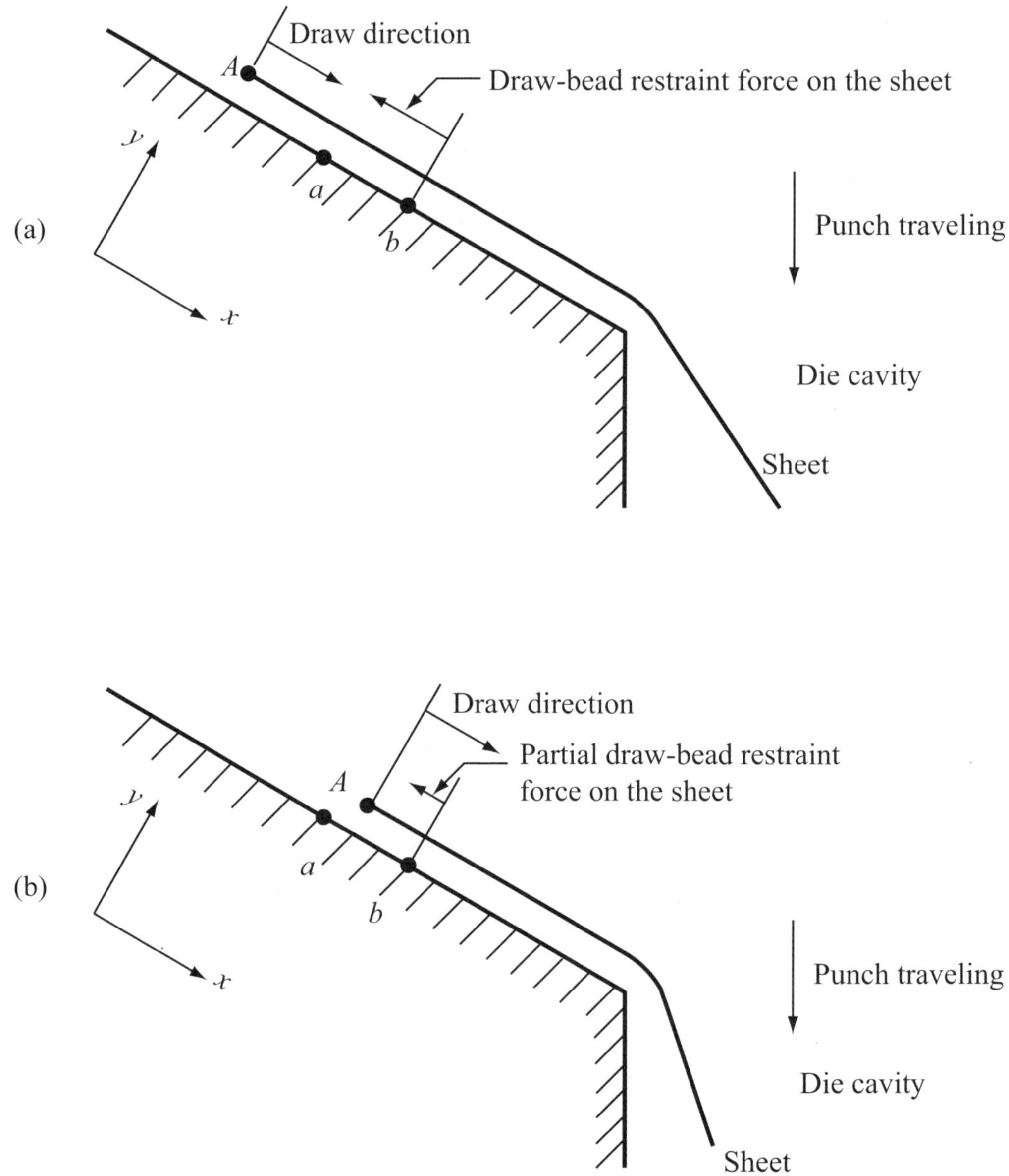

Figure 4.7 Point A at various positions during draw: (a) full draw-bead restraint force acting on the sheet before point A is drawn to the position of point a, and (b) partial draw-bead restraint force acting on the sheet when point A is drawn to a position between points a and b.

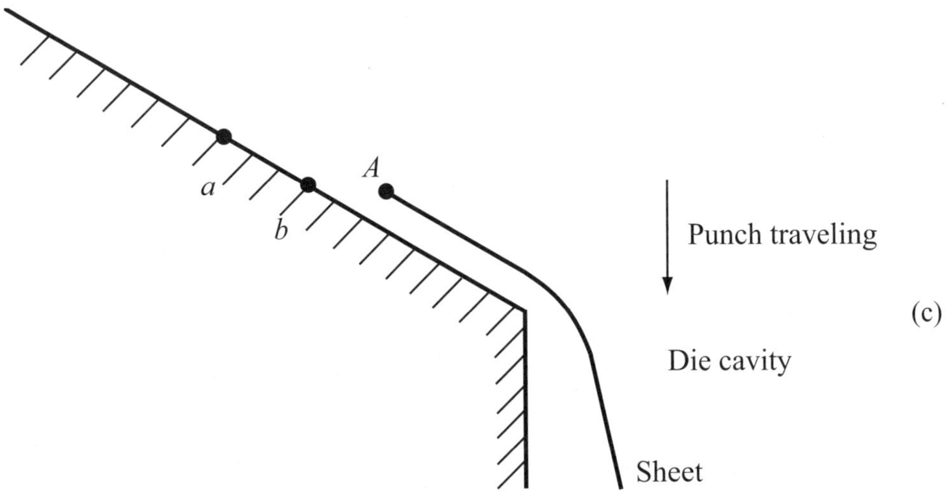

Figure 4.7 *(Continued)* (c) No restraint force acting on the sheet when point A is drawn to a position beyond point b.

Instead of establishing a detailed finite element model, we use the draw-bead restraint force, as shown in Figure 4.6(b), acting at a point on the sheet coinciding with the fixed point b on the binder surface of a die to control the sheet drawn through the groove of the draw-bead in simulations. The draw-bead restraint force may be computed by a separate finite element model or measured from draw-bead tests. This draw-bead model also can be applied to the inverted drawing processes, as shown in Figures 1.3 and 1.4 of Chapter 1. Stamping engineers may set more than one row of draw-beads for adding the strength and steadiness of the restraint force. As an approximation, the draw-bead restraint force for multiple draw-beads is obtained by multiplying the number of draw-beads to the restraint force for a single draw-bead.

4.6.1 Draw-Bead Restraint Force by Computation

Researchers are still trying to compute the draw-bead restraint force for a section as shown in Figure 4.5(c) by means of two-dimensional finite continuum elements. Although some good results have been published, direct measurements of the pulling forces from draw-bead tests are preferred.

4.6.2 Draw-Bead Restraint Force by Measurement

The easiest way to determine the draw forces is by direct measurement using a draw-bead test apparatus. The apparatus can be built with inserts of different corner radii so that different shapes of draw-beads can be tested.

Figure 4.8 shows the shapes of a typical rectangular bead and a typical round bead, respectively, from the Proceedings of a NUMISHEET conference edited by Makinouchi *et al.* [1993]. Their dimensions are listed in Table 4.1, where the radius can be as small as 1.0 mm. Table 4.2 lists the test results for the clamping force and the restraint force or the pulling force per unit length (1 mm) of the bead centerline.

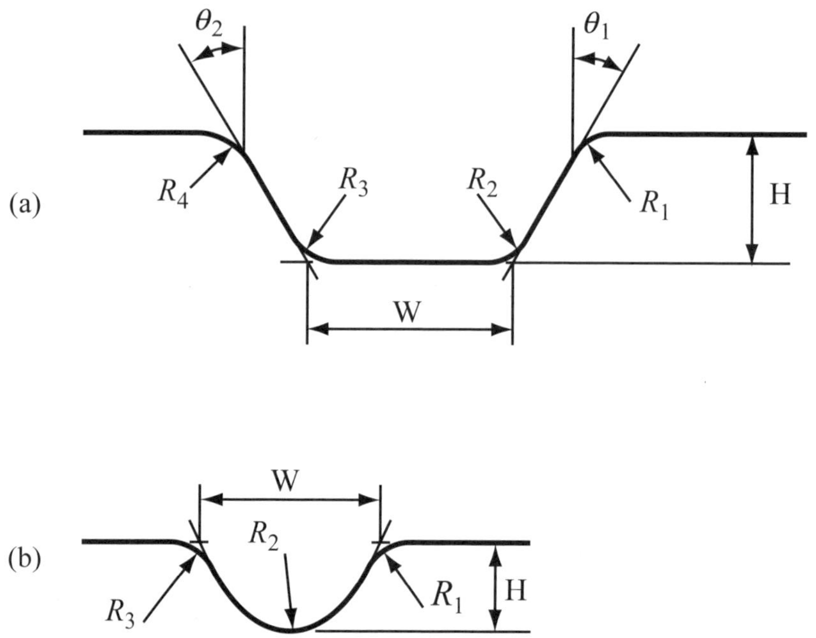

Figure 4.8 Typical draw-bead shapes: (a) rectangular bead, and (b) round bead.

TABLE 4.1
DRAW-BEAD DIMENSIONS

Draw-Beads on the Binder Ring Surface:

Type	Shape	H (mm)	W (mm)	θ_1 (degree)	θ_2 (degree)	R_1 (mm)	R_2 (mm)	R_3 (mm)	R_4 (mm)
A	Rectangle	5.4	6.2	29	40	4.8	3.9	3.7	3.7
C	Rectangle	5.6	10.7	10	30	2.8	1.0	3.7	2.6
B_1	Rectangle	4.9	8.0	25	30	1.3	4.0	4.0	2.1
B_2	Rectangle	5.0	15.7	30	31	2.1	2.1	2.8	3.3
B_3	Round	5.8	12.3			2.1	5.1	1.9	

Draw-Beads on the Binder Surface of a Die:

Type	Shape	H (mm)	W (mm)	θ_1 (degree)	θ_2 (degree)	R_1 (mm)	R_2 (mm)	R_3 (mm)	R_4 (mm)
A	Rectangle	6.0	8.9	29	40	4.5	3.7	4.5	3.2
C	Rectangle	5.8	11.4	13	36	1.9	2.6	4.4	2.8
B_1	Rectangle	5.3	8.7	25	30	2.7	2.4	5.4	3.5
B_2	Rectangle	4.5	13.8	30	31	1.4	1.6	1.9	4.4
B_3	Round	5.9	14.0			1.9	6.1	1.9	

TABLE 4.2
DRAW-BEAD CLAMPING AND PULLING FORCES

Type	Clamping Force (N/mm)	Pulling Force (N/mm)
A	62	65
C	169	189
B_1	113	115
B_2	115	130
B_3	132	128

4.7 References

Bathe, K.J., *Finite Element Procedures*, Prentice Hall, Englewood Cliffs, NJ, 1996.

Bathe, K.J. and Bouzinov, P.A., "On the Constraint Function Methods for Contact Problems," *Computers & Structures*, Vol. 64, No. 5/6, 1997, p. 1069.

Makinouchi, A., Nakamachi, E., Oñate, E., and Wagoner, R.H. (eds.), Proceedings of the 2nd International Conference—"NUMISHEET'93," Isehara, Japan, 1993.

Simo, J.C. and Hughes, T.J.R., *Elastoplasticity and Viscoplasticity: Computational Aspects,* Springer, Berlin, 1991.

Simo, J.C. and Laursen, T.A., "An Augmented Lagrangian Treatment of Contact Problems Involving Friction," *Computers & Structures*, Vol. 42, No. 1, 1992, p. 97.

Wang, N.-M. and Budiansky, B., "Analysis of Sheet Metal Stamping by a Finite-Element Method," *Journal of Applied Mechanics*, American Society of Mechanical Engineers, Vol. 45, 1978, p. 73.

Chapter 5

Thin Plate and Shell Analyses

5.1 Plates and General Shells

For sheet metal products such as beverage cans and parts in aircraft, appliances, and automobiles, the sheet metal used is very thin in comparison to the bending radius in most areas during the forming processes. Therefore, sheet metal may be considered as a thin plate or thin shell for manufacturing or product analysis. When flat sheet metal is undergoing finite or large deformation during the forming processes, the sheet metal behaves as a thin shell. When the sheet metal becomes curved and stretched during the forming processes, both the membrane and bending actions on the sheet metal are significant. In this book, we discuss only the general thin shells. Note that forming processes usually start with initially flat sheet metal, which is called a blank in the stamping industry. A blank is a special case of a general thin shell structure.

5.2 Assumptions and Approximations

To extend the classical thin shell theory to deal with finite or large strains that occurred in a forming process, we relax the restrictions on the thin plate and shell theory such that the following are true:

1. The thickness change of a deformed shell must be considered in a forming process.
2. The ratio of the bending radius to the thickness in a deformed shell can be as small as 5. It is difficult to prove this statement theoretically. However, as demonstrated by Xia and Tang [1999], this statement is valid when the ratio is as small as 3.
3. A normal to the middle surface of a shell does not necessarily remain normal after deformation. However, the material points on a normal to the middle surface of an undeformed shell remain on a straight line after deformation. Therefore, in this book, we call the straight line the director instead of the normal.
4. In addition to in-plane stresses, there are transverse shear stresses and strains in a deformed shell. Consequently, a normal to the middle surface of a shell does not remain normal after deformation.

5.3 Base Vectors and Metric Tensors (Spain [1956])

Let the middle surface of an undeformed shell be the reference surface that is expressed by the equation

$$ {}_0x^i = {}_0x^i\left(\theta^\alpha\right) \qquad (i = 1,2,3;\ \alpha = 1,2) \tag{5.3.1} $$

where ${}_0x^i$ are the rectangular Cartesian coordinates in space, and θ^α are the convected coordinates on the reference surface. Note that in this book, a Latin superscript or subscript has a range of 3, and a Greek superscript or subscript has a range of 2. The vector ${}_0x^i$ also may be interpreted as the position of a material point on the reference surface of an undeformed shell. Note that ${}_0x^i = {}_0x_i$ because both ${}_0x^i$ and ${}_0x_i$ are Cartesian components. Here, θ^α may be called the Gaussian or curvilinear coordinates on a surface. The position vector of a material point with coordinates θ^i $(i = 1, 2, 3)$ with $-1/2 \le \theta^3 \le 1/2$ represents a layer parallel to the middle surface of an undeformed shell, as shown in Figure 5.1. The material point is expressed by the equation

$$ {}_0y^i\left(\theta^\alpha,\theta^3\right) = {}_0x^i\left(\theta^\alpha\right) + h_0\theta^3\,{}_0a_3^i\left(\theta^\alpha\right) \tag{5.3.2} $$

where h_0 is the thickness of the undeformed shell, and ${}_0a_3^i$ is a unit vector to be chosen identical to the normal ${}_0\mathbf{n}$ of the reference surface of the undeformed shell, as shown in Figure 5.2(a). In this case, the normal ${}_0\mathbf{n}$ also is the director.

The corresponding position vectors on the reference surface and a surface parallel to the reference with coordinate θ^3 for the deformed shell, respectively, are

$$ x^i = x^i\left(\theta^\alpha\right) \tag{5.3.3} $$

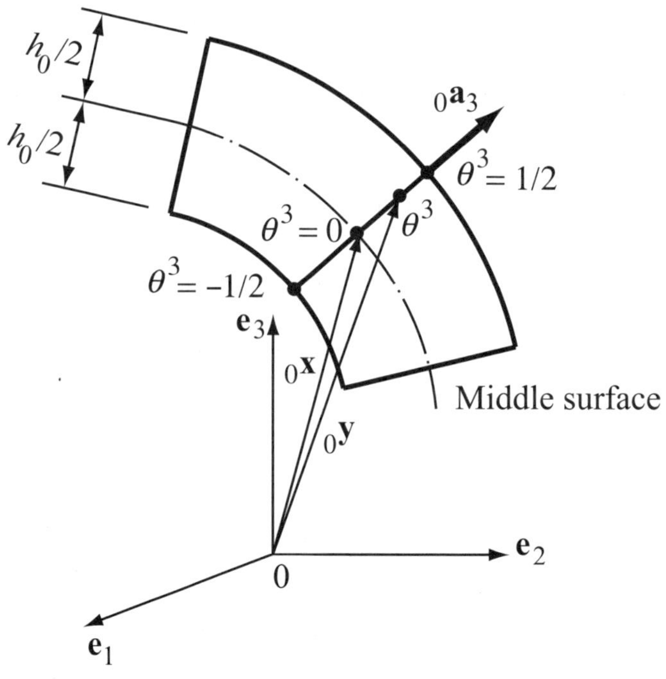

Figure 5.1 Position vectors of material points.

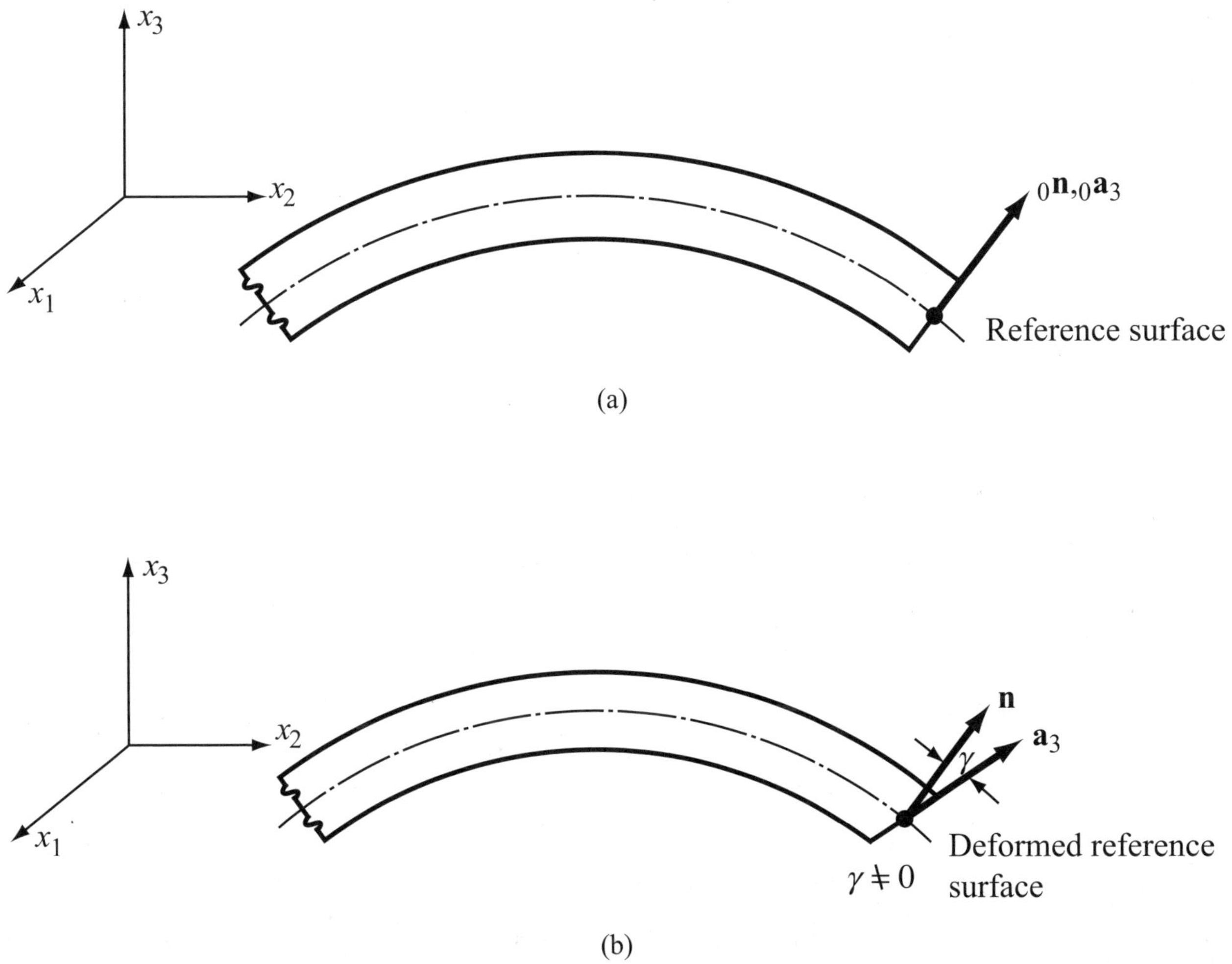

Figure 5.2 A shell with transverse shear deformation: (a) a director parallel to a normal before deformation, and (b) the director no longer parallel to the normal after deformation.

and

$$y^i\left(\theta^\alpha,\theta^3\right) = x^i\left(\theta^\alpha\right) + h\varsigma\left(\theta^3\right)a_3^i\left(\theta^\alpha\right) \tag{5.3.4}$$

where a_3^i is the director, which is a unit vector, of the deformed shell. For the classical shell theory without any transverse shear deformation, the director remains normal to the deformed reference surface after deformation. However, when there is transverse shear deformation, the director is not parallel to the normal **n** to the deformed reference surface, as shown in Figure 5.2(b). Here, h represents the thickness of the deformed shell along the director, and the product $h\varsigma\left(\theta^3\right)$ is the coordinate of the material point in the shell along the director a_3^i.

We use two special cases of cylindrical bending under plane strain conditions with the assumptions of elastic and rigid plastic behaviors for the sheet metal, respectively, to determine the property of the function $\varsigma\left(\theta^3\right)$. As already discussed, the material points on the undeformed director (normal) remain on a straight line, which is the deformed director a_3^i, and the bending is moderate or the ratio h/R_0 is small in comparison to 1 (unity). Here, R_0 represents a measure of the mean radius of the cylindrical shell. Figure 5.3 shows a thin flat sheet of length L and thickness h_0. The dimension of the sheet in the out-of-plane direction to the paper is large. The initially flat plate is bent to a cylindrical shell with the thickness h. The bending first is assumed to be moderate. Therefore, the sheet can be assumed to be in the elastic range.

Seely and Smith [1959] showed that for a curved elastic beam under pure bending without axial force, the neutral axis shifts toward the compression region by

$$z_n = R_0 - R_n = \frac{h}{12} \frac{h}{R_0} \tag{5.3.5}$$

where z_n is the distance between the middle and neutral surfaces, h is the thickness of the curved beam, and R_0 and R_n are the radii of the middle and neutral surfaces of the curved beam, respectively. When h/R_0 is not small, z_n cannot be ignored. The distribution of the bending strain as well as the bending stress through the thickness is nonlinear, as shown in the following paragraph. Note that their analysis of the curved beam is linear.

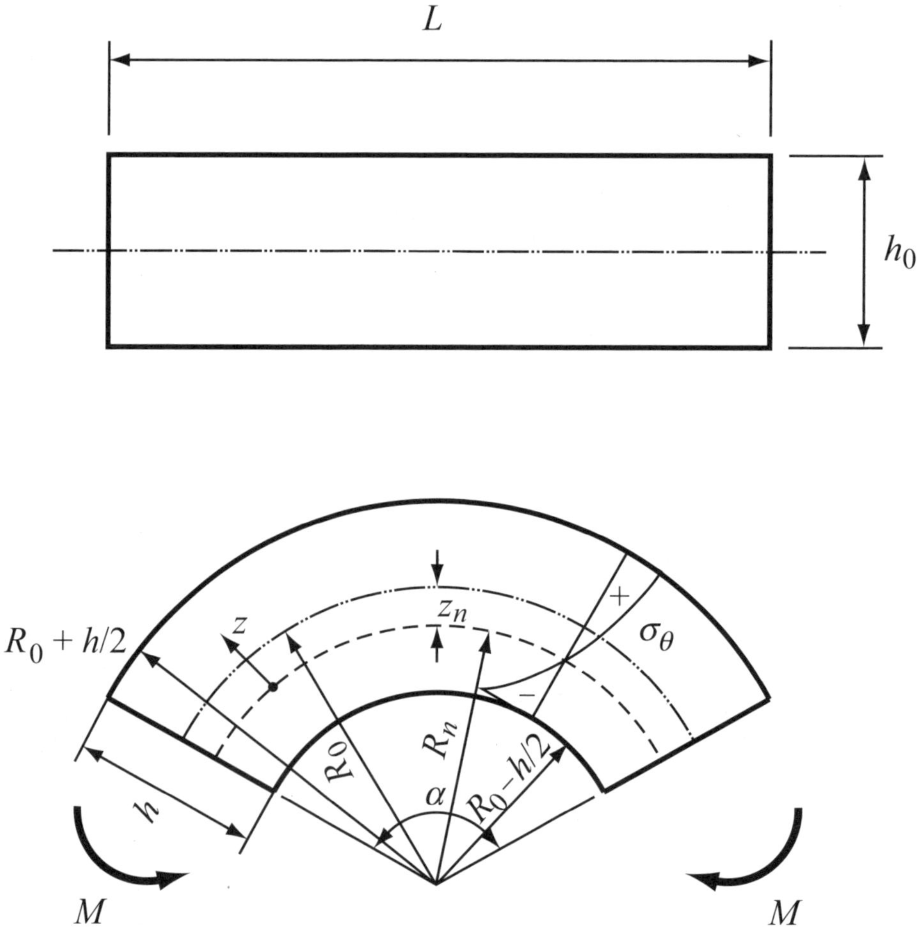

Figure 5.3 Cylindrical bending of an elastic thin flat plate.

We extend the analysis of the curved beam bending of Seely and Smith [1959] to the bending of a flat plate to a cylindrical shell that is no longer under a uniaxial stress state. For finite or large strains, we use the true or natural strain formulation in Eq. 2.8.4 to express the bending strain as

$$\varepsilon_\theta = \ln\frac{(R_n + z)\alpha}{L} \tag{5.3.6a}$$

where z is the distance from the neutral surface, and α is the angle or bend angle extended by the segment of the cylindrical surface, as shown in Figure 5.3. The neutral surface is defined by R_n such that $\alpha = L/R_n$. Therefore, along the neutral surface, the surface has no bending stress or strain. Because of the plane strain assumption in the out-of-plane direction or the longitudinal direction of the long cylinder, the longitudinal strain ε_x is

$$\varepsilon_x = 0 \tag{5.3.6b}$$

Based on the strain components in Eqs. 5.3.6a and 5.3.6b, the assumption of the normal stress component in the direction of the normal of the shell equal to zero, and the generalized Hooke's law in Eqs. 3.1.6 and 3.1.7, the bending stress σ_θ can be derived as

$$\sigma_\theta = \frac{E}{1-\nu^2}\ln\left(1 + z/R_n\right) \tag{5.3.7}$$

This equation indicates that the distribution of the bending stress through the thickness of the cylindrical panel is no longer linear as shown in Figure 5.3. Because the cylindrical surface is under pure bending, the resultant force per unit length in the hoop direction must vanish. Therefore,

$$\int_{-h/2+z_n}^{h/2+z_n} \sigma_\theta \, dz = 0$$

From this equation, we obtain the distance between the middle and the neutral surfaces as

$$z_n = \frac{h}{16}\frac{h}{R_0} \tag{5.3.8}$$

where R_0 is the radius of the middle surface of the cylindrical panel. Note that we assume this middle surface of the cylindrical surface is deformed from the middle surface of the flat plate. Under moderate bending, h/R_0 is small compared to 1, and z_n can be set to zero. In this case, the distribution of the bending stress through the thickness of the elastic cylindrical panel is approximately linear.

If the bending is severe, the sheet can be in the plastic range. For simplicity, we adopt the rigid plastic behavior for the sheet as an illustration. As already shown, when a flat plate is bent to a cylindrical panel, the neutral surface shifts in the bending process. We follow Hill's analysis [1950] of cylindrical bending of a rigid plastic plate. Figure 5.4 shows again a flat sheet of length L and thickness h_0. The flat sheet is bent severely to a cylindrical surface with a bend angle α. Note that the middle plane of the flat sheet is bent to a segment of cylindrical surface with the radius R_0. The arc length for this surface is not equal to the original length L. The deformed middle surface with no change of length can be defined by the radius R under the bend angle α as

$$R\alpha = L \tag{5.3.9}$$

The deformed middle surface with the radius R in this case is not stretched.

The neutral surface can be defined at the location where the bending or the hoop stress changes from tension to compression. The radius of the neutral surface can be derived as

$$R_n = \sqrt{(R + h/2)(R - h/2)} \tag{5.3.10a}$$

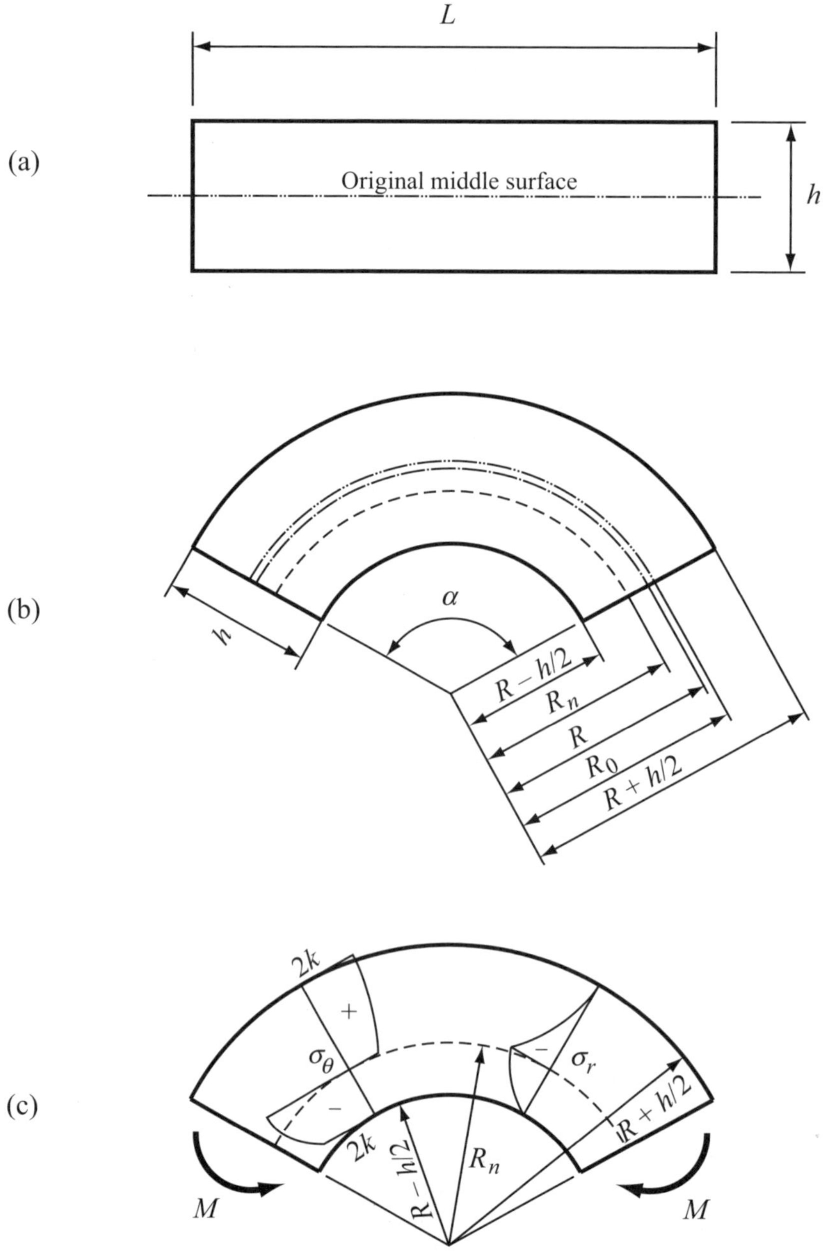

Figure 5.4 Cylindrical bending of a rigid perfect plastic thin flat plate: h_0 = original thickness; h = deformed thickness; R = radius of the deformed middle surface; R_0 = radius of the original middle surface; and R_n = radius of the neutral surface.

Taylor's expansion gives

$$R_n \approx R\left[1 - \frac{1}{8}(h/R)^2\right] \tag{5.3.10b}$$

Based on the assumption of incompressibility for rigid plastic materials, the deformed radius of the original middle surface is

$$R_0 = \sqrt{\frac{(R + h/2)^2 + (R - h/2)^2}{2}} \approx R\left[1 + \frac{1}{8}(h/R)^2\right] \tag{5.3.10c}$$

In the preceding equations, we assume the ratio h/R is much smaller than 1. Therefore, terms that have an order of magnitude equal to or higher than $(h/R)^2$ can be neglected. Hill [1950] also pointed out that the thickness of the shell does not change under cylindrical bending or $h = h_0$.

The deformed middle surface is located between the neutral and the original middle surfaces, as shown in Figure 5.4. Equations 5.3.10a, 5.3.10b, and 5.3.10c give

$$R_0 > R > R_n \tag{5.3.11}$$

Also note that the original position of the deformed middle surface with radius R does not fix on the flat sheet, as shown in Figure 5.4. The original position depends on the bending radius. When the bending is severe, the original position is closer to the bottom of the sheet. Figure 5.4 also shows the distributions of the radial and circumferential (hoop) normal stresses, σ_r and σ_θ. In the figure, k represents the shear yield stress for rigid perfectly plastic materials.

For elastic-plastic work-hardening materials usually assumed in simulations of sheet metal forming processes, the neutral surface is estimated to shift toward the compression region by a distance on the order of h^2/R. For the thin shells in which we are interested, the amount of neutral surface shift from the undeformed middle surface under bending is negligible, as illustrated in the preceding cylindrical bending cases. Therefore, we may set $\varsigma\left(\theta^3\right) = \theta^3$ in Eq. 5.3.4.

For a sheet under membrane forces and bending moments, the thickness may change significantly, and the original h_0 can be quite different from the deformed h. This change is due to the membrane forces but not to bending or twisting moments. Therefore, we may use the stretch ratios on the reference surface of a shell to estimate the thickness change. We must update the thickness to h during forming simulations.

Partial differentiation of the position vectors with respect to θ^α gives the covariant base vectors. Using Eqs. 5.3.1 and 5.3.3, the surface covariant base vectors at a material point with coordinates $\left(\theta^\alpha, 0\right)$ on the reference surface for the undeformed and deformed configurations of a shell, respectively, are

$${}_0a^i_\alpha \equiv \frac{\partial_0 x^i}{\partial \theta^\alpha} = {}_0x^i_{,\alpha} \tag{5.3.12}$$

$$a_\alpha^i \equiv \frac{\partial x^i}{\partial \theta^\alpha} = x,_\alpha^i \tag{5.3.13}$$

Using Eqs. 5.3.2 and 5.3.4 (setting $\varsigma = \theta^3$), the covariant base vectors at a material point with coordinates θ^i ($i = 1, 2, 3$) on a surface parallel to the reference surface of a shell for the undeformed and deformed configurations, respectively, are

$${}_0g_\alpha^i \equiv \frac{\partial {}_0y^i}{\partial \theta^\alpha} = {}_0a_\alpha^i + h_0\theta^3\, {}_0a_{3,\alpha}^i \tag{5.3.14a}$$

$${}_0g_3^i \equiv \frac{\partial {}_0y^i}{\partial \theta^3} = h_0\, {}_0a_3^i \tag{5.3.14b}$$

and

$$g_\alpha^i \equiv \frac{\partial y^i}{\partial \theta^\alpha} = a_\alpha^i + h\theta^3 a_{3,\alpha}^i \tag{5.3.15a}$$

$$g_3^i \equiv \frac{\partial y^i}{\partial \theta^3} = h a_3^i \tag{5.3.15b}$$

where the magnitudes of ${}_0a_{3,\alpha}^i$ and $a_{3,\alpha}^i$ are proportional to the curvatures of the reference surfaces of the undeformed and deformed shells (Spain [1956]), respectively, in the classical thin shell theory as discussed in Section 5.5 of this chapter. Note that ${}_0a_{3,\alpha}^i$ and $a_{3,\alpha}^i$ are parallel to the tangent planes of the undeformed and deformed reference surfaces, respectively. Taking the scalar product of the base vectors in Eq. 5.3.14, we obtain the metric tensors for the undeformed shell as

$$\begin{aligned} {}_0g_{\alpha\beta} &= {}_0g_\alpha^i\, {}_0g_\beta^i \\ &= {}_0a_{\alpha\beta} + h_0\theta^3\left({}_0a_\alpha^i\, {}_0a_{3,\beta}^i + {}_0a_\beta^i\, {}_0a_{3,\alpha}^i\right) + O\left[\left(h_0/R\right)^2\right] \end{aligned} \tag{5.3.16a}$$

where ${}_0a_{\alpha\beta} = {}_0a_\alpha^i\, {}_0a_\beta^i$ is the first fundamental or the metric tensor of the reference surface of an undeformed shell, and

$$\begin{aligned} {}_0g_{\alpha 3} &= {}_0g_\alpha^i\, {}_0g_3^i \\ &= h_0\, {}_0a_3^i\left({}_0a_\alpha^i + h_0\theta^3\, {}_0a_{3,\alpha}^i\right) = 0 \end{aligned} \tag{5.3.16b}$$

when we choose the director ${}_0a_3^i$ as the normal to the reference surface of the undeformed shell, and ${}_0a_\alpha^i$ is tangent to the reference surface. The metric tensor is symmetrical, that is, ${}_0g_{\alpha i} = {}_0g_{i\alpha}$ $(i = 1, 2, 3;\ \alpha = 1, 2)$. Similarly, we obtain the metric tensor for the deformed shell as

$$\begin{aligned} g_{\alpha\beta} &= g^i_\alpha g^i_\beta \\ &= a_{\alpha\beta} + h\theta^3\left(a^i_\alpha a^i_{3,\beta} + a^i_\beta a^i_{3,\alpha}\right) + O\left[(h/R)^2\right] \end{aligned} \tag{5.3.17a}$$

$$\begin{aligned} g_{\alpha 3} &= g^i_\alpha g^i_3 \\ &= h a^i_3\left(a^i_\alpha + h\theta^3 a^i_{3,\alpha}\right) \end{aligned} \tag{5.3.17b}$$

We compute $g_{\alpha 3}$ on the reference surface $(\theta^3 = 0)$ as

$$g_{\alpha 3} = g^i_\alpha g^i_3 = h a^i_\alpha a^i_3 \tag{5.3.17c}$$

In Eq. 5.3.17a, $a_{\alpha\beta} = a^i_\alpha a^i_\beta$ is the first fundamental or the metric tensor of the reference surface of the deformed shell, and R is a reference radius that is smaller than any radius of the undeformed or deformed shell. The metric tensor $g_{\alpha i}$ $(i = 1,2,3;\ \alpha = 1,2)$ for the deformed shell also is symmetrical.

5.4 Lagrangian Strains (Hutchinson [1973])

The Lagrangian strain tensor is expressed by the metric tensors in Eqs. 5.3.16 and 5.3.17 as

$$\varepsilon_{\alpha\beta} = \frac{1}{2}\left(g_{\alpha\beta} - {}_0g_{\alpha\beta}\right) \tag{5.4.1a}$$

and

$$\varepsilon_{\alpha 3} = \frac{1}{2}\left(g_{\alpha 3} - {}_0g_{\alpha 3}\right) \tag{5.4.1b}$$

Note that $\varepsilon_{\alpha 3}$ represent the transverse shear strains. The Lagrangian strain tensor is symmetrical because the metric tensor is symmetrical.

If we choose a director as the normal to the reference surface of the undeformed shell as already mentioned, Eq. 5.4.1b is simplified to

$$\varepsilon_{\alpha 3} = \frac{1}{2} g_{\alpha 3} \tag{5.4.1c}$$

Note that $\varepsilon_{\alpha 3}$ are the transverse shear strains from Mindlin [1951] and from Lin and Morgan [1956].

5.5 Classical Shell Theory

In the classical shell theory, a director ${}_0a_3^i$ chosen to be ${}_0n^i$ normal to the reference surface of an undeformed shell remains as the normal $a_3^i\,(n^i)$ to the reference surface after deformation. Therefore, the transverse shear strains $\varepsilon_{\alpha 3}$ in Eq. 5.4.1c are zero. We cannot use different interpolation functions to interpolate displacements and rotations. Rotations are obtained from the partial derivatives of the transverse displacement with respect to the in-plane coordinates on the reference surface of a deformed shell. Polynomials with higher order and rational functions are required to satisfy C^1 continuity. Not many such interpolation functions exist in the literature. We will choose one and demonstrate it in this book.

5.5.1 Strain-Displacement Relationship

Before we discuss the use of finite element analysis to discretize the equations of motion via the virtual work in Eq. 4.2.1, we need to establish the strain-displacement relationship from the shell geometry. A director is a normal to the reference surface of a thin shell before and after deformation for the classical thin shell theory. Therefore, both ${}_0g_{\alpha 3}$ and $g_{\alpha 3}$ are zero. Consequently, the transverse shear strains $\varepsilon_{\alpha 3}$ from Eq. 5.4.1c vanish. Let us introduce the second fundamental tensors ${}_0b_{\alpha\beta}$ and $b_{\alpha\beta}$ on the reference surface of the undeformed and deformed shell, respectively. Koiter [1966] gave the relationship between $b_{\alpha\beta}$ and the second partial derivatives of the position vector of a point on the reference surface of an undeformed shell as

$$ {}_0x^i_{,\alpha\beta} = {}_0b_{\alpha\beta}\,{}_0a_3^i \tag{5.5.1a} $$

where ${}_0a_3^i$ is the normal to the reference surface at that point. Note that ${}_0b_{\alpha\beta} = {}_0b_{\beta\alpha}$. Multiplying both sides by ${}_0a_3^i$ gives

$$ {}_0b_{\alpha\beta} = {}_0a_3^i\,{}_0x^i_{,\alpha\beta} \tag{5.5.1b} $$

Note that ${}_0x^i_{,\alpha}$ is orthogonal to ${}_0a_3^i$. Therefore,

$$ {}_0x^i_{,\alpha}\,{}_0a_3^i = 0 \tag{5.5.2} $$

Partial differentiation of Eq. 5.5.2 with respect to θ^β gives

$$ {}_0x^i_{,\alpha\beta}\,{}_0a_3^i + {}_0x^i_{,\alpha}\,{}_0a^i_{3,\beta} = 0 \tag{5.5.3} $$

Combining Eqs. 5.5.1a and 5.5.3 and using ${}_0a_3^i\,{}_0a_3^i = 1$ and Eq. 5.3.12, we have the second fundamental tensor of the reference surface of an undeformed shell as

$$ {}_0b_{\alpha\beta} = -{}_0a^i_\alpha\,{}_0a^i_{3,\beta} = -{}_0a^i_\beta\,{}_0a^i_{3,\alpha} \tag{5.5.4a} $$

which is symmetrical. Similarly, the second fundamental tensor $b_{\alpha\beta}$ of the deformed reference surface can be derived as

$$b_{\alpha\beta} = -a^i_\alpha a^i_{3},_\beta = -a^i_\beta a^i_{3},_\alpha \tag{5.5.4b}$$

which likewise is symmetrical. Note that the vectors $a^i_{3},_\alpha$ are parallel to the tangent plane of the deformed reference surface, as mentioned in Section 5.3.

Inserting Eq. 5.5.4a into Eq. 5.3.16a yields the metric tensor ${}_0g_{\alpha\beta}$ as

$${}_0g_{\alpha\beta} = {}_0a_{\alpha\beta} - 2h_0\theta^3\,{}_0b_{\alpha\beta} \tag{5.5.5a}$$

Inserting Eq. 5.5.4b into Eq. 5.3.17a yields the metric tensor $g_{\alpha\beta}$ as

$$g_{\alpha\beta} = a_{\alpha\beta} - 2h\theta^3 b_{\alpha\beta} \tag{5.5.5b}$$

The Lagrangian strain tensor is obtained from one-half of the difference of the metric tensors for the material points on the deformed and undeformed reference surfaces of a thin shell, as shown in Eq. 5.4.1. Using Eq. 5.5.5, Eq. 5.4.1a becomes

$$\varepsilon_{\alpha\beta} = \gamma_{\alpha\beta} + h\theta^3\kappa_{\alpha\beta} \tag{5.5.6a}$$

where the in-plane strain tensor $\gamma_{\alpha\beta}$ is

$$\gamma_{\alpha\beta} = \frac{1}{2}\left(a_{\alpha\beta} - {}_0a_{\alpha\beta}\right) \tag{5.5.6b}$$

and the bending strain tensor $\kappa_{\alpha\beta}$ is

$$\kappa_{\alpha\beta} = -\left(b_{\alpha\beta} - \frac{h_0}{h}\,{}_0b_{\alpha\beta}\right) \tag{5.5.6c}$$

Now, we express the Lagrangian strain tensor at a material point in terms of its initial geometry on the reference surface of an undeformed shell and the displacement derivatives at that point. Let the spatial coordinates on the deformed reference surface of a material point ${}_0x^i$ be

$$x^i = {}_0x^i + u^i \tag{5.5.7}$$

where u^i is the displacement vector. Therefore, the metric tensor of the reference surface of the deformed shell is

$$a_{\alpha\beta} = \left({}_0x^i + u^i\right),_\alpha \left({}_0x^i + u^i\right),_\beta \tag{5.5.8}$$

Inserting $a_{\alpha\beta}$ into Eq. 5.5.6b, we obtain the in-plane strain tensor in terms of the displacement gradients as

$$\gamma_{\alpha\beta} = \frac{1}{2}\left({}_0x,^i_\alpha\, u,^i_\beta + {}_0x,^i_\alpha\, u,^i_\beta\right) + \frac{1}{2} u,^i_\alpha\, u,^i_\beta \tag{5.5.9}$$

Now, we derive the second fundamental tensors of both the undeformed and deformed reference surfaces of a shell so that the bending strain tensor can be expressed in terms of the geometric properties of the undeformed reference surface and the displacement derivatives, $u,^i_\alpha$ and $u,^i_{\alpha\beta}$. Spain [1956] gave the second fundamental tensor of the undeformed reference surface of a shell as

$${}_0b_{\alpha\beta} = \frac{1}{2}\pi^{\gamma\delta} e_{ijk}\, {}_0x,^j_r\, {}_0x,^k_\delta\, {}_0x,^i_{\alpha\beta} \tag{5.5.10}$$

where e_{ijk} are the permutation symbols for the Cartesian coordinate system defined as

$$\begin{aligned} e_{ijk} &= 0 \text{ if the two indices are equal} \\ e_{123} &= e_{231} = e_{312} = 1 \\ e_{132} &= e_{321} = e_{213} = -1 \end{aligned} \tag{5.5.11}$$

and $\pi^{\gamma\delta}$ is the permutation surface tensor in Spain [1956] on the undeformed reference surface of a shell. The permutation surface tensor is equal to

$$\pi^{\gamma\delta} = \frac{e_{\gamma\delta}}{\sqrt{{}_0a}}$$

where $e_{11} = e_{22} = 0$, $e_{12} = -e_{21} = 1$, and ${}_0a$ is the determinant of the surface metric tensor ${}_0a_{\alpha\beta}$. Note that Eq. 5.5.1b and the normal ${}_0\mathbf{a}_3 = {}_0\mathbf{a}_1 \times {}_0\mathbf{a}_2 / \|{}_0\mathbf{a}_1 \times {}_0\mathbf{a}_2\|$ are used to derive Eq. 5.5.10. As an exercise, readers may apply Eq. 5.5.10 to compute ${}_0b_{\alpha\beta}$ for a circular cylindrical surface and find their geometrical meaning. Similar to Eq. 5.5.10, the second fundamental tensor of the deformed reference surface of a shell is

$$b_{\alpha\beta} = \frac{1}{2}\bar{\pi}^{\gamma\delta} e_{ijk}\, x,^j_\gamma\, x,^k_\delta\, x,^i_{\alpha\beta} \tag{5.5.12}$$

where $\bar{\pi}^{\gamma\delta}$ is the permutation surface tensor of the deformed reference surface of a shell. The permutation surface tensor is equal to

$$\bar{\pi}^{\gamma\delta} = \frac{e_{\gamma\delta}}{\sqrt{a}}$$

where a is the determinant of the surface metric tensor $a_{\alpha\beta}$. We want to express $b_{\alpha\beta}$ in terms of ${}_0b_{\alpha\beta}$ and the displacement derivatives $u^i_{,\alpha}$ and $u^i_{,\alpha\beta}$. For simplifying the derivation, we split $b_{\alpha\beta}$ in Eq. 5.5.12 into two groups, $\bar{\pi}^{\gamma\delta}/2$ and $e_{ijk}x^j_{,\gamma}\, x^k_{,\delta}\, x^i_{,\alpha\beta}$. The first group is

$$\frac{1}{2}\bar{\pi}^{\gamma\delta} = \frac{e_{\gamma\delta}}{2\sqrt{a}} \tag{5.5.13}$$

where a is the determinant of the surface metric tensor $a_{\alpha\beta}$, and $a = e_{\lambda\varphi}e_{\mu\omega}a_{\lambda\mu}a_{\varphi\omega}/2$. We leave the proof of the expression for the determinant as an exercise for readers.

We will establish the tangent stiffness tensor due to the initial stresses in the finite element method later in this book. We should take into consideration the nonlinear terms of the displacement gradient in the determinant a. We may assume the maximum displacement gradient due to bending is moderately large. Therefore, we take up to the quadratic terms of the displacement gradients in the determinant a from Eq. 5.5.8 as

$$\begin{aligned} a = \Big[& {}_0a + 2e_{\lambda\varphi}e_{\mu\omega}\,{}_0a_{\lambda\mu}\,{}_0x^i_{,\varphi}\,{}_0u^i_{,\omega} + e_{\lambda\varphi}e_{\mu\omega}\,{}_0a_{\lambda\mu}u^i_{,\varphi}\,u^i_{,\omega} \\ & + \frac{1}{2}e_{\lambda\varphi}e_{\mu\omega}\left(u^i_{,\lambda}\,{}_0x^i_{,\mu} + u^i_{,\mu}\,{}_0x^i_{,\lambda}\right)\left(u^j_{,\varphi}\,{}_0x^j_{,\omega} + u^j_{,\omega}\,{}_0x^j_{,\varphi}\right)\Big] + O\left(U^3\right) \end{aligned} \tag{5.5.14a}$$

where $U = \underset{i,\alpha}{Max}\left(\left|u^i_{,\alpha}\right|\right)$. Taylor's expansion yields

$$\begin{aligned} a^{-1/2} = {}_0a^{-1/2}\Big[& 1 - \frac{1}{{}_0a}e_{\lambda\varphi}e_{\mu\omega}\,{}_0a_{\lambda\mu}\,{}_0x^i_{,\varphi}\,u^i_{,\omega} - \frac{1}{2{}_0a}e_{\lambda\varphi}e_{\mu\omega}\,{}_0a_{\lambda\mu}u^i_{,\varphi}\,u^i_{,\omega} \\ & - \frac{1}{4{}_0a}e_{\lambda\varphi}e_{\mu\omega}\left(u^i_{,\lambda}\,{}_0x^i_{,\mu} + u^i_{,\mu}\,{}_0x^i_{,\lambda}\right)\left(u^j_{,\varphi}\,{}_0x^j_{,\omega} + u^j_{,\omega}\,{}_0x^j_{,\varphi}\right) \\ & + \frac{3}{2({}_0a)^2}\left(e_{\lambda\varphi}e_{\mu\omega}\,{}_0a_{\lambda\mu}\,{}_0x^i_{,\varphi}\,u^i_{,\omega}\right)^2\Big] + O\left(U^3\right) \end{aligned} \tag{5.5.14b}$$

Let

$$\pi^{\gamma\delta}_{(0)} = \frac{1}{2\sqrt{{}_0a}}e_{\gamma\delta} \tag{5.5.15a}$$

$$\pi^{\gamma\delta}_{(1)} = -\frac{1}{2{}_0a^{3/2}}e_{\gamma\delta}e_{\lambda\varphi}e_{\mu\omega}\,{}_0a_{\lambda\mu}\,{}_0x^i_{,\varphi}\,u^i_{,\omega} \tag{5.5.15b}$$

$$\pi_{(2)}^{\gamma\delta} = -\frac{1}{4_0 a^{3/2}} e_{\gamma\delta}\Big[e_{\lambda\varphi}e_{\mu\omega}\,{}_0a_{\lambda\mu}u,_{\varphi}^{i}u,_{\omega}^{i}$$

$$+\frac{1}{2}e_{\lambda\varphi}e_{\mu\omega}\left(u,_{\lambda}^{i}\,{}_0x,_{\mu}^{i}+u,_{\mu}^{i}\,{}_0x,_{\lambda}^{i}\right)\left(u,_{\varphi}^{j}\,{}_0x,_{\omega}^{j}+u,_{\omega}^{j}\,{}_0x,_{\varphi}^{j}\right) \quad (5.5.15c)$$

$$-\frac{3}{{}_0a}\left(e_{\lambda\varphi}e_{\mu\omega}\,{}_0a_{\lambda\mu}\,{}_0x,_{\varphi}^{i}u,_{\omega}^{i}\right)^2\Big]$$

Using Eqs. 5.5.14b and 5.5.15, we derive the first group in Eq. 5.5.13 as

$$\frac{1}{2}\bar{\pi}^{\gamma\delta} = \frac{e_{\gamma\delta}}{2\sqrt{a}} = \pi_{(0)}^{\gamma\delta} + \pi_{(1)}^{\gamma\delta} + \pi_{(2)}^{\gamma\delta} + 0\left(U^3\right) \quad (5.5.16)$$

The second group is

$$e_{ijk}x,_{\gamma}^{j}\,x,_{\delta}^{k}\,x,_{\alpha\beta}^{i} = e_{ijk}\left({}_0x,_{\gamma}^{j}+u,_{\gamma}^{j}\right)\left({}_0x,_{\delta}^{k}+u,_{\delta}^{k}\right)\left({}_0x,_{\alpha\beta}^{i}+u,_{\alpha\beta}^{i}\right) \quad (5.5.17)$$

Let

$$s_{\alpha\beta\gamma\delta}^{(0)} = e_{ijk}\,{}_0x,_{\gamma}^{j}\,{}_0x,_{\delta}^{k}\,{}_0x,_{\alpha\beta}^{i} \quad (5.5.18a)$$

$$s_{\alpha\beta\gamma\delta}^{(1)} = e_{ijk}\left[\left({}_0x,_{\gamma}^{j}u,_{\delta}^{k}+{}_0x,_{\delta}^{k}u,_{\gamma}^{j}\right)x,_{\alpha\beta}^{i}+{}_0x,_{\gamma}^{j}\,{}_0x,_{\delta}^{k}u,_{\alpha\beta}^{i}\right] \quad (5.5.18b)$$

$$s_{\alpha\beta\gamma\delta}^{(2)} = e_{ijk}\left[\left({}_0x,_{\gamma}^{j}u,_{\delta}^{k}+{}_0x,_{\delta}^{k}u,_{\gamma}^{j}\right)u,_{\alpha\beta}^{i}+u,_{\gamma}^{j}u,_{\delta}^{k}\,{}_0x,_{\alpha\beta}^{i}\right] \quad (5.5.18c)$$

Hence, the second group in Eq. 5.5.17 becomes

$$e_{ijk}x,_{\gamma}^{j}\,x,_{\delta}^{k}\,x,_{\alpha\beta}^{i} = s_{\alpha\beta\gamma\delta}^{(0)} + s_{\alpha\beta\gamma\delta}^{(1)} + s_{\alpha\beta\gamma\delta}^{(2)} + O\left(\bar{U}^3\right) \quad (5.5.19)$$

where $\bar{U} = \max\limits_{i,\alpha,\beta}\left(\left|u,_{\alpha}^{i}\right|,\left|u,_{\alpha\beta}^{i}\right|\right)$. Using Eqs. 5.5.16 and 5.5.19, we derive the second fundamental tensor of the deformed reference surface in Eq. 5.5.12 as

$$b_{\alpha\beta} = \frac{1}{2}\bar{\pi}^{\gamma\delta}e_{ijk}x,_{\gamma}^{j}\,x,_{\delta}^{k}\,x,_{\alpha\beta}^{i}$$

$$= {}_0b_{\alpha\beta} + \left[\pi_{(0)}^{\gamma\delta}s_{\alpha\beta\gamma\delta}^{(1)} + \pi_{(1)}^{\gamma\delta}s_{\alpha\beta\gamma\delta}^{(0)}\right] \quad (5.5.20)$$

$$+\left[\pi_{(0)}^{\gamma\delta}s_{\alpha\beta\gamma\delta}^{(2)} + \pi_{(1)}^{\gamma\delta}s_{\alpha\beta\gamma\delta}^{(1)} + \pi_{(2)}^{\gamma\delta}s_{\alpha\beta\gamma\delta}^{(0)}\right] + O\left(\bar{U}^3\right)$$

5.5.2 Principle of Virtual Work

In simulations of sheet metal forming processes, the deformation of the sheet metal can be well into the plastic range. Note that forming tools are contacting the sheet surfaces. Therefore, there are frictional forces. For these reasons, the mechanical system is non-conservative. A potential function does not exist. Instead, we should use the principle of virtual work to establish the equations of motion. Before we discuss the principle of virtual work, we introduce the Kirchhoff stress tensor $\boldsymbol{\tau}$ from Hutchinson [1973] for the plane stress state as

$$\boldsymbol{\tau} = \tau^{\alpha\beta}\mathbf{g}_\alpha\mathbf{g}_\beta \quad (\alpha, \beta = 1,2) \tag{5.5.21}$$

with respect to the deformed base vectors $\mathbf{g}_\alpha$ but referred to the undeformed volume of a shell. For plates and shells, the Kirchhoff stress tensor is convenient to use in the total Lagrangian formulation. Note that $\tau^{i3} = 0$ for the classical thin shell theory.

With a virtual nodal displacement vector $\delta\mathbf{U}$, we apply the principle of virtual work to a shell to establish the equations of motion, as shown in Eq. 4.2.1 in Chapter 4. In the equations of motion, the internal nodal force vector **Fi** may be obtained by the virtual work from the stresses in a shell as in Eq. 4.3.2 in Chapter 4. Because we use the virtual Lagrangian strain $\delta\varepsilon_{\alpha\beta}$ due to $\delta\mathbf{U}$ and the Kirchhoff stress tensor, the virtual work term due to the stresses has the form

$$(\mathbf{Fi})^T\,\delta\mathbf{U} = \int_{v_0} \tau^{\alpha\beta}\,\delta\varepsilon_{\alpha\beta}\,dv \tag{5.5.22}$$

where v_0 is the undeformed volume of a shell. Note that the Kirchhoff stress $\tau^{\alpha\beta}$ and the virtual Lagrangian strain $\delta\varepsilon_{\alpha\beta}$ are work-conjugate.

5.5.3 Constitutive Equation for the Classical Shell Theory

For the classical shell theory, we use the convected coordinates θ^α (Gaussian or curvilinear coordinates) defined earlier in this chapter. There are only three in-plane distinct (independent) stress components ($\tau^{12} = \tau^{21}$) and three in-plane distinct (independent) strain components ($\varepsilon^{12} = \varepsilon^{21}$). Because we use a displacement-based finite element method to discretize the equations of motion, we need to convert the rate of the stress tensor, $\dot{\tau}^{\alpha\beta}$ ($\alpha, \beta = 1,2$), in terms of the rates of the Lagrangian strains $\dot{\varepsilon}_{\gamma\delta}$ that are computed from the rates of the displacement derivatives. Therefore, the only unknowns are the rates of the generalized displacements and displacement derivatives at each node in a finite element model. Applying the flow theory of plasticity, we may write the constitutive equation in the rate form as

$$\dot{\tau}^{\alpha\beta} = D^{\alpha\beta\gamma\delta}\,\dot{\varepsilon}_{\gamma\delta} \tag{5.5.23}$$

where $\dot{\tau}^{\alpha\beta}$ and $\dot{\varepsilon}_{\gamma\delta}$ are the convected rate of the Kirchhoff stress tensor and the rate of Lagrangian strain tensor, respectively, and $D^{\alpha\beta\gamma\delta}$ is the material tangent stiffness tensor of a shell. For rate-independent materials, $D^{\alpha\beta\gamma\delta}$ contains only the history of the stresses and the strains but not any rate quantity. The material tangent stiffness tensor has the form of symmetry as $D^{\alpha\beta\gamma\delta} = D^{\gamma\delta\alpha\beta} = D^{\beta\alpha\gamma\delta} = D^{\alpha\beta\delta\gamma}$ from Hutchinson [1973]. We use the mathematical theory of plasticity and the constitutive relation for hypoelastic materials to derive the material tangent stiffness tensor.

5.5.4 Yield Function and Flow Rule for the Classical Shell Theory

Several essential parts of the mathematical theory of plasticity such as the yield function, flow, and hardening rules are discussed here. We present the modified J_2 flow theory for the isotropic hardening rule first. At the end of Section 5.5.5, we present a formulation based on an anisotropic hardening rule for the material tangent stiffness tensor.

In the following derivation, all stress and strain tensors are symmetrical unless otherwise noted. We occasionally use the mixed or covariant components of the Kirchhoff stress tensor for simplifying the equations. Note that

$$\tau^{\alpha}_{\beta} = g_{\gamma\beta}\tau^{\alpha\gamma} \tag{5.5.24a}$$

$$\tau_{\alpha\beta} = g_{\alpha\gamma}g_{\beta\delta}\tau^{\gamma\delta} \tag{5.5.24b}$$

In general, $\tau^{\alpha}_{\beta} \neq \tau^{\beta}_{\alpha}$ for $\alpha \neq \beta$, but $\tau_{\alpha\beta} = \tau_{\beta\alpha}$. In the derivation of the yield function for transversely anisotropic (but planar isotropic) materials with the anisotropy parameter R, the non-zero generalized deviatoric components of the Kirchhoff stress tensor from Eqs. 3.9.2 through 3.9.5 become

$$s^{\alpha\beta} = \tau^{\alpha\beta} - \frac{R}{1+2R}g^{\alpha\beta}g_{\gamma\delta}\tau^{\gamma\delta} \tag{5.5.25a}$$

$$s^{33} = -\frac{1}{1+2R}g^{33}g_{\gamma\delta}\tau^{\gamma\delta} \tag{5.5.25b}$$

where the metric tensor $g_{\gamma\delta} = g^i_{\gamma}g^i_{\delta}$, the base vectors g^i_{γ} are computed by using Eq. 5.3.15a, and $g^{\alpha\beta}$ are the contravariant components of the metric tensor $g_{\alpha\beta}$. Note that $g^{\alpha\gamma}g_{\gamma\beta} = \delta^{\alpha}_{\beta}$. Here, δ^{α}_{β} represents the generalized Kronecker delta. The mixed and covariant components of $s^{\alpha\beta}$ are

$$s^{\alpha}_{\beta} = g_{\beta\gamma}s^{\alpha\gamma} = g^{\alpha\gamma}s_{\beta\gamma}$$

$$s_{\alpha\beta} = g_{\alpha\gamma}g_{\beta\delta}s^{\gamma\delta}$$

Note that, in general, $s^{\alpha}_{\beta} \neq s^{\beta}_{\alpha}$ for $\alpha \neq \beta$, but $s_{\alpha\beta} = s_{\beta\alpha}$. Also note that $g_{3\alpha} = 0$.

Therefore,

$$s_3^3 = g_{33}s^{33}$$

Using the convected coordinates θ^α by Hutchinson [1973], the modified J_2 in Eq. 3.9.7 in Chapter 3 can be written in the form

$$J_2 = \frac{1}{1+R}\left(s_\beta^\alpha s_\alpha^\beta + Rs_3^3 s_3^3\right) \tag{5.5.26}$$

Note that in Eq. 5.5.26, we use the mixed component s_3^3, which can be computed from Eq. 5.5.25b as

$$s_3^3 = -\frac{1}{1+2R}\tau_\gamma^\gamma \tag{5.5.27}$$

based on $g_{33}g^{33} = 1$. Equation 5.5.27 gives the relationship

$$s_i^i = 0 \quad (i = 1,2,3) \tag{5.5.28a}$$

Therefore,

$$s_3^3 = -s_\alpha^\alpha \tag{5.5.28b}$$

is still satisfied.

The yield function in Eq. 3.9.8 and the effective stress in Eq. 3.9.6 do not change in form for the convected coordinate system. Therefore,

$$f = (1+2R)J_2 - X^2 = 0 \tag{5.5.29}$$

where $X = X\left(\bar{\varepsilon}^p\right)$. The effective stress $\bar{\sigma}$ can be defined as

$$\bar{\sigma} = \sqrt{(1+2R)J_2} \tag{5.5.30}$$

Different definitions of X and $\bar{\sigma}$ are needed for the generalization of the constitutive law later. Based on the flow rule in Eqs. 3.9.9 through 3.9.11, we can write the flow rule referred to the convected coordinates as

$$\dot{\varepsilon}_{\alpha\beta}^p = \dot{\lambda}\frac{\partial f}{\partial \tau^{\alpha\beta}} = 2\dot{\lambda}\frac{1+2R}{1+R}s_{\alpha\beta} \tag{5.5.31}$$

and the proportionality parameter $\dot{\lambda}$ is the same as that in Eq. 3.9.14 as

$$\dot{\lambda} = \frac{\dot{\bar{\varepsilon}}^p}{2X} \tag{5.5.32}$$

Note that $X = \bar{\sigma}$ from Eqs. 5.5.29 and 5.5.30. Multiplying both sides of Eq. 5.5.31 by $g^{\beta\gamma}$, we have

$$g^{\beta\gamma}\dot{\varepsilon}^p_{\alpha\beta} = 2\dot{\lambda}\frac{1+2R}{1+R}s^\gamma_\alpha$$

or

$$s^\gamma_\alpha = \frac{1}{2\dot{\lambda}}\frac{1+R}{1+2R}g^{\beta\gamma}\dot{\varepsilon}^p_{\alpha\beta} \tag{5.5.33}$$

Similarly, we have

$$g^{\alpha\beta}\dot{\varepsilon}^p_{\alpha\beta} = 2\dot{\lambda}\frac{1+2R}{1+R}s^\alpha_\alpha = -2\dot{\lambda}\frac{1+2R}{1+R}s^3_3$$

or

$$s^3_3 = -\frac{1}{2\dot{\lambda}}\frac{1+R}{1+2R}g^{\alpha\beta}\dot{\varepsilon}^p_{\alpha\beta} \tag{5.5.34}$$

Note that we used Eq. 5.5.28b in the preceding derivation. Combining Eqs. 5.5.32 through 5.5.34 inclusive, 5.5.30, and 5.5.26 gives the effective plastic strain rate $\dot{\bar{\varepsilon}}^p$ as

$$\dot{\bar{\varepsilon}}^p = \left[\frac{1+R}{1+2R}\left(g^{\alpha\gamma}g^{\beta\delta}\dot{\varepsilon}^p_{\alpha\beta}\dot{\varepsilon}^p_{\gamma\delta} + Rg^{\alpha\beta}g^{\gamma\delta}\dot{\varepsilon}^p_{\alpha\beta}\dot{\varepsilon}^p_{\gamma\delta}\right)\right]^{1/2} \tag{5.5.35}$$

The total effective plastic strain $\bar{\varepsilon}^p$ is obtained by integrating $\dot{\bar{\varepsilon}}^p$ over the deformation history as that in Eq. 5.5.35. The derivation of Eq. 5.5.35 is valid only under monotonically increasing loading conditions.

5.5.5 Consistent Material Tangent Stiffness Tensor

In implicit finite element methods, an appropriate material tangent stiffness matrix (tensor) is needed, as discussed in Belytschko *et al.* [2000]. Due to the abrupt transition from the elastic to plastic state, the conventional elastic-plastic material tangent stiffness (referred to as the conventional material tangent stiffness later in this book) may cause artificial loading and unloading. To avoid this numerical problem, the consistent material tangent stiffness is used in this book.

The generalized Hooke's law for hypoelastic materials gives the rate relationship between the objective (or frame-independent) stress rates and the strain rate as

$$\breve{\tau}^{\alpha\beta} = D_e^{\alpha\beta\gamma\delta}\dot{\varepsilon}_{\gamma\delta}^{e} \tag{5.5.36}$$

where $\dot{\varepsilon}_{\gamma\delta}^{e}$ is the time rate of the elastic Lagrangian strain tensor, and $\breve{\tau}^{\alpha\beta}$ is the Jaumann rate of the Kirchhoff stress tensor referred to the convected coordinate system. Hutchinson [1973] showed that the Kirchhoff stress tensor in three dimensions can be expressed as

$$\breve{\tau}^{ij} = \dot{\tau}^{ij} + g^{ik}\tau^{jl}\dot{\varepsilon}_{kl} + g^{jk}\tau^{il}\dot{\varepsilon}_{kl} \quad (i,j,k,l = 1,2,3) \tag{5.5.37a}$$

We condense the equation to the plane stress state as

$$\breve{\tau}^{\alpha\beta} = \dot{\tau}^{\alpha\beta} + g^{\alpha\gamma}\tau^{\beta\delta}\dot{\varepsilon}_{\gamma\delta} + g^{\beta\gamma}\tau^{\alpha\delta}\dot{\varepsilon}_{\gamma\delta} \quad (\alpha,\beta,\gamma,\delta = 1,2) \tag{5.5.37b}$$

Note that $\breve{\tau}^{\alpha\beta}$ are objective. The additive theory of Lee [1969] gives

$$\dot{\varepsilon}_{\gamma\delta} = \dot{\varepsilon}_{\gamma\delta}^{e} + \dot{\varepsilon}_{\gamma\delta}^{p}. \tag{5.5.38a}$$

or

$$\dot{\varepsilon}_{\gamma\delta}^{e} = \dot{\varepsilon}_{\gamma\delta} - \dot{\varepsilon}_{\gamma\delta}^{p} \tag{5.5.38b}$$

The hypoelastic material tangent stiffness tensor $D_e^{\alpha\beta\gamma\delta}$ referred to the convected coordinates θ^{α} $(\alpha = 1,2)$ is obtained through condensing the three-dimensional equation (for nine stress and strain components) for hypoelastic materials from Hutchinson [1973] as

$$D_3^{ijkl} = \frac{E}{1+\nu}\left[\frac{1}{2}\left(g^{ik}g^{jl} + g^{il}g^{jk}\right) + \frac{\nu}{1-2\nu}g^{ij}g^{kl}\right] \tag{5.5.39}$$

Under plane stress conditions,

$$\breve{\tau}^{\alpha\beta} = D_3^{\alpha\beta\gamma\delta}\dot{\varepsilon}_{\gamma\delta}^{e} + D_3^{\alpha\beta13}\dot{\varepsilon}_{13}^{e} + D_3^{\alpha\beta23}\dot{\varepsilon}_{23}^{e} + D_3^{\alpha\beta33}\dot{\varepsilon}_{33}^{e} \tag{5.5.40}$$

and

$$\breve{\tau}^{33} = D_3^{33\gamma\delta}\dot{\varepsilon}_{\gamma\delta}^{e} + D_3^{3313}\dot{\varepsilon}_{13}^{e} + D_3^{3323}\dot{\varepsilon}_{23}^{e} + D_3^{3333}\dot{\varepsilon}_{33}^{e} = 0 \tag{5.5.41}$$

Equation 5.5.39 gives

$$D_3^{\alpha\beta\gamma 3} = \frac{E}{1+\nu}\left[\frac{1}{2}\left(g^{\alpha\gamma}g^{\beta 3} + g^{\alpha 3}g^{\beta\gamma}\right) + \frac{\nu}{1-2\nu}g^{\alpha\beta}g^{\gamma 3}\right] = 0 \tag{5.5.42}$$

because $g^{\delta 3} = 0 \ (\delta = 1,2)$ and

$$D_3^{33\gamma 3} = \frac{E}{1+\nu}\left[\frac{1}{2}\left(g^{3\gamma}g^{33} + g^{33}g^{3\gamma}\right) + \frac{\nu}{1-2\nu}g^{33}g^{\gamma 3}\right] = 0 \tag{5.5.43}$$

Substituting Eq. 5.5.43 into Eq. 5.5.41, we can solve for

$$\dot{\varepsilon}_{33}^{e} = -\frac{D_3^{33\gamma\delta}\dot{\varepsilon}_{\gamma\delta}^{e}}{D_3^{3333}}, \quad D_3^{3333} \neq 0 \tag{5.5.44}$$

where

$$D_3^{\alpha\beta 33} = D_3^{33\alpha\beta} = \frac{E}{1+\nu}\frac{\nu}{1-2\nu}g^{\alpha\beta}h^{-2} \tag{5.5.45a}$$

$$D_3^{3333} = \frac{E}{1+\nu}\frac{1-\nu}{1-2\nu}h^{-4} \tag{5.5.45b}$$

from Eq. 5.5.39 with $g_{33} = h^2, g^{33} = h^{-2}$ obtained from Eq. 5.3.15b. Combining Eqs. 5.5.40, 5.5.42, and 5.5.44 gives the hypoelastic material tangent tensor under plane stress conditions as

$$\begin{aligned} D_e^{\alpha\beta\gamma\delta} &= D_3^{\alpha\beta\gamma\delta} - \frac{D_3^{\alpha\beta 33}D_3^{33\gamma\delta}}{D_3^{3333}} \\ &= \frac{E}{1+\nu}\left[\frac{1}{2}\left(g^{\alpha\gamma}g^{\beta\delta} + g^{\alpha\delta}g^{\beta\gamma}\right) + \frac{\nu}{1-\nu}g^{\alpha\beta}g^{\gamma\delta}\right] \end{aligned} \tag{5.5.46}$$

To compute the rate of the plastic strain in Eq. 5.5.38, we apply the flow rule in Eq. 5.5.31, which gives the increments of the plastic strains from time t to $t + \Delta t$ as

$$\Delta\varepsilon_{\gamma\delta}^{p} = \Delta\lambda\frac{\partial f}{\partial\tau^{\gamma\delta}} \tag{5.5.47a}$$

Taking the finite difference approximation of Eq. 5.5.32, we obtain

$$\Delta\lambda = \frac{\Delta\bar{\varepsilon}^{p}}{2X} \tag{5.5.47b}$$

and $\Delta\bar{\varepsilon}^{p}$ is the increment of the effective plastic increment from time t to $t + \Delta t$. Here, X is the effective stress.

Differentiating Eq. 5.5.47a with respect to time t gives the rate at time $t + \Delta t$ as

$$\frac{d}{dt}\Delta\varepsilon^{p}_{\gamma\delta} = \frac{d\Delta\lambda}{dt}\frac{\partial f}{\partial\tau^{\gamma\delta}} + \Delta\lambda\frac{\partial^2 f}{\partial\tau^{\gamma\delta}\partial\tau^{\xi\zeta}}\breve{\tau}^{\xi\zeta} \tag{5.5.48}$$

Combining Eqs. 5.5.36, 5.5.38b, and 5.5.48 yields

$$\breve{\tau}^{\alpha\beta} = D_e^{\alpha\beta\gamma\delta}\left(\dot{\varepsilon}_{\gamma\delta} - \frac{d\Delta\lambda}{dt}\frac{\partial f}{\partial\tau^{\gamma\delta}} - \Delta\lambda\frac{\partial^2 f}{\partial\tau^{\gamma\delta}\partial\tau^{\xi\zeta}}\breve{\tau}^{\xi\zeta}\right) \tag{5.5.49a}$$

Moving the last term on the right side of Eq. 5.5.49a to the left side gives

$$H_{\xi\zeta}^{\alpha\beta}\breve{\tau}^{\xi\zeta} = D_e^{\alpha\beta\gamma\delta}\left(\dot{\varepsilon}_{\gamma\delta} - \frac{d\Delta\lambda}{dt}\frac{\partial f}{\partial\tau^{\gamma\delta}}\right) \tag{5.5.49b}$$

where

$$H_{\xi\zeta}^{\alpha\beta} = \frac{1}{2}\left(\delta_\xi^\alpha\delta_\zeta^\beta + \delta_\xi^\beta\delta_\zeta^\alpha\right) + \Delta\lambda D_e^{\alpha\beta\gamma\delta}\frac{\partial^2 f}{\partial\tau^{\gamma\delta}\partial\tau^{\xi\zeta}} \tag{5.5.49c}$$

The first partial derivatives from Eq. 5.5.29 are

$$\frac{\partial f}{\partial\tau^{\alpha\beta}} = \frac{2(1+2R)}{1+R}s_{\alpha\beta} = \frac{2(1+2R)}{1+R}\left[g_{\alpha\gamma}g_{\beta\delta}\tau^{\gamma\delta} - \frac{R}{1+2R}g_{\alpha\beta}g_{\gamma\delta}\tau^{\gamma\delta}\right] \tag{5.5.49d}$$

and the second derivatives are

$$\frac{\partial^2 f}{\partial\tau^{\xi\zeta}\partial\tau^{\alpha\beta}} = \frac{2(1+2R)}{1+R}\left[g_{\alpha\xi}g_{\beta\zeta} - \frac{R}{1+2R}g_{\alpha\beta}g_{\xi\zeta}\right] \tag{5.5.49e}$$

Let us introduce a tensor $G_{\alpha\beta}^{\phi\psi}$ such that

$$G_{\alpha\beta}^{\phi\psi} H_{\xi\zeta}^{\alpha\beta} = \frac{1}{2}\left(\delta_{\xi}^{\phi}\delta_{\zeta}^{\psi} + \delta_{\zeta}^{\phi}\delta_{\xi}^{\psi}\right) \tag{5.5.50a}$$

and set

$$\bar{D}^{\phi\psi\gamma\delta} = G_{\alpha\beta}^{\phi\psi} D_{e}^{\alpha\beta\gamma\delta} \tag{5.5.50b}$$

Multiplying both sides of Eq. 5.5.49b by $G_{\alpha\beta}^{\phi\psi}$ and using Eqs. 5.5.50a and 5.5.50b, we solve for

$$\breve{\tau}^{\phi\psi} = \bar{D}^{\phi\psi\gamma\delta}\left(\dot{\varepsilon}_{\gamma\delta} - \frac{d\Delta\lambda}{dt}\frac{\partial f}{\partial \tau^{\gamma\delta}}\right)$$

or

$$\breve{\tau}^{\alpha\beta} = \bar{D}^{\alpha\beta\gamma\delta}\left(\dot{\varepsilon}_{\gamma\delta} - \frac{d\Delta\lambda}{dt}\frac{\partial f}{\partial \tau^{\gamma\delta}}\right) \tag{5.5.50c}$$

The consistency condition is obtained by taking the time derivative of the yield function in Eq. 5.5.29 for rate insensitive materials as

$$\frac{df}{dt} = 2\frac{1+2R}{1+R} s_{\alpha\beta}\breve{\tau}^{\alpha\beta} - 2XX'\frac{d}{dt}\Delta\bar{\varepsilon}^{p} = 0 \tag{5.5.51}$$

Here, $X' = \dfrac{dX}{d\bar{\varepsilon}^{p}}$. Now we can write

$$\frac{1+2R}{1+R} s_{\alpha\beta}\breve{\tau}^{\alpha\beta} - XX'\frac{d}{dt}\Delta\bar{\varepsilon}^{p} = 0 \tag{5.5.52}$$

where X and X' are evaluated at time $t + \Delta t$. Differentiating Eq. 5.5.47b with respect to time t gives

$$\frac{d\Delta\lambda}{dt} = \frac{1}{2X}\frac{d\Delta\bar{\varepsilon}^{p}}{dt} \tag{5.5.53}$$

or

$$\frac{d\Delta\bar{\varepsilon}^{p}}{dt} = 2X\frac{d\Delta\lambda}{dt} \tag{5.5.54}$$

Using Eqs. 5.5.31, 5.5.52, 5.5.54, and 5.5.50c, we can solve for $d\Delta\lambda/dt$ in terms of $\dot{\varepsilon}_{\gamma\delta}$, such as

$$\frac{d\Delta\lambda}{dt} = \frac{\dfrac{\partial f}{\partial\tau^{\alpha\beta}}\bar{D}^{\alpha\beta\gamma\delta}\dot{\varepsilon}_{\gamma\delta}}{4X^2X' + \dfrac{\partial f}{\partial\tau^{\alpha\beta}}\bar{D}^{\alpha\beta\gamma\delta}\dfrac{\partial f}{\partial\tau^{\gamma\delta}}} \tag{5.5.55}$$

Inserting Eq. 5.5.55 into Eq. 5.5.50c gives the consistent material tangent stiffness tensor

$$\bar{\tau}^{\alpha\beta} = \left[\bar{D}^{\alpha\beta\gamma\delta} - \frac{\dfrac{\partial f}{\partial\tau^{\xi\zeta}}\bar{D}^{\xi\zeta\gamma\delta}\bar{D}^{\alpha\beta\phi\psi}\dfrac{\partial f}{\partial\tau^{\phi\psi}}}{4X^2X' + \dfrac{\partial f}{\partial\tau^{\xi\zeta}}\bar{D}^{\xi\zeta\phi\psi}\dfrac{\partial f}{\partial\tau^{\phi\psi}}}\right]\dot{\varepsilon}_{\gamma\delta} \tag{5.5.56}$$

In terms of the convected rate of the Kirchhoff stress in Eq. 5.5.37b,

$$\dot{\tau}^{\alpha\beta} = \left[\bar{D}^{\alpha\beta\gamma\delta} - \frac{\dfrac{\partial f}{\partial\tau^{\xi\zeta}}\bar{D}^{\xi\zeta\gamma\delta}\bar{D}^{\alpha\beta\phi\psi}\dfrac{\partial f}{\partial\tau^{\phi\psi}}}{4X^2X' + \dfrac{\partial f}{\partial\tau^{\xi\zeta}}\bar{D}^{\xi\zeta\phi\psi}\dfrac{\partial f}{\partial\tau^{\phi\psi}}} - g^{\alpha\gamma}\tau^{\beta\delta} - g^{\beta\gamma}\tau^{\alpha\delta}\right]\dot{\varepsilon}_{\gamma\delta} \tag{5.5.57}$$

Let $\Delta\tau^{\alpha\beta} \approx \dot{\tau}^{\alpha\beta}\Delta t$ and $\Delta\varepsilon_{\gamma\delta} \approx \dot{\varepsilon}_{\gamma\delta}\Delta t$. We may write the finite difference form as

$$\Delta\tau^{\alpha\beta} = D_{ep}^{\alpha\beta\gamma\delta}\Delta\varepsilon_{\gamma\delta} \tag{5.5.58a}$$

where the consistent elastic-plastic material tangent stiffness tensor (referred to as the consistent material tangent stiffness tensor later in this book) is

$$D_{ep}^{\alpha\beta\gamma\delta} = \bar{D}^{\alpha\beta\gamma\delta} - \frac{\dfrac{\partial f}{\partial\tau^{\xi\zeta}}\bar{D}^{\xi\zeta\gamma\delta}\bar{D}^{\alpha\beta\phi\psi}\dfrac{\partial f}{\partial\tau^{\phi\psi}}}{4X^2X' + \dfrac{\partial f}{\partial\tau^{\xi\zeta}}\bar{D}^{\xi\zeta\phi\psi}\dfrac{\partial f}{\partial\tau^{\phi\psi}}} - g^{\alpha\gamma}\tau^{\beta\delta} - g^{\beta\gamma}\tau^{\alpha\delta} \tag{5.5.58b}$$

We note that as $\Delta\lambda$ or $\Delta\bar{\varepsilon}^p \to 0$, from Eq. 5.5.49c,

$$H_{\xi\zeta}^{\alpha\beta} = \frac{1}{2}\left(\delta_\xi^\alpha\delta_\zeta^\beta + \delta_\xi^\beta\delta_\zeta^\alpha\right) \tag{5.5.59}$$

Therefore, combining Eqs. 5.5.59, 5.5.50a, and 5.5.50b gives

$$\bar{D}^{\alpha\beta\gamma\delta} = D_e^{\alpha\beta\gamma\delta} \tag{5.5.60}$$

Using Eq. 5.5.60, Eq. 5.5.58b is reduced to the material tangent stiffness tensor

$$\begin{aligned} D_{ep}^{\alpha\beta\gamma\delta} = D_e^{\alpha\beta\gamma\delta} - \frac{\dfrac{\partial f}{\partial \tau^{\xi\zeta}} D_e^{\xi\zeta\gamma\delta} D_e^{\alpha\beta\phi\psi} \dfrac{\partial f}{\partial \tau^{\phi\psi}}}{4X^2X' + \dfrac{\partial f}{\partial \tau^{\xi\zeta}} D_e^{\xi\zeta\phi\psi} \dfrac{\partial f}{\partial \tau^{\phi\psi}}} \\ - g^{\alpha\gamma}\tau^{\beta\delta} - g^{\beta\gamma}\tau^{\alpha\delta} \end{aligned} \tag{5.5.61}$$

This is the conventional material tangent stiffness tensor. All of the preceding derivations in this chapter up to this point are for the isotropic hardening rule.

For Mroz's hardening rule, the material tangent stiffness equation is modified accordingly. Because of finite or large strains in simulations of sheet metal forming processes, Chu's constitutive equation [1984] is modified for a curvilinear coordinate system. The modified equation can be applied to a shell analysis of large deformation. The transversely anisotropic material property in terms of the anisotropy parameter R also is included in the yield or loading surface by the proposed hardening rule. Note that because there is transverse anisotropy in plastic deformation, no real advantage can be taken for the yield surface defined in the deviatoric stress space.

We consider the yield surface in the Kirchhoff stress space. Therefore, the yield surface has the same form as that for the isotropic hardening rule, except using

$$r^{\alpha\beta} = \tau^{\alpha\beta} - \alpha^{\alpha\beta} \tag{5.5.62}$$

where $\alpha^{\alpha\beta}$ is the position tensor of the center of the active yield surface. Replacing the Kirchhoff stress tensor by $r^{\alpha\beta}$ in Eq. 5.5.58b or in Eq. 5.5.61, we can establish the consistent material tangent or the conventional material tangent stiffness tensor for Mroz's hardening rule.

5.5.6 Stress Resultant Constitutive Relationship

Computing the term for the virtual work done by the stresses in Eq. 5.5.22 requires numerical integration through the thickness and over the undeformed reference surface (volume integration) of a shell. The numerical integration through the thickness takes significant computing efforts. Researchers have tried to use the stress resultants to compute the virtual work that requires only integration over the undeformed reference surface of a shell in order to save computation time.

Using Eq. 5.5.6a, we may write the virtual strain tensor in the form of

$$\delta\varepsilon_{\alpha\beta} = \delta\gamma_{\alpha\beta} + h\theta^3\delta\kappa_{\alpha\beta} \tag{5.5.63}$$

where $\gamma_{\alpha\beta}$ is the in-plane strain tensor, and $\kappa_{\alpha\beta}$ is the bending strain (or curvature) tensor. Inserting the virtual strain tensor found in Eq. 5.5.63 into Eq. 5.5.22, the virtual work becomes

$$\int_{v_0} \tau^{\alpha\beta}\delta\varepsilon_{\alpha\beta}\, dv = \int_{s_0} h_0 \int_{-1/2}^{1/2} \tau^{\alpha\beta} d\theta^3 \delta\gamma_{\alpha\beta}\, ds + \int_{s_0} h_0 h \int_{-1/2}^{1/2} \tau^{\alpha\beta}\theta^3 d\theta^3 \delta\kappa_{\alpha\beta}\, ds \quad (5.5.64)$$

where $dv = h_0 d\theta^3 ds$ has been applied, and s_0 is the undeformed reference surface of a shell. We introduce two stress resultant tensors, the in-plane force tensor $N^{\alpha\beta}$ as

$$N^{\alpha\beta} = h_0 \int_{-1/2}^{1/2} \tau^{\alpha\beta} d\theta^3 \quad (5.5.65a)$$

and the moment tensor $M^{\alpha\beta}$ as

$$M^{\alpha\beta} = h_0 h \int_{-1/2}^{1/2} \tau^{\alpha\beta}\theta^3\, d\theta^3 \quad (5.5.65b)$$

The virtual work in Eq. 5.5.64 therefore is simplified to

$$\int_{v_0} \tau^{\alpha\beta}\delta\varepsilon_{\alpha\beta}\, dv = \int_{s_0} \left(N^{\alpha\beta}\delta\gamma_{\alpha\beta} + M^{\alpha\beta}\delta\kappa_{\alpha\beta} \right) ds \quad (5.5.66)$$

Therefore, we can avoid the time-consuming integration through the thickness to compute the virtual work discussed in the previous subsection in Eq. 5.5.22.

To be able to do the surface integration, we must establish the constitutive relation between the stress resultant tensors and the generalized strain tensors (in-plane strain and curvature tensors) instead of that for the stresses and strains. Koiter [1966] gave the relation for the elastic deformation. For the plastic deformation, we must establish yield criteria for the in-plane force tensor $N^{\alpha\beta}$ and the moment tensor $M^{\alpha\beta}$. Chou *et al.* [1991, 1993, and 1994] tried to establish the stress resultant constitutive relations and used the relations for plate and shell structures under large plastic deformation. Readers may refer to these references for more details.

5.6 Shell Theory with Transverse Shear Deformation

In the shell theory with consideration of transverse shear deformation, a normal to the reference surface of an undeformed shell no longer remains normal to the reference surface after deformation. Therefore, the transverse shear strains in a shell are not zero as expressed by Eq. 5.4.1b. We use the transverse shear strains on the deformed reference surface of a shell to compute the transverse shear stress resultants. In this shell theory as in the classical one, the normal stress in the thickness direction vanishes.

Following Bathe and Bolourchi [1980], we use the updated Lagrangian method and the natural (isoparametric) coordinate system r, s, and t' (here we use t' to avoid confusion with time t) to formulate the finite element analysis. The coordinate axes r and s are mapped on the reference surface of a shell element. We define a rectangular Cartesian coordinate system with the first two coordinate axes $\overline{r}$ and $\overline{s}$ that are on the tangent plane determined by the axes r and s (Chapter 6 will show the details), while the third axis $\overline{t}$ is normal to the tangent plane. The Cartesian tensors are used in the following subsections.

5.6.1 Constitutive Equation for the Shell Theory with Transverse Shear Deformation

For simplifying the constitutive equations in the updated Lagrangian formulation, we use the rectangular Cartesian coordinate system $\overline{r}$, $\overline{s}$, $\overline{t}$ mentioned in the first paragraph of Section 5.6. In a rectangular Cartesian coordinate system, the metric tensor is the Kronecker delta of

$$\begin{aligned} \delta_{ij} &= 1, \quad i = j \\ \delta_{ij} &= 0, \quad i \neq j \end{aligned} \tag{5.6.1}$$

The constitutive equation for this shell theory involves a total of eight stress and strain components if tensor notation is used. In this case, for example, we consider $\dot{\sigma}_{12}$ and $\dot{\sigma}_{21}$ as two different components, although they are equal. Generally, the constitutive relation (for eight stress and strain components) in the rate form is expressed as

$$\dot{\sigma}_{ij} = D_{ijkl}\dot{\varepsilon}_{kl} \quad \text{(excluding } i = j = 3 \text{ or } k = l = 3) \tag{5.6.2}$$

Here, $1 = \overline{r}$, $2 = \overline{s}$, and $3 = \overline{t}$.

Following Bathe and Bolourchi [1980] for the updated Lagrangian formulation, we use the increment of the second Piola-Kirchhoff stress tensor, add it to the Cauchy stress tensor referred to the current configuration at time t, and transform the sum to the Cauchy stress tensor referred to the updated reference configuration at time $t + \Delta t$. McMeeking and Rice [1975] expressed the constitutive relation between the rate of the second Piola-Kirchhoff stress tensor, $\dot{t}_{ij}$, and the rate of the Lagrangian strain tensor, $\dot{\varepsilon}_{kl}$, for three-dimensional problems as

$$\dot{t}_{ij} = \left(D^{03}_{ijkl} - \sigma_{ik}\delta_{jl} - \sigma_{jk}\delta_{il} \right)\dot{\varepsilon}_{kl} \tag{5.6.3}$$

where D^{03}_{ijkl} is the three-dimensional elastic-plastic material tangent stiffness tensor for small strains, and σ_{ij} is the Cauchy stress tensor. For the eight stress and strain components ($\sigma_{33} = 0$), we may condense it to

$$\dot{t}_{ij} = \left(D^0_{ijkl} - \sigma_{ik}\delta_{jl} - \sigma_{jk}\delta_{il} \right)\dot{\varepsilon}_{kl} - \left(\sigma_{i3}\delta_{j3} + \sigma_{j3}\delta_{i3} \right)\dot{\varepsilon}_{33} \quad \text{(excluding } i = j = 3 \text{ or } k = l = 3) \tag{5.6.4a}$$

Because the plastic strain rate usually is significantly larger than the elastic strain rate under so-called plastic loading conditions in typical metal forming processes, we may use the approximation of the volume incompressibility for the total strain rate tensor to simplify Eq. 5.6.4a. Otherwise, the governing equations for numerical implementations would be much more complicated. Based on the approximation of the incompressibility, we can take $\dot{\varepsilon}_{33} = -\dot{\varepsilon}_{\alpha\alpha}$. Then Eq. 5.6.4a becomes

$$\dot{t}_{ij} = \left[D^0_{ijkl} - \left(\sigma_{ik}\delta_{jl} + \sigma_{jk}\delta_{il} \right) + \left(\sigma_{i3}\delta_{j3} + \sigma_{j3}\delta_{i3} \right)\delta_{k\alpha}\delta_{l\alpha} \right]\dot{\varepsilon}_{kl} \quad \text{(excluding } i = j = 3 \text{ or } k = l = 3) \tag{5.6.4b}$$

where D^0_{ijkl} is the elastic-plastic material tangent stiffness tensor for eight stress and strain components within the context of the small strain approach.

5.6.2 Consistent Material Tangent Stiffness Tensor with Transverse Shear Deformation

Following the method presented in Section 5.5.5, we now derive the consistent material tangent stiffness as shown in Eq. 5.6.4b. First, we write the rate form for eight stress and strain components as

$$\dot{\sigma}_{ij} = D^e_{ijkl}\dot{e}^e_{kl} \quad \text{(excluding } i = j = 3 \text{ or } k = l = 3) \tag{5.6.5}$$

where D^e_{ijkl} is the modified elastic material tangent stiffness tensor, $\dot{e}^e_{kl}$ is the rate of the small elastic strain tensor (with eight components), and $\dot{\sigma}_{ij}$ is the rate of the Cauchy stress tensor (with eight components). Equation 3.1.28 in Chapter 3 gives the elastic material tangent stiffness tensor for three-dimensional problems (nine stress components and nine strain components) as

$$D^3_{ijkl} = \frac{E}{1+\nu}\left[\frac{1}{2}\left(\delta_{ik}\delta_{jl} + \delta_{il}\delta_{jk} \right) + \frac{\nu}{1-2\nu}\delta_{ij}\delta_{kl} \right]$$

We now condense the above equation for the rates of the eight stress and strain components by setting $\dot{\sigma}_{33} = 0$. Following the procedure to derive Eqs. 5.5.44 through 5.5.46 inclusive, we have

$$\dot{e}^e_{33} = -\frac{D^3_{33ij}\dot{e}^e_{ij}}{D^{3333}_3}, \quad D^{3333}_3 \neq 0 \quad \text{(excluding } i = j = 3) \tag{5.6.6a}$$

$$D^e_{ijkl} = D^3_{ijkl} - \frac{D^3_{ij33} D^3_{33kl}}{D^3_{3333}} \quad \text{(excluding } i = j = 3 \text{ or } k = l = 3) \tag{5.6.6b}$$

where the non-zero components of the elastic material tangent stiffness tensor for eight stress and strain components are

$$D^e_{1111} = D^e_{2222} = \frac{E}{1 - \nu^2}$$

$$D^e_{1122} = \frac{\nu E}{1 - \nu^2}$$

$$D^e_{1212} = \frac{E}{2(1 + \nu)}$$

$$D^e_{2323} = D^e_{3131} = \frac{\kappa E}{2(1 + \nu)}$$

and κ is a correction factor for computing the average transverse shear strains across the thickness based on the transverse shear strains (using Eqs. 5.3.17c and 5.4.1c) on the deformed reference surface of a shell, and κ is set to $5/6$. Note that $1 = \bar{r}$, $2 = \bar{s}$, and $3 = \bar{t}$ in these equations. Invoking the symmetry conditions, we have other non-zero components of the tensor as

$$\begin{aligned} D^e_{2211} &= D^e_{1122} \\ D^e_{2121} &= D^e_{2112} = D^e_{1212} \\ D^e_{3232} &= D^e_{2332} = D^e_{2323} \\ D^e_{1313} &= D^e_{1331} = D^e_{3131} \end{aligned} \tag{5.6.7}$$

Next, we consider the effect of the rate of plastic strain, $\dot{e}^p_{ij}$. The additive theory of Lee [1969] gives the total rate of the elastic and plastic strain tensors as

$$\dot{e}_{ij} = \dot{e}^e_{ij} + \dot{e}^p_{ij} \quad \text{(excluding } i = j = 3) \tag{5.6.8a}$$

or

$$\dot{e}^e_{ij} = \dot{e}_{ij} - \dot{e}^p_{ij} \quad \text{(excluding } i = j = 3) \tag{5.6.8b}$$

To compute the rate of plastic strain based on the mathematical theory of plasticity, we may use the yield function in Eq. 3.9.15 of Chapter 3 and the effective plastic strain rate computed by Eq. 3.9.23 of Chapter 3, referred to the rectangular Cartesian coordinate system. Tensor notation is used in this chapter. Therefore, we rewrite the yield function in Eq. 3.9.15 as

$$\begin{aligned} f &= \sigma_{11}^2 + \sigma_{22}^2 - \frac{2R}{1+R}\sigma_{11}\sigma_{22} + \frac{(1+2R)}{1+R}\left(\sigma_{12}^2 + \sigma_{21}^2\right) \\ &+ \frac{3}{1+R}\left(\sigma_{23}^2 + \sigma_{32}^2 + \sigma_{31}^2 + \sigma_{13}^2\right) - X^2 = 0 \end{aligned} \tag{5.6.9}$$

where X is the flow stress of the shell material to be determined by the effective plastic strain. The effective stress in Eq. 3.9.22 of Chapter 3 is rewritten in terms of the tensor components in the following form as

$$\begin{aligned} \bar{\sigma} = \Bigg[&\sigma_{11}^2 + \sigma_{22}^2 - \frac{2R}{1+R}\sigma_{11}\sigma_{22} + \frac{1+2R}{1+R}\left(\sigma_{12}^2 + \sigma_{21}^2\right) \\ &+ \frac{3}{1+R}\left(\sigma_{23}^2 + \sigma_{32}^2 + \sigma_{31}^2 + \sigma_{13}^2\right)\Bigg]^{1/2} \end{aligned} \tag{5.6.10}$$

In the following derivations, we exclude the subscripts $i = j = 3$. The flow rule gives the rates of the plastic strain components as

$$\dot{e}_{ij}^p = \dot{\lambda}\frac{\partial f}{\partial \sigma_{ij}} \tag{5.6.11a}$$

We write Eq. 5.6.11a in the finite difference form for the increments of the plastic strains from time t to $t + \Delta t$ as

$$\Delta e_{ij}^p = \Delta\lambda\frac{\partial f}{\partial \sigma_{ij}} \tag{5.6.11b}$$

where the proportionality parameter $\Delta\lambda$ can be obtained by the finite difference approximation in Eq. 5.5.47b as

$$\Delta\lambda = \frac{\Delta\bar{e}^p}{2X} \tag{5.6.12}$$

Here $\Delta\bar{e}^p$ is the increment of the effective plastic strain.

In the following derivations, we exclude the subscripts $i = j = 3$, $k = l = 3$, $m = n = 3$, or $p = q = 3$. Differentiating Eq. 5.6.11b with respect to time t gives the rate at time $t + \Delta t$ as

$$\frac{d}{dt}\Delta e_{ij}^p = \frac{d\Delta\lambda}{dt}\frac{\partial f}{\partial \sigma_{ij}} + \Delta\lambda\frac{\partial^2 f}{\partial \sigma_{ij}\partial \sigma_{kl}}\dot{\sigma}_{kl} \tag{5.6.13}$$

Combining Eqs. 5.6.5, 5.6.8b, and 5.6.13 yields

$$\dot{\sigma}_{ij} = D^e_{ijkl}\left(\dot{e}_{kl} - \frac{d\Delta\lambda}{dt}\frac{\partial f}{\partial \sigma_{kl}} - \Delta\lambda \frac{\partial^2 f}{\partial \sigma_{kl}\partial \sigma_{mn}}\dot{\sigma}_{mn} \right) \tag{5.6.14a}$$

or

$$H_{ijkl}\dot{\sigma}_{kl} = D^e_{ijkl}\left(\dot{e}_{kl} - \frac{d\Delta\lambda}{dt}\frac{\partial f}{\partial \sigma_{kl}} \right) \tag{5.6.14b}$$

where the tensor **H** can be simplified from that in Eq. 5.5.49c (referred to as the rectangular Cartesian coordinate system in this section) as

$$H_{ijkl} = \frac{1}{2}\left(\delta_{ik}\delta_{jl} + \delta_{jk}\delta_{il}\right) + \Delta\lambda D^e_{ijmn}\frac{\partial^2 f}{\partial \sigma_{mn}\partial \sigma_{kl}} \tag{5.6.14c}$$

Therefore, Eqs. 5.5.50a and 5.5.50b become, respectively,

$$G_{ijkl}H_{mnij} = \frac{1}{2}\left(\delta_{mk}\delta_{nl} + \delta_{nk}\delta_{ml}\right) \tag{5.6.15}$$

$$\bar{D}_{ijkl} = G_{ijmn}D^e_{mnkl} \tag{5.6.16}$$

Similar to those in Eqs. 5.5.49d and 5.5.49e, we can obtain the first and the second partial derivatives of the yield function f in Eq. 5.6.9 with respect to the Cauchy stress components σ_{ij}. Following the procedure to use the consistency condition to derive Eq. 5.5.55, we can obtain the time rate of $\Delta\lambda$ as

$$\frac{d\Delta\lambda}{dt} = \frac{\dfrac{\partial f}{\partial \tau_{ij}}\bar{D}_{ijkl}\dot{\varepsilon}_{kl}}{4X^2X' + \dfrac{\partial f}{\partial \tau_{ij}}\bar{D}_{ijkl}\dfrac{\partial f}{\partial \tau_{kl}}} \tag{5.6.17}$$

Inserting the partial derivatives of the yield function and Eq. 5.6.17 into Eq. 5.6.14a and using Eqs. 5.6.14b through 5.6.16, we obtain the consistent material tangent stiffness tensor for eight stress and strain components within the context of the small strain approach as

$$D^0_{ijkl} = \bar{D}_{ijkl} - \frac{\dfrac{\partial f}{\partial \sigma_{mn}}\bar{D}_{mnkl}\bar{D}_{ijpq}\dfrac{\partial f}{\partial \sigma_{pq}}}{4X^2X' + \dfrac{\partial f}{\partial \sigma_{mn}}\bar{D}_{mnpq}\dfrac{\partial f}{\partial \sigma_{pq}}} \tag{5.6.18}$$

Inserting this equation into Eq. 5.6.4b yields the constitutive equation for the large or finite elastic-plastic strains (eight components) in the finite difference form as

$$\Delta t_{ij} = D^{ep}_{ijkl}\Delta\varepsilon_{kl} \quad (\text{excluding } i = j = 3 \text{ or } k = l = 3) \tag{5.6.19a}$$

where

$$D^{ep}_{ijkl} = D^{0}_{ijkl} - \left(\sigma_{ik}\delta_{jl} + \sigma_{jk}\delta_{il}\right) + \left(\sigma_{i3}\delta_{j3} + \sigma_{j3}\delta_{i3}\right)\delta_{k\alpha}\delta_{l\alpha} \tag{5.6.19b}$$

Note again that $1 = \overline{r}$, $2 = \overline{s}$, and $3 = \overline{t}$.

Similar to the classical shell case, $\Delta\lambda$ or $\Delta\overline{\varepsilon}_p \to 0$, Eq. 5.6.14c is reduced to

$$H_{ijkl} = \frac{1}{2}\left(\delta_{ik}\delta_{jl} + \delta_{jl}\delta_{ik}\right) \tag{5.6.20}$$

Therefore, combining Eqs. 5.6.15, 5.6.20, and 5.6.16 gives

$$\overline{D}_{ijkl} = D^{e}_{ijkl} \tag{5.6.21}$$

Using Eq. 5.6.21, Eq. 5.6.18 is reduced to the material tangent stiffness tensor (not consistent) for small strains

$$D^{0}_{ijkl} = D^{e}_{ijkl} - \frac{\dfrac{\partial f}{\partial \sigma_{mn}} D^{e}_{mnkl} D^{e}_{ijpq} \dfrac{\partial f}{\partial \sigma_{pq}}}{4X^2X' + \dfrac{\partial f}{\partial \sigma_{mn}} D^{e}_{mnpq} \dfrac{\partial f}{\partial \sigma_{pq}}} \tag{5.6.22}$$

and the material tangent stiffness tensor for large strains is

$$D^{ep}_{ijkl} = D^{0}_{ijkl} - \left(\sigma_{ik}\delta_{jl} + \sigma_{jk}\delta_{il}\right) + \left(\sigma_{i3}\delta_{j3} + \sigma_{j3}\delta_{i3}\right)\delta_{k\alpha}\delta_{l\alpha} \tag{5.6.23}$$

If we need to use Mroz's hardening rule, we replace the Cauchy tensor $\sigma_{\alpha i}$ by $\sigma_{\alpha i} - \alpha_{\alpha i}$ in Eq. 5.6.19 or Eq. 5.6.23, where $\alpha_{\alpha i}$ is the center of the active yield surface.

5.7 References

Bathe, K.-J. and Bolourchi, S., "A Geometric and Material Nonlinear Plate and Shell Element," *Computers and Structures*, Vol. 11, 1980, p. 23.

Belytschko, T., Liu, W.K., and Moran, B., *Nonlinear Finite Elements for Continua and Structures*, John Wiley, New York, 2000.

Chou, C.H., Pan, J., and Tang, S.C., "Analysis of Sheet Metal Operations by a Stress Resultant Constitutive Law for Thin Plates of Power-Law Hardening Material," *International Journal for Numerical Methods in Engineering*, Vol. 37, 1994, p. 717.

Chou, C.H., Pan, J., and Tang, S.C., "Constitutive Law for Thin Plates of Power-Law Materials," *International Journal of Solids and Structures*, Vol. 27, No. 11, 1991, p. 1387.

Chou, C.H., Pan, J., and Tang, S.C., "A Hardening Rule Between Stress Resultants and Generalized Plastic Strains for Thin Plates of Power-Law Hardening Material," *Journal of Applied Mechanics*, American Society of Mechanical Engineers, Vol. 60, 1993, p. 548.

Chu, C.C., "A Three Dimensional Model of Anisotropic Hardening in Metals and Its Application to the Analysis of Sheet Metal Formability," *Journal of Mechanics and Physics of Solids*, Vol. 2, 1984, p. 197.

Hill, R., *The Mathematical Theory of Plasticity*, Clarendon Press, Oxford, 1950.

Hutchinson, J.W., "Finite Strain Analysis of Elastic-Plastic Solids and Structures," *Numerical Solution of Nonlinear Structural Problems*, Applied Mechanics Division, American Society of Mechanical Engineers, Vol. 6, 1973, p. 77.

Koiter, W.T., "On the Nonlinear Theory of Thin Elastic Shells, I–III," Proc. Kon. Ned. Ak. Wet., B69, 1966, p. 1.

Lee, E.H., "Elastic-Plastic Deformation at Finite Strains," *Journal of Applied Mechanics*, American Society of Mechanical Engineers, Vol. 36, 1969, p. 1.

Lin, T.C. and Morgan, G.W., "A Study of Axisymmetric Vibrations of Cylindrical Shells as Affected by Rotary Inertia and Transverse Shear," *Journal of Applied Mechanics*, American Society of Mechanical Engineers, Vol. 23, 1956, p. 255.

McMeeking, R.M. and Rice, J.R., "Finite-Element Formulations for Problems of Large Elastic-Plastic Deformation," *International Journal of Solids and Structures*, Vol. 11, 1975, p. 601.

Mindlin, R.D., "Influence of Rotary Inertia and Shear on Flexural Motion of Isotropic Elastic Plates," *Journal of Applied Mechanics*, American Society of Mechanical Engineers, Vol. 18, 1951, p. 31.

Seely, F.B. and Smith, J.O., *Advanced Mechanics of Materials*, John Wiley, New York, 1959.

Spain, B., *Tensor Calculus*, Oliver and Boyd, Edinburgh and London, 1956.

Xia, Z.C. and Tang, S.C., "Locking Phenomena in the Use of Solid Elements for Sheet Metal Forming Simulation," SAE Paper No. 1999-01-1005, Society of Automotive Engineers, Warrendale, PA, 1999.

CHAPTER 6

Finite Element Methods for Thin Shells

6.1 Introduction

In this book, we use the finite element method to discretize the spatial variables in a thin shell, and the finite difference method (to be discussed in Chapter 7) to approximate the velocities and accelerations in terms of the discrete spatial variables and time increments. Therefore, a set of nonlinear algebraic equations in terms of unknown variables at each discrete point instead of the partial differential equations governing the motion of a thin shell can be obtained. The nonlinear algebraic equations can be solved incrementally in time by Newton-Raphson's iteration. Although the finite difference method may be applied to discretize the spatial variables directly in the governing partial differential equations of motion for a thin shell, there are some shortcomings of the finite difference method, particularly for application to simulations of sheet metal forming processes. First, the governing partial differential equations for a thin shell under large deflections, rotations, and strains are lacking in the literature, and these equations are high orders even for a shell of arbitrary shape with linear elastic material. Second, it is difficult to establish the boundary conditions automatically for a sheet (or blank) with a complex shape. To minimize the material to be trimmed off after a part is formed, the shape of a blank might be irregular for a complex sheet metal part, such as an automotive inner or outer panel. It also is difficult to establish a finite difference mesh automatically for a blank with an irregular geometry. On the other hand, it is not required to establish the equations of motion in the form of partial differential equations for a thin shell to use the finite element method. In the finite element method, we need only to establish the strain and displacement relationship and then apply the principle of virtual work to the equations of motion in terms of the discrete spatial variables. There are quite a few commercial software packages for establishing a finite element mesh on the reference surface of a shell automatically in spite of the irregular geometry of a sheet. Furthermore, techniques for adaptive meshing are available. We will discuss the features for adaptive meshing in Chapter 7.

In this chapter, we use both tensor and matrix notations for convenience of presentation. When matrix notations are used, the stress and strain vectors or their increments contain distinct components. For example, only the shear stress component σ_{12} (but not σ_{21}) is contained in the stress vector.

6.1.1 Computer-Aided Engineering (CAE) Requirements for Shell Elements

Following Bathe and Dvorkin [1986], several requirements should be satisfied for a shell element for use in production. These conditions are as follows:

- The element should be formulated based on a general shell theory for shells of any shape.
- The theoretical formulation should come strictly from the mechanics principles, with assumptions and approximations in the finite element discretization that are mechanically clear and well-founded.
- The shell element should be numerically sound and should not contain any spurious zero energy modes. It must be free of any lock and should not be based on numerically adjusted factors to obtain accurate results.
- The shell element should be simple to use for engineers with strong backgrounds in mechanics.
- The predictive capability of the element should be high and relatively insensitive to element distortions.

6.1.2 Displacement Method

There are two basic methods—the displacement method and the force method—for obtaining finite element solutions. The displacement method is more systematic; therefore, it is suitable for automation. Most commercial software for structural analyses are based on this method. The reference surface of a thin shell is divided into a set of curved or flat triangular and/or quadrilateral elements. For each element, we interpolate variables inside the elements by the values at nodes (at vertices, on sides, or at some assigned points inside the elements). If we use the same interpolation functions to interpolate the reference surface of a shell as those for the motion from the values at the nodes, this is the parametric method. If the parametric method is used for interpolation, the rigid body condition is satisfied automatically, and no strains are generated under rigid body motion. This is a necessary condition for convergence of the finite element method.

Another condition for monotonic convergence of the displacement method for thin shell analyses is that the interpolation function is C^1 continuous where the interpolation function and its first derivatives with respect to the spatial variables are continuous. For C^1 continuity, not only the displacement but also the edge rotation (slope) must be continuous at the interface of two neighboring elements. This is desirable for the classical shell theory. In the classical shell theory, there are no transverse shear strains because a normal to the reference surface of a thin shell remains normal after deformation. The rotations at a node are derived from the differentiations of the transverse displacement function with respect to the in-plane coordinates on the reference surface of a deformed shell. Hence, the interpolation function for transverse displacement usually requires a polynomial function higher than a cubic function.

If we relax the C^1 continuity requirement to the C^0 continuity, where the interpolation function is continuous but the derivatives are not, we allow only the material points on the original normal to remain on a straight line that is called a director. To do so, the displacements are continuous, but the edge rotations at the interface between two neighboring elements are not continuous. Because a director is no longer normal to the deformed reference surface, there are transverse shear strains as shown in Eq. 5.4.1c in Chapter 5. C^0 continuous elements normally do not require high-order interpolation functions. We may interpolate the displacements and rotations independently. The differences among the slopes, which are the partial derivatives of the transverse displacement with respect to the in-plane coordinates on the reference surface of a deformed shell, and the rotations generate the transverse shear strains on the reference surface.

Because a sheet may contact tool surfaces with very sharp radii, an extremely fine mesh for this contact area is required to obtain accurate results. To use simple flat elements in simulations of sheet metal forming is not a bad idea. For elements of this kind, we may superimpose membrane forces on the plate bending elements.

6.2 Finite Element Method for the Classical Shell Theory—Total Lagrangian Formulation

We write the virtual work term due to the stresses in Eq. 5.5.22 from Chapter 5 in incremental form, so that we are able solve them incrementally for a specified number of steps. We note that the transverse shear strains in these equations are omitted for the classical shell theory. We use the finite element method to discretize the spatial variables in a finite element model with shell elements such that the virtual work involves only unknowns of nodal variables, which can be solved iteratively by the Newton-Raphson method in a load step from time t to $t + \Delta t$.

6.2.1 Strain-Displacement Relationship in Incremental Forms

Let us review the strain-displacement relation in Eq. 5.5.6 from Chapter 5

$$\varepsilon_{\alpha\beta} = \gamma_{\alpha\beta} + h\theta^3\kappa_{\alpha\beta} \tag{6.2.1}$$

where the membrane (in-plane) strains are

$$\gamma_{\alpha\beta} = \frac{1}{2}\left(a_{\alpha\beta} - {}_0a_{\alpha\beta}\right) \tag{6.2.2}$$

and the bending strains are

$$\kappa_{\alpha\beta} = -\left(b_{\alpha\beta} - \frac{h_0}{h}\,{}_0b_{\alpha\beta}\right) \tag{6.2.3}$$

The incremental forms are

$$\Delta\gamma_{\alpha\beta} = \frac{1}{2}\left(\bar{a}_{\alpha\beta} - a_{\alpha\beta}\right) \tag{6.2.4a}$$

and

$$\Delta b_{\alpha\beta} = \bar{b}_{\alpha\beta} - b_{\alpha\beta} \tag{6.2.4b}$$

The fundamental tensors $\bar{a}_{\alpha\beta}$ and $\bar{b}_{\alpha\beta}$ are computed at the deformed coordinates $(x^i + \Delta u^i)$ at time $t + \Delta t$. Note that x^i are the coordinates at time t, and Δu^i that we are solving for are the increments of displacements from time t to $t + \Delta t$. In terms of x^i and Δu^i, Eq. 6.2.4a becomes

$$\Delta\gamma_{\alpha\beta} = \frac{1}{2}\left(x,^i_\alpha\, \Delta u,^i_\beta + x,^i_\beta\, \Delta u,^i_\alpha\right) + \frac{1}{2}\Delta u,^i_\alpha\, \Delta u,^i_\beta \tag{6.2.5}$$

Assuming the thickness of a shell is a slow variation function in Eq. 6.2.3 for sheet forming processes, we may freeze h in establishing the incremental form of the bending strain measurement as

$$\Delta\kappa_{\alpha\beta} = -\Delta b_{\alpha\beta} \tag{6.2.6}$$

6.2.2 Virtual Work Due to the Internal Nodal Force Vector

Equation 5.5.22 in Chapter 5 gives the virtual work done by all stresses from the finite element model of a shell

$$(\mathbf{Fi})^T\, \delta\mathbf{U} = \int_{v_0} \tau^{\alpha\beta}\, \delta\varepsilon_{\alpha\beta}\, dv \tag{6.2.7}$$

This equation represents the virtual work done by the internal nodal force vector **Fi**. Here, the virtual strain tensor $\delta\varepsilon_{\alpha\beta}$ is generated by the virtual nodal displacement vector $\delta\mathbf{U}$. Now we write Eq. 6.2.7 for a generic element $P_rP_sP_t$ in terms of the displacement increments. Note that for all other elements of the model, we repeat this procedure, and the total virtual work is obtained by adding (also called assembling) those for all elements. The virtual work for the element $P_rP_sP_t$ at time $t + \Delta t$ is

$$\delta w_j = \left(\mathbf{Fi}_j\right)^T \delta\mathbf{U}^j = \int_{v_{0j}} \bar{\tau}^{\alpha\beta}\, \delta\bar{\varepsilon}_{\alpha\beta}\, dv \quad \text{(nso } j\text{; nso means no sum on)} \tag{6.2.8}$$

where the subscript and superscript j represent for the variables of the shell element $P_rP_sP_t$, v_{0j} is the undeformed volume of the shell element, δw_j is the virtual work done by the internal nodal force vector on a virtual displacement $\delta\mathbf{U}^j$ in the element, and the bars mean the values at time $t + \Delta t$. Hence, at time $t + \Delta t$, the stress and strain tensors are

$$\bar{\tau}^{\alpha\beta} = \tau^{\alpha\beta} + \Delta\tau^{\alpha\beta} \quad (6.2.9a)$$

$$\bar{\varepsilon}_{\varepsilon\beta} = \varepsilon_{\alpha\beta} + \Delta\varepsilon_{\alpha\beta} \quad (6.2.9b)$$

Combining Eqs. 6.2.5 and 6.2.6 and using Eq. 6.2.1 gives the total strain increments from time t and $t + \Delta t$ (freezing h during this time interval) as

$$\Delta\varepsilon_{\alpha\beta} = \Delta\gamma_{\alpha\beta} + h\theta^3\Delta\kappa_{\alpha\beta} \quad (6.2.10)$$

Equation 6.2.5 gives the membrane strain increments as

$$\Delta\gamma_{\alpha\beta} = \Delta e_{\alpha\beta} + \Delta\eta_{\alpha\beta} \quad (6.2.11a)$$

where the linear part is

$$\Delta e_{\alpha\beta} = \frac{1}{2}\left(x,^i_\alpha \Delta u,^i_\beta + x,^i_\beta \Delta u,^i_\alpha\right) \quad (6.2.11b)$$

and the nonlinear part is

$$\Delta\eta_{\alpha\beta} = \frac{1}{2}\Delta u,^i_\alpha \Delta u,^i_\beta \quad (6.2.11c)$$

We will neglect the nonlinear terms (the reason will be explained later in Section 6.2.4) in the increment of the second fundamental tensor in Eq. 6.2.6 to compute the bending strains. Because the strain tensor $\varepsilon_{\alpha\beta}$ at time t is given, only the increment terms in Eq. 6.2.9b can be varied, and the variation of $\bar{\varepsilon}_{\alpha\beta}$ is

$$\delta\bar{\varepsilon}_{\alpha\beta} = \delta\Delta\varepsilon_{\alpha\beta} = \delta\Delta e_{\alpha\beta} + h\theta^3\delta\Delta\kappa_{\alpha\beta} + \delta\Delta\eta_{\alpha\beta} \quad (6.2.12)$$

Using Eqs. 6.2.12 and 6.2.9, we have the additional virtual work from time t to $t + \Delta t$ as

$$\begin{aligned} \delta\Delta w_j &= \Delta(\mathbf{Fi}_j)^T \delta\Delta\mathbf{U}^j \\ &= \int_{v_{0j}} \Delta\tau^{\alpha\beta}\delta\Delta\varepsilon_{\alpha\beta}\, dv + \int_{v_{0j}} \tau^{\alpha\beta}\delta\Delta\eta_{\alpha\beta} dv \quad (\text{nso } j) \end{aligned} \quad (6.2.13)$$

Note that the virtual work is

$$\delta w_j(t) = \int_{v_{0j}} \Delta\tau^{\alpha\beta}\left(\delta\Delta e_{\alpha\beta} + h\theta^3\delta\Delta\kappa_{\alpha\beta}\right)dv \quad (6.2.14)$$

at time t due to the virtual strain, and $\delta\Delta e_{\alpha\beta} + h\theta^3\delta\Delta\kappa_{\alpha\beta}$ is invoked to derive Eq. 6.2.13.

Inserting the constitutive relationship of Eq. 5.5.58 from Chapter 5 into Eq. 6.2.13 and taking the linear terms in Eqs. 6.2.10 and 6.2.12 to approximate $\Delta\varepsilon_{\alpha\beta}$ and $\Delta\delta\varepsilon_{\alpha\beta}$, respectively, we obtain the final form of the additional virtual work from the internal nodal forces in a generic element $P_rP_sP_t$ as

$$\begin{aligned}\delta\Delta w_j &= \Delta\left(\mathbf{Fi}_j\right)^T \delta\Delta\mathbf{U}^j \\ &= \int_{v_{0j}} D_{ep}^{\alpha\beta\gamma\delta}\left(\Delta e_{\gamma\delta} + h\theta^3\Delta\kappa_{\gamma\delta}\right)\left(\delta\Delta e_{\alpha\beta} + h\theta^3\delta\Delta\kappa_{\alpha\beta}\right)dv \\ &+ \int_{v_{0j}} \tau^{\alpha\beta}\,\delta\Delta\eta_{\alpha\beta}\,dv \quad (\text{nso } j)\end{aligned} \tag{6.2.15}$$

where $D_{ep}^{\alpha\beta\gamma\delta}$ is the consistent elastic-plastic material tangent stiffness tensor. We may sum over the virtual work for each element in the finite element model to obtain the total virtual work done by all internal nodal forces.

6.2.3 Discretization of Spatial Variables in a Curved Triangular Shell Element

We now express the strain increments in terms of the displacement and its gradient increments such that the element nodal force vector and the element tangent stiffness matrix can be established.

A curved triangle $P_rP_sP_t$ is shown in Figure 6.1. Given the values z^i (ith component in the spatial coordinate system) and its first partial derivatives $z^i_{,\alpha}$ $(i = 1,2,3; \alpha = 1,2)$ with respect to θ^α at the three nodal points P_r, P_s, and P_t, we compute the values z^i and the first and second partial derivatives $z^i_{,\alpha}$ and $z^i_{,\alpha\beta}$ anywhere inside the triangle $P_rP_sP_t$ by the interpolation function matrix $\boldsymbol{\varphi}$ from Goël [1968], which is C^1 continuous. Therefore, the curved triangular element is conforming. We write the interpolated values in terms of those nodal values and their first derivatives as follows:

$$\begin{bmatrix} z^i \\ z^i_{,1} \\ z^i_{,2} \\ z^i_{,11} \\ z^i_{,22} \\ z^i_{,12} \end{bmatrix} = \begin{bmatrix} \varphi_r & \varphi_s & \varphi_t \\ \varphi_{r,1} & \varphi_{s,1} & \varphi_{t,1} \\ \varphi_{r,2} & \varphi_{s,2} & \varphi_{t,2} \\ \varphi_{r,11} & \varphi_{s,11} & \varphi_{t,11} \\ \varphi_{r,22} & \varphi_{s,22} & \varphi_{t,22} \\ \varphi_{r,12} & \varphi_{s,12} & \varphi_{t,12} \end{bmatrix} \begin{bmatrix} \mathbf{z}^i_r \\ \mathbf{z}^i_s \\ \mathbf{z}^i_t \end{bmatrix} \tag{6.2.16}$$

where $\boldsymbol{\varphi}_r$ is a 1 by 3 matrix that contains the interpolation functions for the node P_r of the curved triangle $P_rP_sP_t$. Here, the vector $\mathbf{z}^i_r$ is

$$\left[\mathbf{z}^i_r\right] = \begin{bmatrix} z^i_r \\ z^i_{r,1} \\ z^i_{r,2} \end{bmatrix} \tag{6.2.17}$$

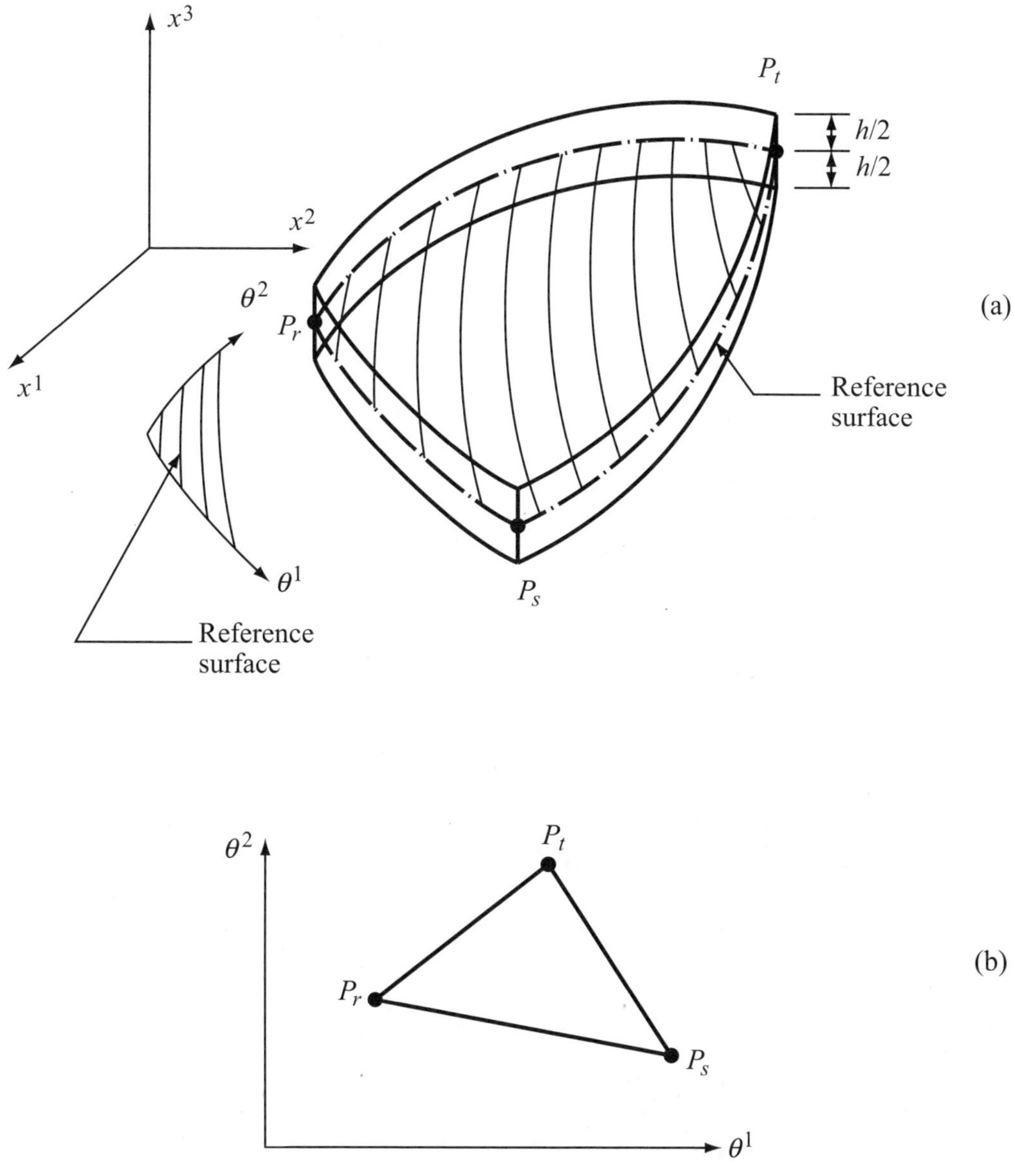

Figure 6.1 Curved triangular thin shell element: (a) shell element in the Cartesian space, and (b) reference surface of the element on the θ-plane.

where z_r^i are the values at the node P_r, and $z_{r,\alpha}^i$ are the first derivatives with respect to the convected coordinates θ^α, but $\mathbf{z}_s^i$ and $\mathbf{z}_t^i$ (Eq. 6.2.17) with the subscripts s and t denote the vectors at nodes P_s and P_t, respectively. Note again that z^i may represent the ith component of a position vector of a material point on the reference surface of the undeformed shell, the ith component of a displacement vector, or the ith component of an incremental displacement vector. Based on the interpolation functions and their partial derivatives in Eq. 6.2.16, we may express the strain-displacement increment relationship in the matrix form.

6.2.4 Increments of the Strain Field in Terms of Nodal Displacement Increments

Here, we establish the relation between the strain increments and the nodal displacement increments for a curved triangular shell element. With Eqs. 6.2.16 and 6.2.11b, the linear part of the membrane strains in the matrix form in Dupuis [1971] is

$$\begin{bmatrix} \Delta e_{11} \\ \Delta e_{22} \\ 2\Delta e_{12} \end{bmatrix} = \begin{bmatrix} \mathbf{A}^1 & \mathbf{A}^2 & \mathbf{A}^3 \end{bmatrix} \begin{bmatrix} \Delta \mathbf{u}^1 \\ \Delta \mathbf{u}^2 \\ \Delta \mathbf{u}^3 \end{bmatrix} \tag{6.2.18a}$$

where

$$\mathbf{A}^i = \begin{bmatrix} x_{,1}^i \varphi_{r,1} & x_{,1}^i \varphi_{s,1} & x_{,1}^i \varphi_{t,1} \\ x_{,2}^i \varphi_{r,2} & x_{,2}^i \varphi_{s,2} & x_{,2}^i \varphi_{t,2} \\ \left(x_{,2}^i \varphi_{r,1} + x_{,1}^i \varphi_{r,2}\right) & \left(x_{,2}^i \varphi_{s,1} + x_{,1}^i \varphi_{s,2}\right) & \left(x_{,2}^i \varphi_{t,1} + x_{,1}^i \varphi_{t,2}\right) \end{bmatrix} \tag{6.2.18b}$$

$$\Delta \mathbf{u}^i = \begin{bmatrix} \Delta u_r^i & \Delta u_{r,1}^i & \Delta u_{r,2}^i & \Delta u_s^i & \Delta u_{s,1}^i & \Delta u_{s,2}^i & \Delta u_t^i & \Delta u_{t,1}^i & \Delta u_{t,2}^i \end{bmatrix}^T \tag{6.2.18c}$$

Here, the superscript i denotes the component i in the spatial (global) coordinate system. We have

$$x_{,\alpha}^i = {}_0x_{,\alpha}^i + u_{,\alpha}^i \tag{6.2.19}$$

For the bending strains, we take only the linear part of the (generalized) displacement increments into consideration. Furthermore, as already mentioned in this chapter, we assume that the thickness varies slowly in establishing the incremental form in Eq. 6.2.6. We neglect the nonlinear parts of the bending strains $\kappa_{\alpha\beta}$ ($\kappa_{\alpha\beta} = -b_{\alpha\beta} + \left(h_0/h\right){}_0b_{\alpha\beta}$), expressed in terms of the second fundamental tensor $b_{\alpha\beta}$ in Eq. 5.5.20, which is rather complicated to express in the matrix form. From our past experience, neglecting the nonlinear parts due to the initial bending moments in establishing the tangent stiffness has negligible effect on the convergence of the Newton-Raphson iteration. However, we must use the updated coordinates $\overline{x}^i = x^i + \Delta u^i$ where $x^i = {}_0x^i + u^i$ to compute the first and second fundamental tensors $\overline{a}_{\alpha\beta}$ and $\overline{b}_{\alpha\beta}$. The linear terms of the increments of the bending strains from Eqs. 5.5.6c and 5.5.20 from Chapter 5 become

$$\left(\Delta\kappa_{\alpha\beta}\right)_l = -\left(\Delta b_{\alpha\beta}\right)_l = -\overline{\pi}_{(0)}^{\gamma\delta}\overline{s}_{\alpha\beta\gamma\delta}^{(1)} - \overline{\pi}_{(1)}^{\gamma\delta}\overline{s}_{\alpha\beta\gamma\delta}^{(0)} \tag{6.2.20}$$

where the subscript l denotes the linear terms from Eqs. 5.5.15 to 5.5.18 inclusive from Chapter 5 and

$$\overline{\pi}_{(0)}^{\gamma\delta} = \frac{1}{2\sqrt{a}} e_{\gamma\delta} \tag{6.2.21a}$$

$$\bar{\pi}^{\gamma\delta}_{(1)} = -\frac{1}{2a^{3/2}} e_{\gamma\delta} e_{\lambda\varphi} e_{\mu\omega} a_{\lambda\mu} x,^i_\varphi \Delta u,^i_\omega \tag{6.2.21b}$$

$$\bar{s}^{(0)}_{\alpha\beta\gamma\delta} = e_{ijk} x,^j_\gamma x,^k_\delta x,^i_{\alpha\beta} \tag{6.2.22a}$$

$$\bar{s}^{(1)}_{\alpha\beta\gamma\delta} = e_{ijk}\left[\left(x,^j_\gamma \Delta u,^k_\delta + x,^k_\delta \Delta u,^j_\gamma\right) x,^i_{\alpha\beta} + x,^j_\gamma x,^k_\delta \Delta u,^i_{\alpha\beta}\right] \tag{6.2.22b}$$

Note that we replace ${}_0a$ by a, ${}_0x$ by x, and u by Δu in Eqs. 5.5.15 through 5.5.18 from Chapter 5 to obtain Eqs. 6.2.21 and 6.2.22. Using Eqs. 6.2.21 and 6.2.22, we have the expressions for establishing the matrix relation as

$$\begin{aligned}\bar{\pi}^{\gamma\delta}_{(0)}\bar{s}^{(1)}_{\alpha\beta\gamma\delta} = \frac{1}{\sqrt{a}}\Big[&\left(x,^2_2 x,^3_{\alpha\beta} - x,^3_2 x,^2_{\alpha\beta}\right)\Delta u,^1_1 + \left(x,^3_1 x,^2_{\alpha\beta} - x,^2_1 x,^3_{\alpha\beta}\right)\Delta u,^1_2 \\ &+ \left(x,^3_2 x,^1_{\alpha\beta} - x,^1_2 x,^3_{\alpha\beta}\right)\Delta u,^2_1 + \left(x,^1_1 x,^3_{\alpha\beta} - x,^3_1 x,^1_{\alpha\beta}\right)\Delta u,^2_2 \\ &+ \left(x,^1_2 x,^2_{\alpha\beta} - x,^2_2 x,^1_{\alpha\beta}\right)\Delta u,^3_1 + \left(x,^2_1 x,^1_{\alpha\beta} - x,^1_1 x,^2_{\alpha\beta}\right)\Delta u,^3_2 \\ &+ \left(x,^2_1 x,^3_2 - x,^3_1 x,^2_2\right)\Delta u,^1_{\alpha\beta} + \left(x,^3_1 x,^1_2 - x,^1_1 x,^3_2\right)\Delta u,^2_{\alpha\beta} \\ &+ \left(x,^2_2 x,^1_1 - x,^2_1 x,^1_2\right)\Delta u,^3_{\alpha\beta}\Big]\end{aligned} \tag{6.2.23a}$$

$$\begin{aligned}\bar{\pi}^{\gamma\delta}_{(1)}\bar{s}^{(0)}_{\alpha\beta\gamma\delta} = -\frac{1}{a^{3/2}}\Big[&\left(x,^2_1 x,^3_2 - x,^3_1 x,^2_2\right)x,^1_{\alpha\beta} + \left(x^3_1 x,^1_2 - x,^1_1 x,^3_2\right)x,^2_{\alpha\beta} \\ &+ \left(x,^2_2 x,^1_1 - x,^2_1 x,^1_2\right)x,^3_{\alpha\beta}\Big]\cdot\Big[\left(a_{22}x,^m_1 - a_{12}x,^m_2\right)\Delta u,^m_1 \\ &+ \left(a_{11}x,^m_2 - a_{12}x,^m_1\right)\Delta u,^m_2\Big]\end{aligned} \tag{6.2.23b}$$

Summing these two terms and using Eq. 6.2.20, we have the linear term in the matrix form for the bending strains as

$$\begin{bmatrix}\Delta\kappa_{11}\\ \Delta\kappa_{22}\\ 2\Delta\kappa_{12}\end{bmatrix} = -\begin{bmatrix}\mathbf{B}^1\mathbf{\Phi} & \mathbf{B}^2\mathbf{\Phi} & \mathbf{B}^3\mathbf{\Phi}\end{bmatrix}\begin{bmatrix}\Delta\mathbf{u}^1\\ \Delta\mathbf{u}^2\\ \Delta\mathbf{u}^3\end{bmatrix} \tag{6.2.24}$$

where

$$\mathbf{B}^1 = \frac{1}{\sqrt{a}}\begin{bmatrix} \left(x_{,2}^{2}x_{,11}^{3} - x_{,2}^{3}x_{,11}^{2} - s_1\alpha_1\right) & \left(x_{,1}^{3}x_{,11}^{2} - x_{,1}^{2}x_{,11}^{3} - s_1\beta_1\right) & c_1 & 0 & 0 \\ \left(x_{,2}^{2}x_{,22}^{3} - x_{,2}^{3}x_{,22}^{2} - s_2\alpha_1\right) & \left(x_{,1}^{3}x_{,22}^{2} - x_{,1}^{2}x_{,22}^{3} - s_2\beta_1\right) & 0 & c_1 & 0 \\ 2\left(x_{,2}^{2}x_{,12}^{3} - x_{,2}^{3}x_{,12}^{2} - s_3\alpha_1\right) & 2\left(x_{,1}^{3}x_{,12}^{2} - x_{,1}^{2}x_{,12}^{3} - s_3\beta_1\right) & 0 & 0 & 2c_1 \end{bmatrix}$$

$$\mathbf{B}^2 = \frac{1}{\sqrt{a}}\begin{bmatrix} \left(x_{,2}^{3}x_{,11}^{1} - x_{,2}^{1}x_{,11}^{3} - s_1\alpha_2\right) & \left(x_{,1}^{1}x_{,11}^{3} - x_{,1}^{3}x_{,11}^{1} - s_1\beta_2\right) & c_2 & 0 & 0 \\ \left(x_{,2}^{3}x_{,22}^{1} - x_{,2}^{1}x_{,22}^{3} - s_2\alpha_2\right) & \left(x_{,1}^{1}x_{,22}^{3} - x_{,1}^{3}x_{,22}^{1} - s_2\beta_2\right) & 0 & c_2 & 0 \\ 2\left(x_{,2}^{3}x_{,12}^{1} - x_{,2}^{1}x_{,12}^{3} - s_3\alpha_2\right) & 2\left(x_{,1}^{1}x_{,12}^{3} - x_{,1}^{3}x_{,12}^{1} - s_3\beta_2\right) & 0 & 0 & 2c_2 \end{bmatrix}$$

$$\mathbf{B}^3 = \frac{1}{\sqrt{a}}\begin{bmatrix} \left(x_{,2}^{1}x_{,11}^{2} - x_{,2}^{2}x_{,11}^{1} - s_1\alpha_1\right) & \left(x_{,1}^{2}x_{,11}^{1} - x_{,1}^{1}x_{,11}^{2} - s_1\beta_3\right) & c_3 & 0 & 0 \\ \left(x_{,2}^{2}x_{,22}^{3} - x_{,2}^{3}x_{,22}^{2} - s_2\alpha_1\right) & \left(x_{,1}^{3}x_{,22}^{2} - x_{,1}^{2}x_{,22}^{3} - s_2\beta_3\right) & 0 & c_3 & 0 \\ 2\left(x_{,2}^{1}x_{,12}^{2} - x_{,2}^{2}x_{,12}^{1} - s_3\alpha_1\right) & 2\left(x_{,1}^{2}x_{,12}^{1} - x_{,1}^{1}x_{,12}^{2} - s_3\beta_3\right) & 0 & 0 & 2c_3 \end{bmatrix}$$

$$\alpha_i = a_{22}x_{,1}^{i} - a_{12}x_{,2}^{i}$$

$$\beta_i = a_{11}x_{,2}^{i} - a_{12}x_{,1}^{i}$$

$$c_1 = x_{,1}^{2}x_{,2}^{3} - x_{,1}^{3}x_{,2}^{2}$$

$$c_2 = x_{,1}^{3}x_{,2}^{1} - x_{,1}^{1}x_{,2}^{3}$$

$$c_3 = x_{,1}^{1}x_{,2}^{2} - x_{,1}^{2}x_{,2}^{1}$$

$$s_1 = c_i x_{,11}^{i} / a$$

$$s_2 = c_i x_{,22}^{i} / a$$

$$s_3 = c_i x_{,12}^{i} / a$$

$$\mathbf{\Phi} = \begin{bmatrix} \varphi_{r,1} & \varphi_{s,1} & \varphi_{t,1} \\ \varphi_{r,2} & \varphi_{s,2} & \varphi_{t,2} \\ \varphi_{r,11} & \varphi_{s,11} & \varphi_{t,11} \\ \varphi_{r,22} & \varphi_{s,22} & \varphi_{t,22} \\ \varphi_{r,12} & \varphi_{s,12} & \varphi_{t,12} \end{bmatrix} \tag{6.2.25}$$

The relationship between the increments of the linear strains $\left(\Delta\varepsilon_{\alpha\beta}\right)_l = \Delta e_{\alpha\beta} + h\theta^3 \Delta\kappa_{\alpha\beta}$ and the increments of displacements $\Delta\mathbf{u}^i$ can be written in the matrix form as

$$\begin{bmatrix} \left(\Delta\varepsilon_{11}\right)_l \\ \left(\Delta\varepsilon_{22}\right)_l \\ 2\left(\Delta\varepsilon_{12}\right)_l \end{bmatrix} = \left[\mathbf{B}_l\right] \begin{bmatrix} \Delta\mathbf{u}^1 \\ \Delta\mathbf{u}^2 \\ \Delta\mathbf{u}^3 \end{bmatrix} \tag{6.2.26a}$$

where the transformation matrix from the displacement increments to the linear strain increments is

$$\mathbf{B}_l = \left[\left(\mathbf{A}^1 - h\theta^3\mathbf{B}^1\mathbf{\Phi}\right) \ \left(\mathbf{A}^2 - h\theta^3\mathbf{B}^2\mathbf{\Phi}\right) \ \left(\mathbf{A}^3 - h\theta^3\mathbf{B}^3\mathbf{\Phi}\right)\right] \tag{6.2.26b}$$

The virtual strain increments due to the virtual displacement increments in Eq. 6.2.26b are

$$\begin{bmatrix} \delta\left(\Delta\varepsilon_{11}\right)_l \\ \delta\left(\Delta\varepsilon_{22}\right)_l \\ 2\delta\left(\Delta\varepsilon_{12}\right)_l \end{bmatrix} = \left[\mathbf{B}_l\right] \begin{bmatrix} \delta\Delta\mathbf{u}^1 \\ \delta\Delta\mathbf{u}^2 \\ \delta\Delta\mathbf{u}^3 \end{bmatrix} \tag{6.2.26c}$$

The variation of the nonlinear terms in Eq. 6.2.11c is

$$\delta\Delta\eta_{\alpha\beta} = \frac{1}{2}\left(\delta\Delta u,^i_\alpha \Delta u,^i_\beta + \Delta u,^i_\alpha \delta\Delta u,^i_\beta\right) \tag{6.2.27a}$$

By interpolation, we express

$$\Delta u,^i_\alpha = \begin{bmatrix} \Delta u,^i_1 \\ \Delta u,^i_2 \end{bmatrix} = \mathbf{B}_{nl} \begin{bmatrix} \Delta\mathbf{u}^i_r \\ \Delta\mathbf{u}^i_s \\ \Delta\mathbf{u}^i_t \end{bmatrix} \tag{6.2.27b}$$

$$\delta\Delta u,^i_\beta = \begin{bmatrix} \delta\Delta u,^i_1 \\ \delta\Delta u,^i_2 \end{bmatrix} = \mathbf{B}_{nl} \begin{bmatrix} \delta\Delta\mathbf{u}^i_r \\ \delta\Delta\mathbf{u}^i_s \\ \delta\Delta\mathbf{u}^i_t \end{bmatrix} \tag{6.2.27c}$$

where the transformation matrix from the displacement increments to the nonlinear strain increments is

$$\mathbf{B}_{nl} = \begin{bmatrix} \varphi_{r,1} & \varphi_{s,1} & \varphi_{t,1} \\ \varphi_{r,2} & \varphi_{s,2} & \varphi_{t,2} \end{bmatrix} \tag{6.2.27d}$$

Note that $\Delta u,^{i}_{\alpha}$ are the increments of the displacement gradient vector field, while $\Delta \mathbf{u}^{i}_{r} = \left[\Delta u^{i}_{r}\ \Delta u^{i}_{r,1}\ \Delta u^{i}_{r,2} \right]$ are the increments of the generalized displacement (including displacement gradient) vectors at the node r.

6.2.5 Element Tangent Stiffness Matrix and Nodal Force Vector

We can insert the consistent material tangent stiffness matrix from Eq. 5.5.58 or the material tangent stiffness matrix from Eq. 5.5.61 from Chapter 5 into the additional virtual work by the internal nodal forces in Eq. 6.2.15 to establish the element tangent stiffness matrix for a generic shell element j.

The last term due to the nonlinear terms in Eq. 6.2.15 can be expressed in the matrix form as

$$\tau^{\alpha\beta}\delta\Delta\eta_{\alpha\beta} = \sum_{i=1}^{3} \left[\delta\Delta\mathbf{u}^{iT}_{r} \quad \delta\Delta\mathbf{u}^{iT}_{s} \quad \delta\Delta\mathbf{u}^{iT}_{t} \right] \mathbf{B}^{T}_{nl}\, \boldsymbol{\tau}\, \mathbf{B}_{nl} \begin{bmatrix} \Delta\mathbf{u}^{i}_{r} \\ \Delta\mathbf{u}^{i}_{s} \\ \Delta\mathbf{u}^{i}_{t} \end{bmatrix} \tag{6.2.28a}$$

where

$$\boldsymbol{\tau} = \begin{bmatrix} \tau^{11} & \tau^{12} \\ \tau^{21} & \tau^{22} \end{bmatrix} \tag{6.2.28b}$$

Inserting Eqs. 6.2.26 through 6.2.28 inclusive into Eq. 6.2.15, we have the additional virtual work in the matrix form as

$$\begin{aligned} \delta\Delta w_j &= \Delta\left(\mathbf{Fi}_j\right)^T \delta\Delta\mathbf{U}^j \\ &= \left(\mathbf{K}_l \Delta\mathbf{U}^j\right)^T \delta\Delta\mathbf{U}^j + \left(\mathbf{K}_{nl} \Delta\mathbf{U}^j\right)^T \delta\Delta\mathbf{U}^j \end{aligned} \qquad (\text{nso } j) \tag{6.2.29}$$

where

$$\mathbf{K}_l = \int_{v_{0j}} \mathbf{B}_l^{\,T} \mathbf{D}\, \mathbf{B}_l \, dv \tag{6.2.30}$$

Here, **D** is a 3 by 3 symmetrical material tangent matrix in Eq. 5.5.58 or 5.5.61 from Chapter 5, expressing the relation between the increments of the Kirchhoff stress and the Lagrangian strain vectors. Here,

$$\mathbf{D} \equiv \begin{bmatrix} D_{11} & D_{12} & D_{13} \\ D_{21} & D_{22} & D_{23} \\ D_{31} & D_{32} & D_{33} \end{bmatrix} = \begin{bmatrix} D_{ep}^{1111} & D_{ep}^{1122} & D_{ep}^{1112} \\ D_{ep}^{1122} & D_{ep}^{2222} & D_{ep}^{2212} \\ D_{ep}^{1112} & D_{ep}^{2212} & D_{ep}^{1212} \end{bmatrix} \tag{6.2.31}$$

and

$$\mathbf{K}_{nl} = \int_{v_{0j}} \begin{bmatrix} \mathbf{S} & \mathbf{0} & \mathbf{0} \\ \mathbf{0} & \mathbf{S} & \mathbf{0} \\ \mathbf{0} & \mathbf{0} & \mathbf{S} \end{bmatrix} dv \tag{6.2.32a}$$

where

$$\mathbf{S} = \mathbf{B}_{nl}^T \boldsymbol{\tau} \mathbf{B}_{nl} \tag{6.2.32b}$$

Because $\delta\Delta\mathbf{U}^j$ is arbitrary in Eq. 6.2.29, we have

$$\Delta\left(\mathbf{Fi}_j\right)^T = \left(\mathbf{K}_l \Delta\mathbf{U}^j\right)^T + \left(\mathbf{K}_{nl}\Delta\mathbf{U}^j\right)^T \tag{6.2.33a}$$

or

$$\Delta\mathbf{Fi}_j = \mathbf{K}\Delta\mathbf{U}^j \tag{6.2.33b}$$

where the tangent stiffness matrix for an element is the sum of the linear and nonlinear parts as

$$\mathbf{K} = \mathbf{K}_l + \mathbf{K}_{nl} \tag{6.2.34}$$

The nonlinear part that is due to the initial stress is called the geometric stiffness matrix. Dividing both sides of Eq. 6.2.33b by $\Delta\mathbf{U}^j$ and setting $\Delta\mathbf{U}^j$ approaching zero, we obtain the partial derivative of

$$\frac{\partial \mathbf{Fi}_j}{\partial \mathbf{U}^j} = \mathbf{K} \quad (\text{nso } j) \tag{6.2.35}$$

We then can assemble all element tangent stiffness matrices of the shell elements in the finite element model to obtain the global tangent stiffness matrix.

The internal nodal forces for the element $P_r P_s P_t$ are obtained by

$$\left(\mathbf{Fi}_j\right)^T \delta\mathbf{U}^j = \int_{v_{0j}} \overline{\tau}^{\alpha\beta} \delta\left(\overline{\varepsilon}_{\alpha\beta}\right)_l dv$$

in Eq. 6.2.8, where the subscript l means the linear terms in the variation of the strain and displacement relationship in Eq. 6.2.12 evaluated at the configuration at time $t + \Delta t$ (replace $x^i_{,\alpha}$ by $\overline{x}^i_{,\alpha}$). Therefore,

$$\mathbf{Fi} = \int_{v_{0j}} \mathbf{B}_l^T \begin{bmatrix} \overline{\tau}^{11} \\ \overline{\tau}^{22} \\ \overline{\tau}^{12} \end{bmatrix} dv \tag{6.2.36}$$

where the matrix $\mathbf{B}_l$ is from Eq. 6.2.26b. We also can assemble all element internal nodal forces of the shell elements in the finite element model to obtain the global internal nodal force vector.

6.2.6 Basic and Shape (Interpolation) Functions

To define the shape functions $\boldsymbol{\varphi}\left(\theta^1,\theta^2\right)$, we consider the linear mapping that maps the triangle $P_1P_2P_3$ on the plane $\left(\theta^1,\theta^2\right)$ as shown in Figure 6.2(a) to a unit triangle on the plane $(\varsigma_1,\varsigma_2)$, whose vertices are $(0,0)$, $(1,0)$, and $(0,1)$, as shown in Figure 6.2(b). This mapping is defined by the relation

$$\begin{bmatrix} \theta^1 \\ \theta^2 \end{bmatrix} = \begin{bmatrix} \theta_2^1 - \theta_1^1 & \theta_3^1 - \theta_1^1 \\ \theta_2^2 - \theta_1^2 & \theta_3^2 - \theta_1^2 \end{bmatrix} \begin{bmatrix} \varsigma_1 \\ \varsigma_2 \end{bmatrix} + \begin{bmatrix} \theta_1^1 \\ \theta_1^2 \end{bmatrix} \tag{6.2.37a}$$

or

$$\begin{bmatrix} \varsigma_1 \\ \varsigma_2 \end{bmatrix} = \begin{bmatrix} \theta_2^1 - \theta_1^1 & \theta_3^1 - \theta_1^1 \\ \theta_2^2 - \theta_1^2 & \theta_3^2 - \theta_1^2 \end{bmatrix}^{-1} \left(\begin{bmatrix} \theta^1 \\ \theta^2 \end{bmatrix} - \begin{bmatrix} \theta_1^1 \\ \theta_1^2 \end{bmatrix} \right) \tag{6.2.37b}$$

where the subscripts 1, 2, and 3 for θ denote for the vertices P_1, P_2, and P_3, respectively.

We define the basic functions $\overline{\boldsymbol{\varphi}}(\varsigma_1,\varsigma_2)$ for the unit triangle on the plane $(\varsigma_1,\varsigma_2)$ and then transform them to the triangle $P_1P_2P_3$ on the plane $\left(\theta^1,\theta^2\right)$ as

$$\boldsymbol{\varphi}\left(\theta^1,\theta^2\right) = \mathbf{T}\overline{\boldsymbol{\varphi}}(\varsigma_1,\varsigma_2) \tag{6.2.38a}$$

where the transformation matrix from Dupuis [1971] is

$$\mathbf{T} = \begin{bmatrix} \mathbf{A}_{123} & \mathbf{0} & \mathbf{0} & \mathbf{0} & \mathbf{b}_{123} & \mathbf{c}_{123} \\ \mathbf{0} & \mathbf{A}_{231} & \mathbf{0} & \mathbf{c}_{231} & \mathbf{0} & \mathbf{b}_{231} \\ \mathbf{0} & \mathbf{0} & \mathbf{A}_{312} & \mathbf{b}_{312} & \mathbf{c}_{312} & \mathbf{0} \end{bmatrix} \tag{6.2.38b}$$

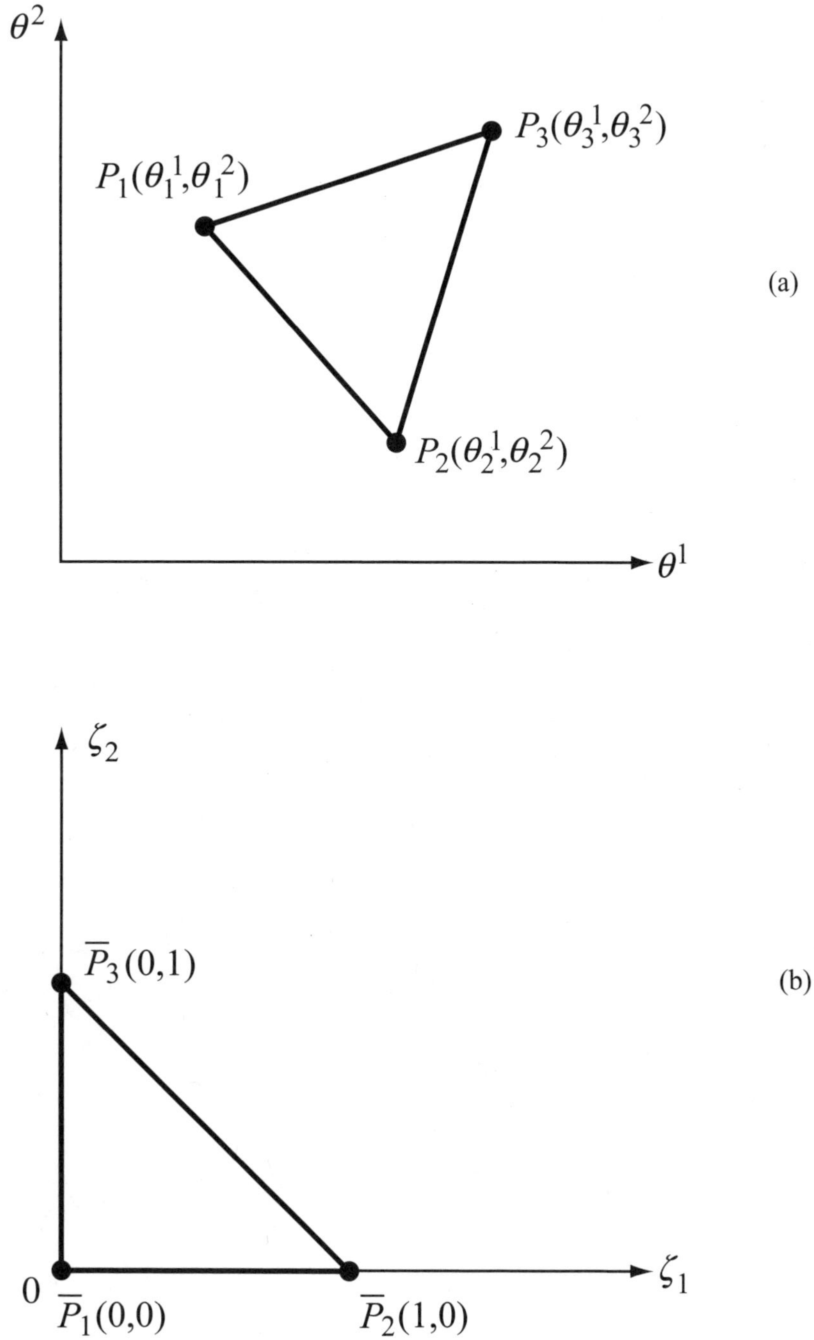

Figure 6.2 Mapping of a triangle $P_1P_2P_3$, to a unit triangle: (a) triangle on the θ-plane, and (b) unit triangle on the ς-plane.

where

$$\mathbf{A}_{ijk} = \begin{bmatrix} 1 & 0 & 0 \\ 0 & \theta_j^1 - \theta_i^1 & \theta_k^1 - \theta_i^1 \\ 0 & \theta_j^2 - \theta_i^2 & \theta_k^2 - \theta_i^2 \end{bmatrix}$$

$$\mathbf{b}_{ijk} = \begin{bmatrix} \dfrac{c_i}{l_j^2} \\ -\dfrac{\left(\theta_k^2 - \theta_i^2\right)J}{2l_j^2} \\ \dfrac{\left(\theta_k^1 - \theta_i^1\right)J}{2l_j^2} \end{bmatrix} \tag{6.2.38c}$$

$$\mathbf{c}_{ijk} = \begin{bmatrix} \dfrac{c_i}{l_k^2} \\ \dfrac{\left(\theta_j^2 - \theta_i^2\right)J}{2l_k^2} \\ -\dfrac{\left(\theta_j^1 - \theta_i^1\right)J}{2l_k^2} \end{bmatrix}$$

Here, l_1, l_2, and l_3 are the lengths of the sides P_2P_3, P_3P_1, and P_1P_2, respectively,and

$$c_i = \frac{1}{2}\sum_{j=1}^{3} l_j^2 - l_i^2 \quad (i = 1, 2, 3)$$

$$J = \left(\theta_2^1 - \theta_1^1\right)\left(\theta_3^2 - \theta_1^2\right) - \left(\theta_3^1 - \theta_1^1\right)\left(\theta_2^2 - \theta_1^2\right) \tag{6.2.38d}$$

The basic functions defined for the unit triangle on the plane $(\varsigma_1, \varsigma_2)$ are

$$
\overline{\varphi}(\varsigma_1, \varsigma_2) = \begin{bmatrix}
1 & 0 & 0 & -3 & -4 & -3 & 2 & 4 & 4 & 2 & \frac{1}{3} & -\frac{2}{3} & -\frac{2}{3} \\
0 & 1 & 0 & -2 & -1.5 & 0 & 1 & 1.5 & 0.5 & 0 & \frac{1}{12} & \frac{5}{12} & -\frac{1}{4} \\
0 & 0 & 1 & 0 & -1.5 & -2 & 0 & 0.5 & 1.5 & 1 & \frac{1}{12} & -\frac{1}{4} & \frac{5}{12} \\
0 & 0 & 0 & 3 & 2 & 0 & -2 & -2 & -2 & 0 & -\frac{2}{3} & \frac{1}{3} & -\frac{2}{3} \\
0 & 0 & 0 & 0 & 0.5 & 0 & 0 & 0.5 & -0.5 & 0 & -\frac{1}{4} & \frac{1}{12} & \frac{5}{12} \\
0 & 0 & 0 & 1 & 0.5 & 0 & -1 & -1.5 & -0.5 & 0 & \frac{5}{12} & \frac{1}{12} & -\frac{1}{4} \\
0 & 0 & 0 & 0 & 2 & 3 & 0 & -2 & -2 & -2 & -\frac{2}{3} & -\frac{2}{3} & \frac{1}{3} \\
0 & 0 & 0 & 0 & 0.5 & 1 & 0 & -0.5 & -1.5 & -1 & \frac{5}{12} & -\frac{1}{4} & \frac{1}{12} \\
0 & 0 & 0 & 0 & 0.5 & 0 & 0 & -0.5 & 0.5 & 0 & -\frac{1}{4} & \frac{5}{12} & \frac{1}{12} \\
0 & 0 & 0 & 0 & 0 & 0 & 0 & 0 & 0 & 0 & 1 & 0 & 0 \\
0 & 0 & 0 & 0 & 0 & 0 & 0 & 0 & 0 & 0 & 0 & 1 & 0 \\
0 & 0 & 0 & 0 & 0 & 0 & 0 & 0 & 0 & 0 & 0 & 0 & 1
\end{bmatrix}
\begin{bmatrix}
1 \\ \varsigma_1 \\ \varsigma_2 \\ \varsigma_1^2 \\ \varsigma_1\varsigma_2 \\ \varsigma_2^2 \\ \varsigma_1^3 \\ \varsigma_1^2\varsigma_2 \\ \varsigma_1\varsigma_2^2 \\ \varsigma_2^3 \\ \rho_1 \\ \rho_2 \\ \rho_3
\end{bmatrix} \tag{6.2.39a}
$$

where

$$
\rho_1 = -\frac{6\varsigma_1^2\varsigma_2^2(1-\varsigma_1-\varsigma_2)}{(1-\varsigma_1)(1-\varsigma_2)} - \frac{4\varsigma_1^2\varsigma_2^2(1-\varsigma_1-\varsigma_2)^2}{(1-\varsigma_1)(1-\varsigma_2)(\varsigma_1+\varsigma_2)} \tag{6.2.39b}
$$

$$
\rho_2 = -\frac{6\varsigma_2^2(1-\varsigma_1-\varsigma_2)^2\varsigma_1}{(1-\varsigma_2)(\varsigma_1+\varsigma_2)} - \frac{4\varsigma_1^2\varsigma_2^2(1-\varsigma_1-\varsigma_2)^2}{(1-\varsigma_1)(1-\varsigma_2)(\varsigma_1+\varsigma_2)} \tag{6.2.39c}
$$

$$
\rho_3 = -\frac{6(1-\varsigma_1-\varsigma_2)^2\varsigma_1^2\varsigma_2}{(\varsigma_1+\varsigma_2)(1-\varsigma_1)} - \frac{4\varsigma_1^2\varsigma_2^2(1-\varsigma_1-\varsigma_2)^2}{(1-\varsigma_1)(1-\varsigma_2)(\varsigma_1+\varsigma_2)} \tag{6.2.39d}
$$

After using the transformation in Eq. 6.2.38a, we have the interpolation functions

$$
\varphi^T = \begin{bmatrix} \varphi_r & \varphi_s & \varphi_t \end{bmatrix} \tag{6.2.40a}
$$

and we may write

$$\begin{aligned} \boldsymbol{\varphi}_r &= \begin{bmatrix} \phi_1 & \phi_2 & \phi_3 \end{bmatrix} \\ \boldsymbol{\varphi}_s &= \begin{bmatrix} \phi_4 & \phi_5 & \phi_6 \end{bmatrix} \\ \boldsymbol{\varphi}_t &= \begin{bmatrix} \phi_7 & \phi_8 & \phi_9 \end{bmatrix} \end{aligned} \tag{6.2.40b}$$

for the element $P_r P_s P_t$.

We must use not only the shape functions $\boldsymbol{\varphi}$ but also the partial derivatives of the shape functions $\boldsymbol{\varphi}$ with respect to θ^1 and θ^2. In Eq. 6.2.38a, these are expressed in terms of ς_1 and ς_2. Applying the chain rule of differentiation and using the mapping in Eq. 6.2.37, we can establish the partial derivatives of the functions $\boldsymbol{\varphi}$ in Eq. 6.2.40 with respect to θ^1 and θ^2 as

$$\begin{bmatrix} \boldsymbol{\varphi}^T \\ (\boldsymbol{\varphi}_{,1})^T \\ (\boldsymbol{\varphi}_{,2})^T \\ (\boldsymbol{\varphi}_{,11})^T \\ (\boldsymbol{\varphi}_{,22})^T \\ (\boldsymbol{\varphi}_{,12})^T \end{bmatrix} = \begin{bmatrix} 1 & 0 & 0 & 0 & 0 & 0 \\ 0 & t_1 & t_3 & 0 & 0 & 0 \\ 0 & t_2 & t_4 & 0 & 0 & 0 \\ 0 & 0 & 0 & t_1^2 & t_3^2 & 2t_1t_3 \\ 0 & 0 & 0 & t_2^2 & t_4^2 & 2t_2t_4 \\ 0 & 0 & 0 & t_1t_2 & t_3t_4 & t_1t_4 + t_2t_4 \end{bmatrix} \begin{bmatrix} \overline{\boldsymbol{\varphi}}^T \\ (\overline{\boldsymbol{\varphi}}_{,1})^T \\ (\overline{\boldsymbol{\varphi}}_{,2})^T \\ (\overline{\boldsymbol{\varphi}}_{,11})^T \\ (\overline{\boldsymbol{\varphi}}_{,22})^T \\ (\overline{\boldsymbol{\varphi}}_{,12})^T \end{bmatrix} \mathrm{T}^T \tag{6.2.41}$$

where

$$\boldsymbol{\varphi}_{,\alpha} = \frac{\partial \boldsymbol{\varphi}}{\partial \theta^\alpha}$$

$$\boldsymbol{\varphi}_{,\alpha\beta} = \frac{\partial^2 \boldsymbol{\varphi}}{\partial \theta^\alpha \partial \theta^\beta}$$

$$\overline{\boldsymbol{\varphi}}_{,\alpha} = \frac{\partial \overline{\boldsymbol{\varphi}}}{\partial \varsigma_\alpha}$$

$$\overline{\boldsymbol{\varphi}}_{,\alpha\beta} = \frac{\partial^2 \overline{\boldsymbol{\varphi}}}{\partial \varsigma_\alpha \partial \varsigma_\beta}$$

$$t_1 = \left(\theta_3^2 - \theta_1^2\right) / J$$

$$t_2 = -\left(\theta_3^1 - \theta_1^1\right) / J$$

$$t_3 = -\left(\theta_2^2 - \theta_1^2\right)/J$$

$$t_4 = \left(\theta_2^1 - \theta_1^1\right)/J$$

We now summarize the procedure to compute the interpolated values as well as the derivatives for a point θ^α in a curved triangular element $P_r P_s P_t$ in Table 6.1. Given the values z^i and its first derivatives $z,_\alpha^i$ $(i = 1,2,3; \alpha = 1,2)$ at the three nodal points P_r, P_s, and $_\mathbf{r}$, we compute the values z^i, and the first and the second derivatives $z,_\alpha^i$ and $z,_{\alpha\beta}^i$ anywhere inside the triangle $P_r P_s P_t$ by the interpolation function in the vector form, $\boldsymbol{\varphi}$, which is C^1 continuous.

TABLE 6.1
PROCEDURE FOR THE COMPUTATION OF INTERPOLATION (SHAPE) FUNCTIONS FOR THE CURVED TRIANGULAR THIN SHELL

1. Map the point θ^α to the point ς_α on the unit triangle on the ς-plane by means of Eq. 6.2.37.
2. Establish the basic functions in the vector form, $\overline{\boldsymbol{\varphi}}(\varsigma_1, \varsigma_2)$, for the unit triangle on the ς-plane, as shown in Eq. 6.2.39.
3. Map $\overline{\boldsymbol{\varphi}}(\varsigma_1, \varsigma_2)$ to $\boldsymbol{\varphi}\left(\theta^1, \theta^2\right)$ by means of the transformation matrix $\mathbf{T}$ in Eq. 6.2.38.
4. Compute $\boldsymbol{\varphi}$, $\boldsymbol{\varphi},_\alpha$, and $\boldsymbol{\varphi},_{\alpha\beta}$ from Eq. 6.2.41 and then the interpolated values and the first and second derivatives from Eq. 6.2.16. Note that $\boldsymbol{\varphi}^T = \left[\varphi_1 \varphi_2 \cdots \varphi_9\right]$ in Eq. 6.2.40.

6.2.7 Numerical Integration for a Curved Triangular Shell Element

There are integrations involved in computing the element tangent stiffness matrix and nodal force vector. We use the numerical procedures to compute these integrals. A typical integral is written as

$$I = \int_{v_0} f\left(\theta^1, \theta^2, \theta^3\right) dv \tag{6.2.42}$$

where

$$dv = h_0 \sqrt{{}_0a}\, d\theta^1 d\theta^2 d\theta^3$$

Therefore,

$$I = \int_{-1/2}^{1/2} d\theta^3 \iint_{\Delta} F\left(\theta^1, \theta^2, \theta^3\right) d\theta^1 d\theta^2 \tag{6.2.43}$$

where Δ is the area of the element $P_r P_s P_t$ on the θ-plane (Figure 6.1(b)) and

$$F\left(\theta^1, \theta^2, \theta^3\right) = f\left(\theta^1, \theta^2, \theta^3\right) h_0 \sqrt{{}_0 a}$$

The numerical integration scheme gives

$$I = \Delta \sum_{j=1}^{m} w_h^j \sum_{k=1}^{n} w_s^k F\left(\theta_k^1, \theta_k^2, \theta_j^3\right) \tag{6.2.44}$$

The sampling points $\left(\theta_k^1, \theta_k^2, \theta_j^3\right)$ are along the normal (Figure 6.3) passing through the point $\left(\theta_k^1, \theta_k^2, 0\right)$ on the undeformed reference surface of a shell for the element, and the index j means the point $\left(\theta_k^1, \theta_k^2, \theta_j^3\right)$ at the jth integration point through the undeformed thickness of the shell element. We use Simpson's rule for the numerical integration with respect to θ^3. Because we need to analyze loading as well as unloading, the seven or more points for Simpson's rule for numerical integration with respect to θ^3 are recommended. We equally divide the coordinate θ^3 from $-1/2$ to $1/2$ into six divisions. Therefore, there are seven equally spaced integration points (m = 7) as shown in Figure 6.3, and the weights w_h^j are $[1\ 4\ 2\ 4\ 2\ 4\ 1]/18$. As shown in Tang [1987],

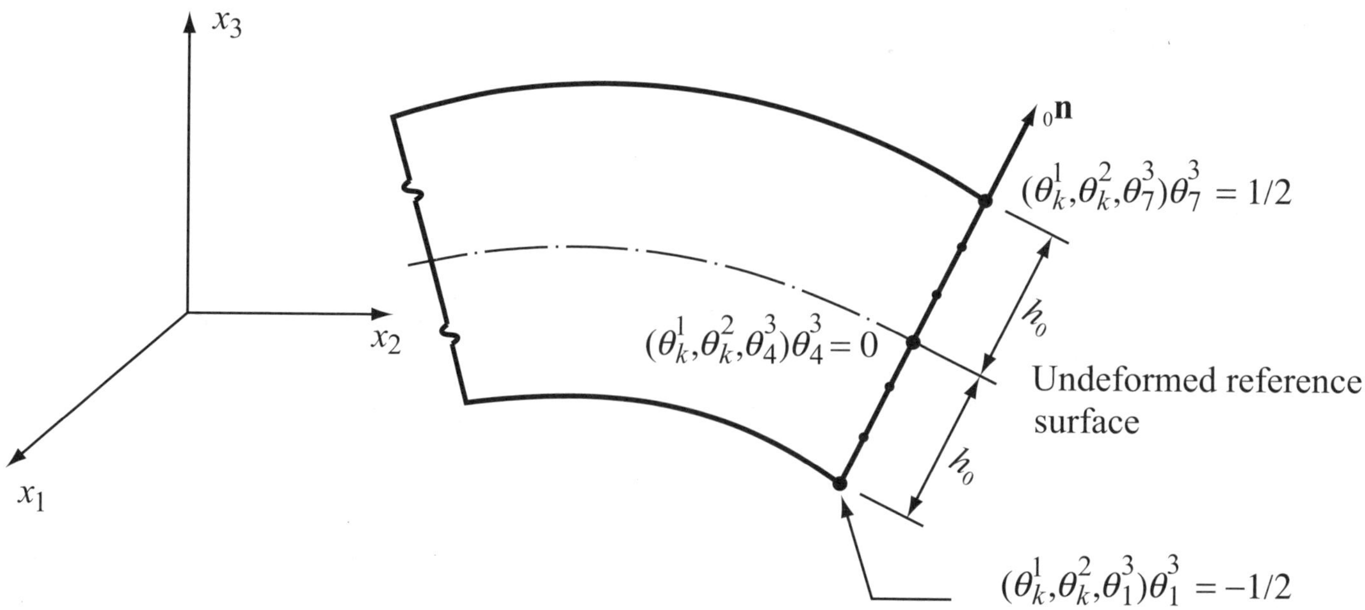

Figure 6.3 Seven integration points through the thickness of a triangular shell element.

the residual stress distribution along a normal section is very irregular after a sheet is released from the forming tools. Based on our experience, numerical integration with seven sampling points satisfies the minimum requirement for accuracy.

The quintic accuracy method in Zienkiewicz and Taylor [1988] with a degree of precision of 5 is applied for the numerical integration over the undeformed reference surface for better accuracy. Table 6.2 shows the locations and weight factors of the integration points (with $n = 7$ also)

TABLE 6.2
NUMERICAL INTEGRATION OVER A TRIANGULAR ELEMENT $P_1P_2P_3$ (OR $P_rP_sP_t$) AS SHOWN IN FIGURE 6.2(a)

Point	Area Coordinates			Weight Factor
k	L_1,	L_2,	L_3	w_s^k
1	1/3,	1/3,	1/3	0.22500 00000 000
2	α_1,	β_1,	β_1	0.13239 41527 885
3	β_1,	α_1,	β_1	0.13239 41527 885
4	β_1,	β_1,	α_1	0.13239 41527 885
5	α_2,	β_2,	β_2	0.12593 91805 448
6	β_2,	α_2,	β_2	0.12593 91805 448
7	β_2,	β_2,	α_2	0.12593 91805 448

where

$$\alpha_1 = 0.05971\,58717\,898$$

$$\beta_1 = 0.47014\,20641\,051$$

$$\alpha_2 = 0.79742\,69853\,531$$

$$\beta_2 = 0.10128\,65073\,235$$

Here, L_i ($i = 1, 2, 3$) are the area coordinates of point a as shown in Figure 6.4, and the locations of points number 1 through 7 are shown in Figure 6.5. From the area coordinates (L_1, L_2, L_3) of an integration point, we convert them to the coordinates (θ^1, θ^2) on the θ-plane by the formula

$$\theta^\alpha = \sum_{i=1}^{3} L_i \theta_i^\alpha \quad \left(\alpha = 1,2;\ \theta_1^\alpha = \theta_r^\alpha,\ \theta_2^\alpha = \theta_s^\alpha,\ \theta_3^\alpha = \theta_t^\alpha\right) \tag{6.2.45}$$

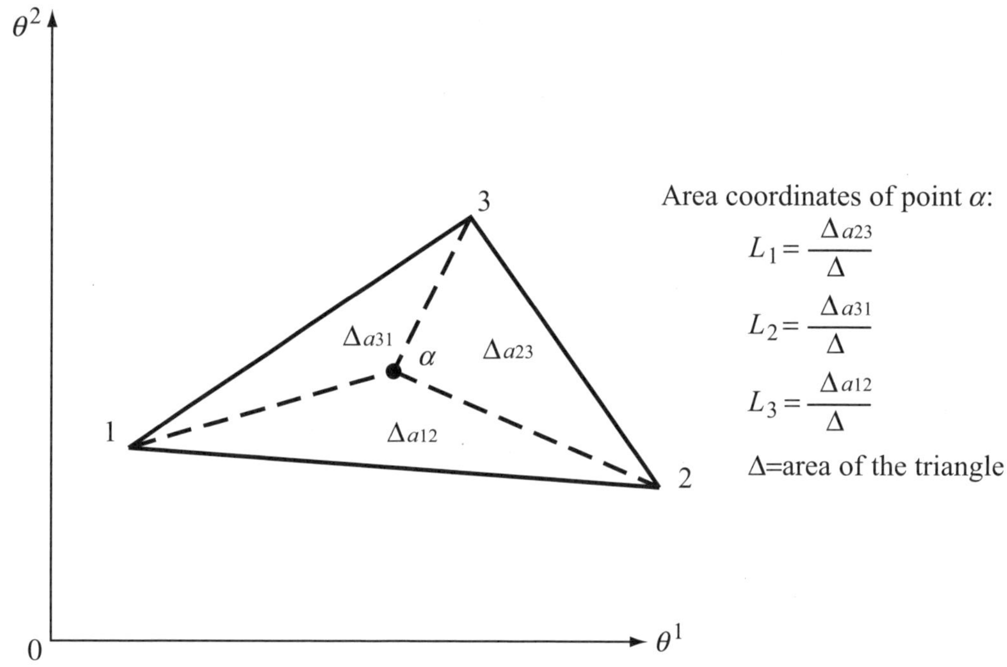

Figure 6.4 Area coordinates of point a in a triangle.

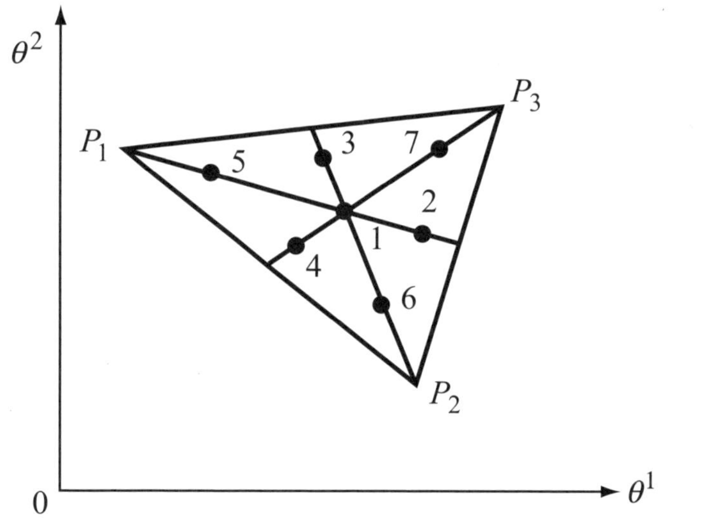

Figure 6.5 Seven integration points in a triangle $P_1P_2P_3$ on the θ-plane.

Because the basic functions $\overline{\varphi}$ are defined on a unit triangle and then are transformed to the shape functions φ by Eq. 6.2.38a, we need to map the seven integration points on the unit triangle. Therefore, we can compute the basic functions at the corresponding seven points in the unit triangle. Because the area coordinates are invariants, Eq. 6.2.45 is still valid for finding the corresponding locations of the seven points in the unit triangle on the ς-plane. For the unit triangle, Eq. 6.2.45 takes the form

$$\varsigma_\alpha = \sum_{i=1}^{3} L_i \varsigma_\alpha^i$$

$$\alpha = 1,2$$

$$\left(\varsigma_1^1, \varsigma_2^1\right) = (0,0) \tag{6.2.46}$$

$$\left(\varsigma_1^2, \varsigma_2^2\right) = (1,0)$$

$$\left(\varsigma_1^3, \varsigma_2^3\right) = (0,1)$$

6.2.8 Updating Configurations, Strains, and Stresses

For updating the geometric properties of a deformed shell, the first and second fundamental tensors $a_{\alpha\beta}$ and $b_{\alpha\beta}$, we need to update only the (generalized) displacements and the coordinates of each node by simple addition of the increments. Using the interpolation functions to compute the coordinates and their first and second derivatives with respect to the convected coordinates θ^α at each integration point, we can obtain from Chapter 5 the first fundament tensor $a_{\alpha\beta}$ from Eq. 5.5.8 and the second fundamental tensor $b_{\alpha\beta}$ from Eq. 5.5.12. For updating strains, we still can use the operation of simple addition because the Lagrangian strains refer to the undeformed configuration. We add the increment of the Kirchhoff stress vector to the Kirchhoff stress vector at time t to update the Kirchhoff stress vector at time $t + \Delta t$ as an approximation to update stresses. The increment of the Kirchhoff stress vector $\Delta\boldsymbol{\tau}$ can be obtained from multiplying the material tangent stiffness matrix in Eq. 6.2.31 by the computed increment of the Lagrangian strain vector $\Delta\boldsymbol{\varepsilon}$. To do so, the updated stress vector and the metric tensor do not necessarily satisfy the yield function in Eq. 5.5.29, even if we subdivide the strain increments into 200 sub-intervals. In Chapter 7, we discuss the radial return method to update the stresses that can satisfy the yield function.

The Lagrangian strain tensor and the Kirchhoff stress tensor that referred to the convected coordinate system (generally not rectangular) are not physically meaningful to engineers. We should convert them to the principal values that are meaningful. The appendix in a paper by Tang *et al.* [1988] shows the equations to compute the principal values of the Lagrangian strains. They are computed by solving a quadratic equation for ε as

$$\det\left|\varepsilon_{\alpha\beta} - \varepsilon\ {}_0 g_{\alpha\beta}\right| = 0 \tag{6.2.47}$$

Two roots ε_α for ε of Eq. 6.2.4.7 are the principal components of the Lagrangian strains. Tang *et al.* [1988] also gave the stretch ratios as

$$\lambda_\alpha = \sqrt{1 + 2\varepsilon_\alpha} \tag{6.2.48}$$

The natural strains are given by

$$\varepsilon_\alpha^l = \ln \lambda_\alpha \tag{6.2.49a}$$

and the principal engineering strains are

$$e_\alpha = \lambda_\alpha - 1 \tag{6.2.49b}$$

Using the stretch ratios at an integration point on the reference surface of an undeformed shell element, we can compute the updated thickness of the shell. Note that the bending strains do not significantly affect the thickness change, as mentioned in Chapter 5. The deformation of a shell element during forming processes is mostly plastic that is incompressible (no volume change). Therefore,

$$\lambda_1 \lambda_2 \lambda_3 \approx 1 \tag{6.2.50}$$

where λ_3 is the stretch ratio in the thickness direction. Thus, the deformed thickness is updated approximately as

$$h = \lambda_3 h_0 \tag{6.2.51}$$

We may use the weighted average of the thickness over all integration points as the element thickness instead of a non-uniform thickness distribution in an element.

Using a similar method, we can compute the principal stresses for a given Kirchhoff stress tensor. The principal stresses are computed by solving a quadratic equation for σ as

$$\det\left|\tau^{\alpha\beta} - \sigma g^{\alpha\beta}\right| = 0 \tag{6.2.52}$$

Two roots σ_α for σ in Eq. 6.2.52 are the principal components of the Cauchy stresses (true stresses). We use an approximation, the Cauchy stress tensor $\sigma^{\alpha\beta} \approx \tau^{\alpha\beta}$ based on the work of Needleman [1972], which referred to the convected coordinate system. Using these principal values, Eq. 3.9.1 from Chapter 3 gives the effective stress in a simple form as

$$\bar{\sigma} = \left(\sigma_1^2 + \sigma_2^2 - \frac{2R}{1+R}\sigma_1\sigma_2\right)^{1/2} \tag{6.3.53}$$

Note that the metric tensors ${}_0g_{\alpha\beta}$ and $g^{\alpha\beta}$ are on the original and deformed reference surfaces of a shell element, respectively.

6.3 Finite Element Method for a Shell with Transverse Shear Deformation—Updated Lagrangian Formulation

In Section 6.2, we realized how complicated the C^1 continuous interpolation functions are. We may consider much simpler interpolation functions for only C^0 continuous interpolation functions. In addition, we may ignore one of the Love-Kirchhoff hypotheses, where a normal to the reference surface of an undeformed shell remains normal to the reference surface after deformation. To do so, lower-order interpolation functions can be used for simplicity. In this section, a quadrilateral element with four nodes is illustrated. Because a normal does not remain normal after deformation, transverse shear strains are generated. Figure 5.2(b) shows a director that deviates from the normal after deformation. This is similar to the cylindrical shell theory by Lin and Morgan [1956], or the plate theory by Reissner [1945] and Mindlin [1951]. The formulation of the elements of this kind is degenerated from that of the three-dimensional brick elements. In fact, the full six components of the stress and stain vectors are used in the processes of computation.

Following Bathe and Bolourch [1980], we use the updated Lagrangian formulation for this C^0 continuous element. In this section, we use the Cartesian coordinate system with Cartesian tensors and a full three-dimensional analysis in the updated Lagrangian formulation. We use both the Cauchy stress tensor and the second Piola-Kirchhoff stress tensor (in the configuration at time t during computation for the next time step $(t + \Delta t)$) in the formulation.

6.3.1 Strain-Displacement Relationship in Incremental Form

We model a sheet with a set of quadrilateral shell elements. In sheet metal forming, the sheet generally starts from a flat or developable surface (not often) with a uniform thickness. We use the isoparametric coordinates (the natural coordinates), r, s, and t', to express the geometry and the displacements of an element. Figure 6.6(a) shows a four-node quadrilateral shell element initially at time $t = 0$. The geometry of the four-node shell element can be interpolated by the nodal coordinates ${}_0x_i^k$ as

$$ {}_0x_i = \sum_{k=1}^{4} \phi_k \, {}_0x_i^k + \frac{t'}{2} h_0 \sum_{k=1}^{4} \phi_k \, {}_0V_{ni}^k \tag{6.3.1} $$

where k (= 1, 2, 3, 4) represents the node number; ${}_0x_i$ represent the Cartesian coordinates of any point in the element at time $t = 0$; $\phi_k (= \phi_k(r,s)$ defined in Section 6.3.6) represent isoparametric interpolation functions; ${}_0V_{ni}^k$ represents the ith component of the unit vector ${}_0\mathbf{V}_n^k$, the director ${}_0\mathbf{a}_3$, or the normal vector ${}_0\mathbf{n}$ at node k; and h_0 represents the undeformed thickness of a shell element along the director ${}_0\mathbf{V}_n^k$. Note that Figure 6.6(a) shows ${}_0\mathbf{V}_n$ for node 4 because we assume the sheet initially is flat.

We denote three unit vectors $\mathbf{e}_i$ parallel to the global coordinate axes x_i, respectively. We establish two mutually orthogonal unit vectors ${}_0\mathbf{V}_\alpha^k$ at time $t = 0$, as shown in Figure 6.6(a) for $k = 4$. These two vectors are orthogonal to the director ${}_0\mathbf{a}_3$ (or the normal ${}_0\mathbf{n}$) denoted by ${}_0\mathbf{V}_n^k$. These two vectors are defined as

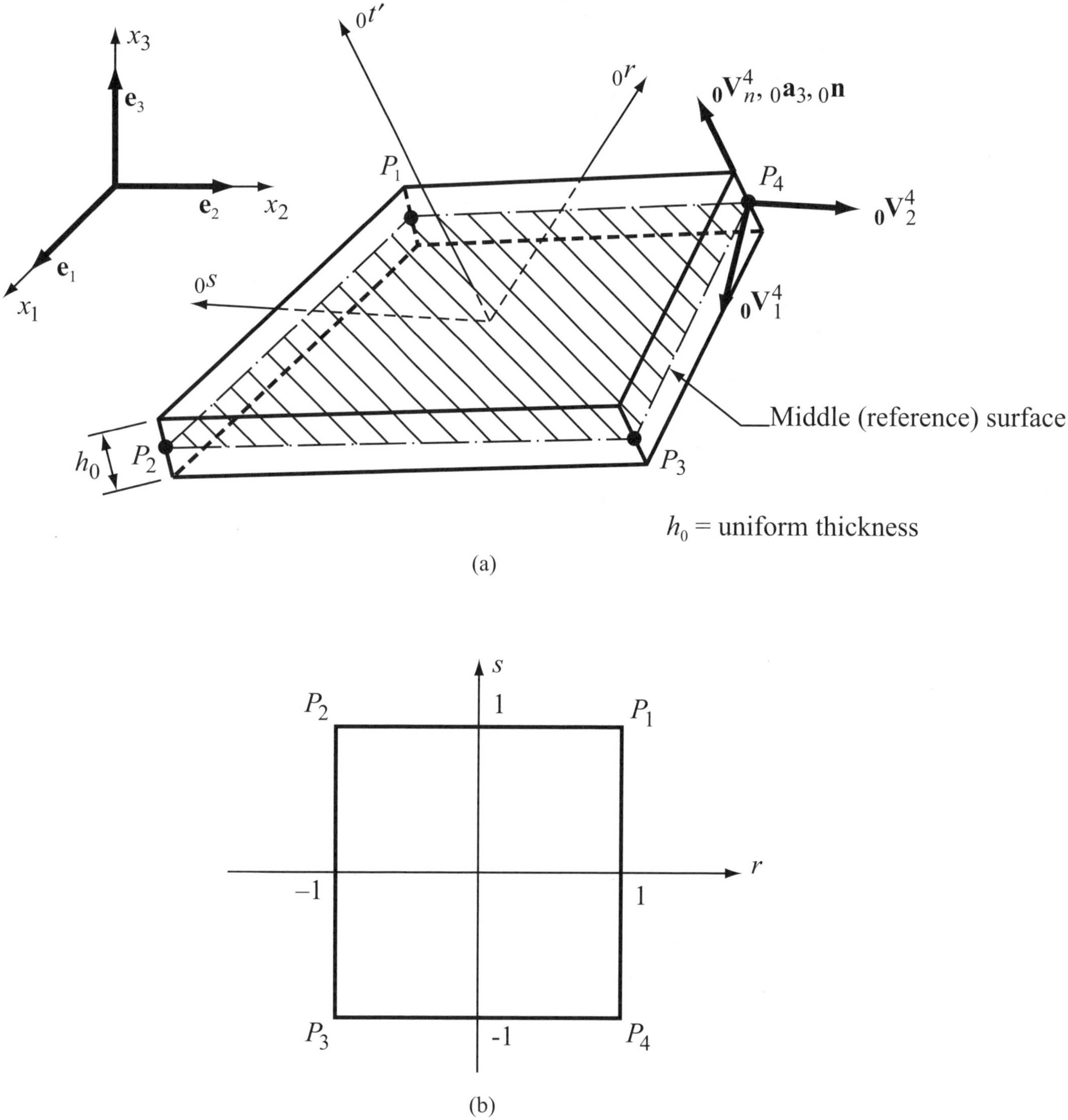

Figure 6.6 Undeformed four-node quadrilateral shell element with a uniform thickness at time $t = 0$: (a) shell element in the Cartesian space, and (b) reference surface on the r–s plane.

$$ {}_0\mathbf{V}_1^k = \mathbf{e}_2 \times {}_0\mathbf{V}_n^k \Big/ \left\| \mathbf{e}_2 \times {}_0\mathbf{V}_n^k \right\| \tag{6.3.2a} $$

If ${}_0\mathbf{V}_n^k$ is parallel to $\mathbf{e}_2$, we set ${}_0\mathbf{V}_1^k$ equal to $\mathbf{e}_3$, and

$$ {}_0\mathbf{V}_2^k = {}_0\mathbf{V}_n^k \times {}_0\mathbf{V}_1^k \tag{6.3.2b} $$

We update this orthonormal system by the amount of nodal rotation at each time step.

The natural coordinate axes r, s, and t' in Figure 6.6(b) can be mapped onto the undeformed reference surface of the shell element as ${}_0\mathbf{r} = \partial_0\mathbf{x}/\partial r$, ${}_0\mathbf{s} = \partial_0\mathbf{x}/\partial s$, and ${}_0\mathbf{t}' = \partial_0\mathbf{x}/\partial t'$ and in Figure 6.6(a) and the director

$$ {}_0\mathbf{V}_n^k = \frac{\dfrac{\partial_0\mathbf{x}}{\partial t'}}{\left\| \dfrac{\partial_0\mathbf{x}}{\partial t'} \right\|} \tag{6.3.3} $$

where ${}_0\mathbf{x}\,({}_0x^i)$ is a position vector, and $\left\|\partial_0\mathbf{x}/\partial t'\right\| = h_0/2$ for a flat element with a uniform thickness.

We now omit the left subscript 0 in Eq. 6.3.1. Then the geometry of the four-node shell element at the current time t can be interpolated by the nodal values x_i^k as

$$ x_i = \sum_{k=1}^{4} \phi_k x_i^k + \frac{t'}{2} \sum_{k=1}^{4} h_k \phi_k V_{ni}^k \tag{6.3.4} $$

where k (=1, 2, 3, 4) denotes the node number; x_i represent the Cartesian coordinates of any point in the element at time t; ϕ_k $(= \phi_k(r,s))$ represent isoparametric interpolation functions; V_{ni}^k represents the ith component of the unit vector $\mathbf{V}_n^k$, or the director $\mathbf{a}_3$, at node k; and h_k represents the thickness of the shell along the director $\mathbf{V}_n^k$ at node k (= 1,2,3,4) in Figure 6.7. Note that $\mathbf{V}_n^k$ initially is a normal to the undeformed reference surface of a shell. However, it does not necessarily remain normal to the deformed reference surface due to the transverse shear deformation.

We denote the displacement increments

$$ \Delta u_i = \overline{x}_i - x_i \tag{6.3.5} $$

where $\overline{x}_i$ are the Cartesian coordinates of a material point at time $t + \Delta t$. Replacing the variables at time t with those at time $t + \Delta t$ in Eq. 6.3.4, we obtain

$$ \overline{x}_i = \sum_{k=1}^{4} \phi_k \overline{x}_i^k + \frac{t'}{2} \sum_{k=1}^{4} h_k \phi_k \overline{V}_{ni}^k \tag{6.3.6} $$

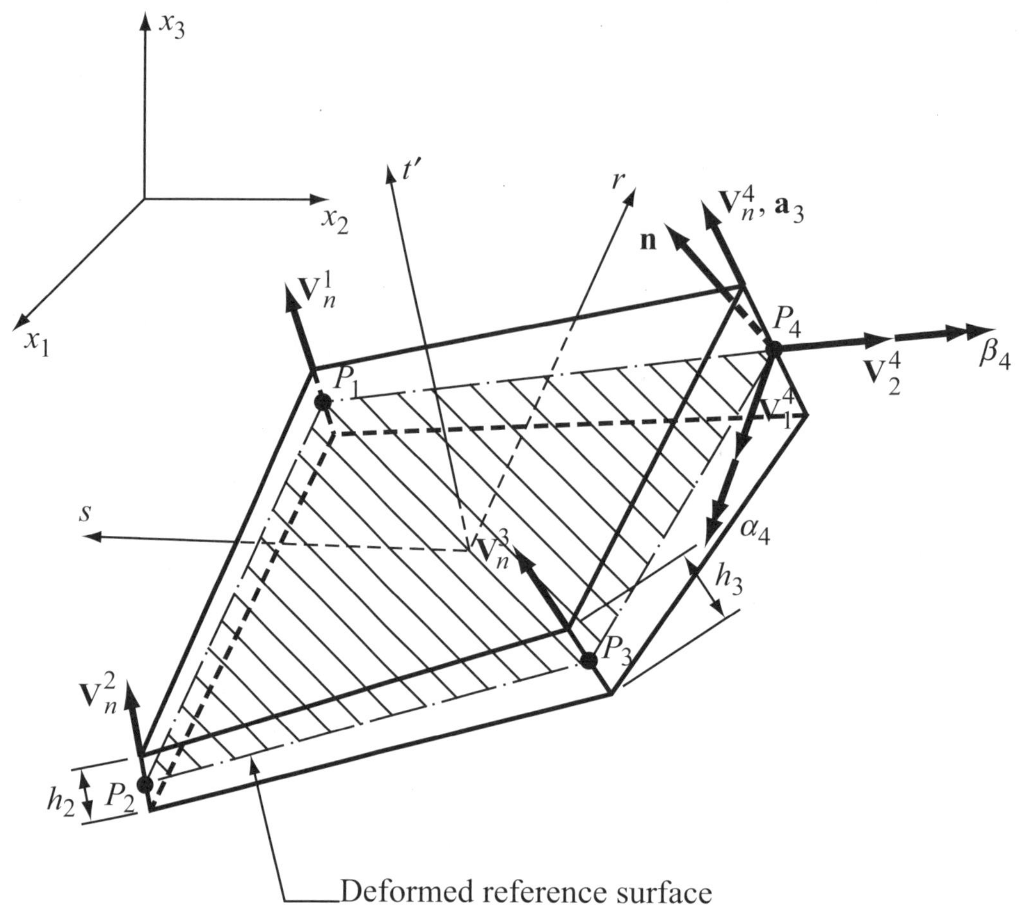

Figure 6.7 Deformed four-node quadrilateral shell element at time t, where h_k ($k = 1,2,3,4$) is the thickness at node P_k, and $\left({}_0\mathbf{V}_1^k, {}_0\mathbf{V}_2^k, {}_0\mathbf{V}_n^k\right)$ rotates to $\left(\mathbf{V}_1^k, \mathbf{V}_2^k, \mathbf{V}_n^k\right)$.

where $\bar{V}_{ni}^k$ represents the ith component of the unit vector $\bar{\mathbf{V}}_n^k$, or the director, of the deformed reference surface at time $t + \Delta t$. In Eq. 6.3.6, we assume that the thickness h_k changes slowly from time t to $t + \Delta t$. However, we update the thickness as well as the configuration of the shell after the solution procedure is convergent for the time step $t + \Delta t$.

Substituting Eqs. 6.3.4 and 6.3.6 into Eq. 6.3.5 gives

$$\Delta u_i = \sum_{k=1}^{4} \phi_k \Delta u_i^k + \frac{t'}{2} \sum_{k=1}^{4} h_k \phi_k \Delta V_{ni}^k \tag{6.3.7a}$$

$$\Delta V_{ni}^k = \bar{V}_{ni}^k - V_{ni}^k \tag{6.3.7b}$$

We express the change of the director ΔV_{ni}^k in terms of the increments of rotation, α_k and β_k, at node k as shown in Figure 6.7 (for $k = 4$). Because we do not consider the in-plane rotation (about $\mathbf{V}_n^k$) and if the rotation angles of the director are small from time t to $t + \Delta t$, we can express

$$\Delta \mathbf{V}_n^k = -\mathbf{V}_2^k \alpha_k + \mathbf{V}_1^k \beta_k \qquad (6.3.8)$$

where the orthonormal system $\mathbf{V}_1^k$, $\mathbf{V}_2^k$, and $\mathbf{V}_n^k$ is obtained by updating step by step from time 0 to t. Substituting Eq. 6.3.8 into Eq. 6.3.7, we thus obtain the increments of the displacement vector in terms of the nodal incremental displacements and rotations for an element

$$\Delta u_i = \sum_{k=1}^{4} \phi_k \Delta u_i^k + \frac{t'}{2} \sum_{k=1}^{4} h_k \phi_k \left(-V_{2i}^k \alpha^k + V_{1i}^k \beta^k \right) \qquad (6.3.9)$$

Note that we neglect the nonlinear terms due to the increments of rotation. Once the increments of rotation at node k are known, we can update the orthonormal system $\mathbf{V}_1^k$, $\mathbf{V}_2^k$, and $\mathbf{V}_n^k$ for time $t + \Delta t$ by rigid body rotation, which will be discussed in Section 6.3.9. The natural coordinate axes r, s, and t' on the deformed reference surface of the shell element at time t are $\mathbf{r} = \partial \mathbf{x}/\partial r$, $\mathbf{s} = \partial \mathbf{x}/\partial s$, and $\mathbf{t}' = \partial \mathbf{x}/\partial t'$, as shown in Figure 6.7 and the director

$$\mathbf{V}_n^k = \frac{\dfrac{\partial \mathbf{x}}{\partial t'}}{\left\| \dfrac{\partial \mathbf{x}}{\partial t'} \right\|} \qquad (\mathbf{a}_3 \text{ at node } k \text{ parallel to } \mathbf{V}_n^k) \qquad (6.3.10)$$

where $\mathbf{x}\left(x^i\right)$ is the position vector of a point on the deformed reference surface at time t. The thickness of the deformed shell at that point is equal to

$$h = 2\left\| \frac{\partial \mathbf{x}}{\partial t'} \right\| . \qquad (6.3.11)$$

6.3.2 Virtual Work Due to the Internal Nodal Force Vector

For the updated Lagrangian formulation, the virtual work term due to the internal nodal forces at time $t + \Delta t$ in Eq. 5.5.22 in Chapter 5 is modified for Cartesian tensors in three dimensions to integrate over the deformed volume of a shell element m at time t. The second Piola-Kirchhoff stress tensor is used in conjugating with the virtual increment of the Lagrangian strain tensor. Hence, the virtual work done by the internal nodal force vector with a virtual increment of displacement vector $\delta \Delta \mathbf{U}_m$ at time $t + \Delta t$ is

$$\delta w_m \left(t + \Delta t\right) = \left(\mathbf{Fi}_m\right)^T \delta \Delta \mathbf{U}_m = \int_{v_m} \bar{t}_{ij} \delta \Delta \varepsilon_{ij} dv \qquad (\text{nso } m) \qquad (6.3.12)$$

where

$$\bar{t}_{ij} = \sigma_{ij} + \Delta t_{ij} \tag{6.3.13}$$

Here, σ_{ij} represents the Cauchy stress tensor referred to the configuration of the element at time t, Δt_{ij} represents the increment of the second Piola-Kirchhoff stress tensor referred to the configuration of shell element m at time t, $\Delta\varepsilon_{ij}$ represents the increment of the Lagrangian strain tensor from time t to $t + \Delta t$ referred to the configuration of the element at time t, and v_m represents the volume of shell element m at time t. Note that $\bar{\varepsilon}_{ij} = \varepsilon_{ij} + \Delta\varepsilon_{ij}$, but $\delta\bar{\varepsilon}_{ij} = \delta\Delta\varepsilon_{ij}$ is used to derive Eq. 6.3.12 because ε_{ij} is given at time t.

Similar to the total Lagrangian formulation, we decompose the strain increment into the linear and nonlinear parts as

$$\Delta\varepsilon_{ij} = \Delta e_{ij} + \Delta\eta_{ij} \tag{6.3.14a}$$

where the linear part is

$$\Delta e_{ij} = \frac{1}{2}\left(\Delta u_{i,j} + \Delta u_{j,i}\right) \tag{6.3.14b}$$

and the nonlinear part is

$$\Delta\eta_{ij} = \frac{1}{2}\Delta u_{k,i}\Delta u_{k,j} \tag{6.3.14c}$$

Because the virtual work at time t due to the virtual incremental strain is

$$\delta w_m(t) = \int_{v_m} \sigma_{ij}\delta\Delta e_{ij}dv$$

the additional virtual work from time t to $t + \Delta t$ is $\delta\Delta w_m = \delta w_m(t + \Delta t) - \delta w_m(t)$. Using Eqs. 6.3.12 and 6.3.14a, we have the additional virtual work (neglecting the higher-order terms) as

$$\delta\Delta w_m = \Delta(\mathbf{Fi}_m)^T \delta\Delta\mathbf{U}_m = \int_{v_m} \Delta t_{ij}\delta\Delta\varepsilon_{ij}dv + \int_{v_m} \sigma_{ij}\Delta\eta_{ij}dv \quad \text{(nso } m\text{)} \tag{6.3.15}$$

We use an approximation to replace $\delta\Delta\varepsilon_{ij}$ by $\delta\Delta e_{ij}$ in Eq. 6.3.15 and

$$\Delta t_{ij} \approx D_{ijkl}\Delta e_{kl} \tag{6.3.16}$$

where D_{ijkl} is the consistent or conventional material tangent stiffness tensor transformed from a material tangent stiffness tensor D_{ijkl}^{ep} (excluding $i = j = 3$ or $k = l = 3$) in Eq. 5.6.19 or Eq. 5.6.23. The transformation will be discussed in Section 6.3.5. Consequently, Eq. 6.3.15 is changed to

$$\Delta(\mathbf{Fi}_m)^T \delta\Delta\mathbf{U}_m = \int_{v_m} D_{ijkl}\Delta e_{kl}\delta\Delta e_{ij}dv + \int_v \sigma_{ij}\Delta\eta_{ij}dv \tag{6.3.17}$$

6.3.3 Discretization of Spatial Variables in a Quadrilateral Shell Element

Because we use the displacement method of finite element analysis, we need to establish a matrix $\mathbf{B}_l$ to transform the (generalized) displacement increments at all nodes to the increments of the linear strain field, and the matrix $\mathbf{B}_{nl}$ to transform the displacement increments to the increments to the nonlinear strain field, for a generic shell element $P_1P_2P_3P_4$ with four corner nodes on the reference plane (Figure 6.7). To establish these matrices at time t, we make partial differentiations of Eq. 6.3.9 with respect to the natural coordinates r, s, and t' (see Figure 6.7). Because the interpolation functions ϕ_k are functions of r, s, and t',

$$\begin{bmatrix} \Delta u_{i,r} \\ \Delta u_{i,s} \\ \Delta u_{i,t'} \end{bmatrix} = \sum_{k=1}^{4} \begin{bmatrix} \phi_{k,r} & \phi_{k,r}t'g_{1i}^k & \varphi_{k,r}t'g_{2i}^k \\ \phi_{k,s} & \phi_{k,s}t'g_{1i}^k & \phi_{k,s}t'g_{2i}^k \\ 0 & \phi_k g_{1i}^k & \phi_k g_{2i}^k \end{bmatrix} \begin{bmatrix} \Delta u_i^k \\ \alpha^k \\ \beta^k \end{bmatrix} \tag{6.3.18}$$

where

$$g_{1i}^k = -\frac{1}{2}h_k V_{2i}^k$$

$$g_{2i}^k = \frac{1}{2}h_k V_{1i}^k$$

Because we need to compute the strains referred to the three-dimensional Cartesian coordinate system, we change the independent variables r, s, and t' to x_i by means of the transformation

$$\begin{bmatrix} \dfrac{\partial}{\partial r} \\ \dfrac{\partial}{\partial s} \\ \dfrac{\partial}{\partial t'} \end{bmatrix} = \mathbf{J} \begin{bmatrix} \dfrac{\partial}{\partial x_1} \\ \dfrac{\partial}{\partial x_2} \\ \dfrac{\partial}{\partial x_3} \end{bmatrix} \tag{6.3.19a}$$

or

$$\begin{bmatrix} \dfrac{\partial}{\partial x_1} \\ \dfrac{\partial}{\partial x_2} \\ \dfrac{\partial}{\partial x_3} \end{bmatrix} = \mathbf{J}^{-1} \begin{bmatrix} \dfrac{\partial}{\partial r} \\ \dfrac{\partial}{\partial s} \\ \dfrac{\partial}{\partial t'} \end{bmatrix} \tag{6.3.19b}$$

where the Jacobian matrix is

$$\mathbf{J} = \begin{bmatrix} \dfrac{\partial x_1}{\partial r} & \dfrac{\partial x_2}{\partial r} & \dfrac{\partial x_3}{\partial r} \\ \dfrac{\partial x_1}{\partial s} & \dfrac{\partial x_2}{\partial s} & \dfrac{\partial x_3}{\partial s} \\ \dfrac{\partial x_1}{\partial t'} & \dfrac{\partial x_2}{\partial t'} & \dfrac{\partial x_3}{\partial t'} \end{bmatrix} \tag{6.3.19c}$$

Using Eq. 6.3.19b, we transform Eq. 6.3.18 into

$$\begin{bmatrix} \dfrac{\partial \Delta u_i}{\partial x_1} \\ \dfrac{\partial \Delta u_i}{\partial x_2} \\ \dfrac{\partial \Delta u_i}{\partial x_3} \end{bmatrix} = \sum_{k=1}^{4} \begin{bmatrix} \phi_{k,1} & g_{1i}^k G_1^k & g_{2i}^k G_1^k \\ \phi_{k,2} & g_{1i}^k G_2^k & g_{2i}^k G_2^k \\ \phi_{k,3} & g_{1i}^k G_3^k & g_{2i}^k G_3^k \end{bmatrix} \begin{bmatrix} \Delta u_i^k \\ \alpha^k \\ \beta^k \end{bmatrix} \tag{6.3.20}$$

where

$$\phi_{k,i} = J_{i1}^{-1} \phi_{k,r} + J_{i2}^{-1} \phi_{k,s}$$

$$G_i^k = t' \phi_{k,i} + J_{i3}^{-1} \phi_k$$

and J_{ij}^{-1} is the element (i, j) of the matrix $\mathbf{J}^{-1}$ in Eq. 6.3.19b.

6.3.4 Increment of the Strain Field in Terms of Nodal Displacement Increments

Using Eq. 6.3.20, we express the linear part of the strain increments in Eq. 6.3.14b in the matrix form as

$$\Delta \mathbf{e} = \mathbf{B}_l \Delta \mathbf{U} \tag{6.3.21a}$$

where

$$\Delta \mathbf{e} = \left[\Delta e_{11} \quad \Delta e_{22} \quad \Delta e_{33} \quad 2\Delta e_{12} \quad 2\Delta e_{23} \quad 2\Delta e_{31} \right]^T$$

$$\Delta \mathbf{U} = \left[\Delta u_1^1 \ \Delta u_2^1 \ \Delta u_3^1 \ \alpha^1 \ \beta^1 \cdots \Delta u_1^4 \ \Delta u_2^4 \ \Delta u_3^4 \ \alpha^4 \ \beta^4 \right]^T$$

$$\mathbf{B}_l = \begin{bmatrix} & \phi_{k,1} & 0 & 0 & g_{11}^k G_1^k & g_{21}^k G_1^k & \\ & 0 & \phi_{k,2} & 0 & g_{12}^k G_2^k & g_{22}^k G_2^k & \\ \cdots & 0 & 0 & \phi_{k,3} & g_{13}^k G_3^k & g_{23}^k G_3^k & \cdots \\ & \phi_{k,2} & \phi_{k,1} & 0 & (g_{12}^k G_1^k + g_{11}^k G_2^k) & (g_{22}^k G_1^k + g_{21}^k G_2^k) & \\ & 0 & \phi_{k,3} & \phi_{k,2} & (g_{13}^k G_2^k + g_{12}^k G_3^k) & (g_{23}^k G_2^k + g_{22}^k G_3^k) & \\ & \phi_{k,3} & 0 & \phi_{k,1} & (g_{11}^k G_3^k + g_{13}^k G_1^k) & (g_{21}^k G_3^k + g_{23}^k G_1^k) & \end{bmatrix} \quad (6.3.21b)$$

For $\mathbf{B}_l$ in Eq. 6.3.21b, the matrix should be repeated for $k = 1, 2, 3$, and 4.

For the nonlinear part expressed by Eq. 6.3.14c, we express the increments of the displacement gradient field $\Delta u_{i,j}$ in terms of the interpolation functions and increments of the generalized displacements at all nodes of the element, such as

$$\begin{bmatrix} \Delta \mathbf{u}_{,1} \\ \Delta \mathbf{u}_{,2} \\ \Delta \mathbf{u}_{,3} \end{bmatrix} = \mathbf{B}_{nl} \Delta \mathbf{U} \quad (6.3.21c)$$

where

$$\mathbf{B}_{nl} = \begin{bmatrix} & \phi_{k,1} & 0 & 0 & g_{11}^k G_1^k & g_{21}^k G_1^k & \\ & 0 & \phi_{k,1} & 0 & g_{12}^k G_1^k & g_{22}^k G_1^k & \\ & 0 & 0 & \phi_{k,1} & g_{13}^k G_1^k & g_{23}^k G_1^k & \\ & \phi_{k,2} & 0 & 0 & g_{11}^k G_2^k & g_{21}^k G_2^k & \\ \cdots & 0 & \phi_{k,2} & 0 & g_{12}^k G_2^k & g_{22}^k G_2^k & \cdots \\ & 0 & 0 & \phi_{k,2} & g_{13}^k G_2^k & g_{23}^k G_2^k & \\ & \phi_{k,3} & 0 & 0 & g_{11}^k G_3^k & g_{21}^k G_3^k & \\ & 0 & \phi_{k,3} & 0 & g_{12}^k G_3^k & g_{22}^k G_3^k & \\ & 0 & 0 & \phi_{k,3} & g_{13}^k G_3^k & g_{23}^k G_3^k & \end{bmatrix} \quad (6.3.21d)$$

For $\mathbf{B}_{nl}$ in Eq. 6.3.21d, the matrix should be repeated for $k = 1, 2, 3$, and 4.

6.3.5 Element Tangent Stiffness Matrix and Nodal Force Vector

To establish the element tangent stiffness matrix, we require a material tangent stiffness matrix. Based on Eq. 5.6.19 or Eq. 5.6.23 from Chapter 5, we convert the material tangent stiffness tensor to the matrix form in a local coordinate system. The coordinate axes

r and s are mapped on the deformed reference surface of a shell element (Figure 6.7). We define a rectangular Cartesian coordinate system with the first two coordinate axes $\bar{r}$ and $\bar{s}$ located on the tangent plane, while the third axis $\bar{t}$ is normal to the tangent plane. We define an orthonormal system with $\mathbf{e}_{\bar{r}}$, $\mathbf{e}_{\bar{s}}$, and $\mathbf{e}_{\bar{t}}$ along the coordinate axes $\bar{r}$, $\bar{s}$, and $\bar{t}$, respectively, as shown in Figure 6.8. In this figure, α is the angle between the axes r and s, and $\varphi = \pi/4 - \alpha/2$ if $\alpha < \pi/2$, or $\varphi = \alpha/2 - \pi/4$ if $\alpha > \pi/2$. Note that the first two coordinate axes r and s of the natural coordinate system are on the reference surface of the deformed shell element, but the third axis t' is not necessarily normal to the reference due to the transverse shear deformation.

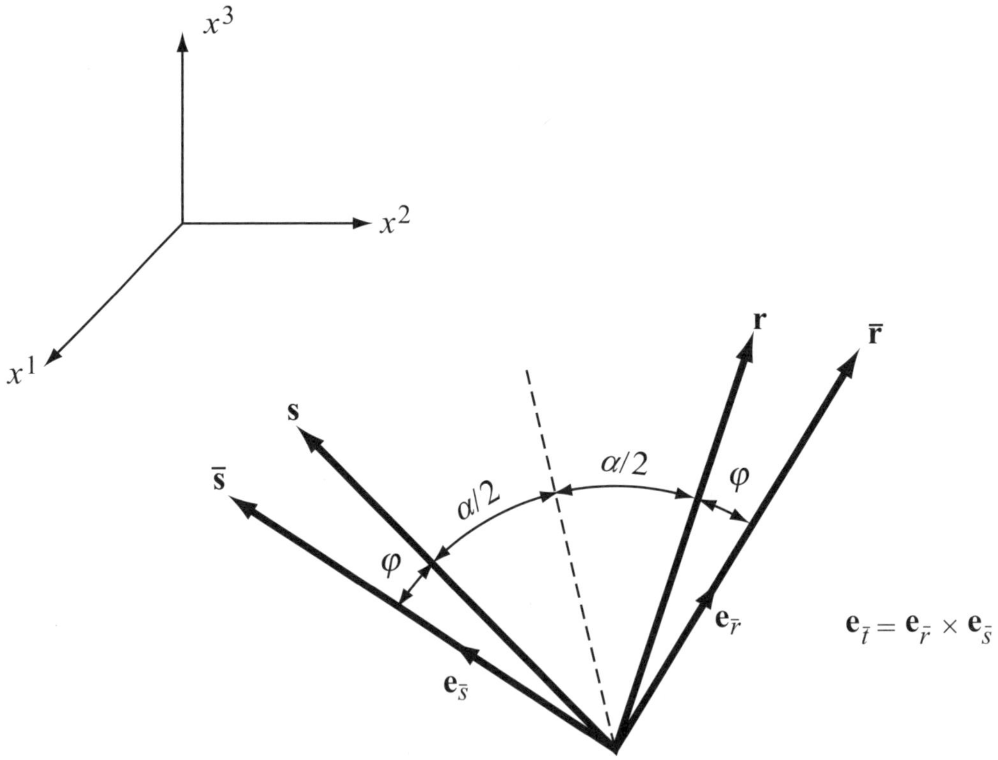

Figure 6.8 A rectangular Cartesian coordinate system $\bar{r}$, $\bar{s}$, and $\bar{t}$, with $\bar{r}$ and $\bar{s}$ on the r–s plane. Note that $\mathbf{e}_{\bar{r}}$, $\mathbf{e}_{\bar{s}}$, and $\mathbf{e}_{\bar{t}}$ form an orthonormal system.

Because of transversely anisotropic characteristics of sheet materials for plastic deformation, the material tangent stiffness matrix must be referred to the local coordinate system for simplicity. Because the computation is based on the global Cartesian coordinate system such as the virtual work in Eqs. 6.3.12 and 6.3.17, we must transform the material tangent stiffness matrix from the local to the global coordinate system. First, we transform the increments of the strain and stress vectors referred to the local coordinate system (five distinct components) to the global Cartesian coordinate system, where both vectors have six distinct components. The transformation matrix from five to six components is singular. To make the transformation matrix regular, we change the increments of both the strain and stress vectors to six components referred to the local coordinate system. Similar to the total Lagrangian formulation, we use only the linear part of the strain increment, $\Delta\mathbf{e}$, in Eq. 6.3.14 to establish the constitutive equation as

$$\Delta \mathbf{t}_{loc}^{6} = \mathbf{D}_{loc}^{6} \Delta \mathbf{e}_{loc}^{6} \tag{6.3.22}$$

where

$$\Delta \mathbf{e}_{loc}^{6} = \begin{bmatrix} \Delta e_{\overline{rr}} & \Delta e_{\overline{ss}} & \Delta e_{\overline{tt}} & 2\Delta e_{\overline{rs}} & 2\Delta e_{\overline{st}} & 2\Delta e_{\overline{tr}} \end{bmatrix}^{T} \tag{6.3.23a}$$

$$\Delta \mathbf{t}_{loc}^{6} = \begin{bmatrix} \Delta t_{\overline{rr}} & \Delta t_{\overline{ss}} & 0 & \Delta t_{\overline{rs}} & \Delta t_{\overline{st}} & \Delta t_{\overline{tr}} \end{bmatrix}^{T} \tag{6.3.23b}$$

Note that $\Delta e_{\overline{tt}}$ is not independent because of the incompressibility assumed in this book, and $t_{\overline{tt}}$ and $\Delta t_{\overline{tt}}$ are zero for the thin shell theory. The material tangent stiffness matrix is modified accordingly as

$$\mathbf{D}_{loc}^{6} = \begin{bmatrix} D_{1111}^{ep} & D_{1122}^{ep} & 0 & D_{1112}^{ep} & D_{1123}^{ep} & D_{1131}^{ep} \\ & D_{2222}^{ep} & 0 & D_{2212}^{ep} & D_{2223}^{ep} & D_{2231}^{ep} \\ & & 0 & 0 & 0 & 0 \\ & & & D_{1212}^{ep} & D_{1223}^{ep} & D_{1231}^{ep} \\ & & & & D_{1323}^{ep} & D_{1331}^{ep} \\ sym & & & & & D_{2331}^{ep} \end{bmatrix} \tag{6.3.23c}$$

where D_{ijkl}^{ep} $(1 = \overline{r}, 2 = \overline{s}, 3 = \overline{t}$, excluding $i = j = 3$ or $k = l = 3)$ are from the consistent material tangent stiffness tensor (referred to the local coordinate system) in Eq. 5.6.19 or the conventional material tangent stiffness tensor in Eq. 5.6.23 of Chapter 5.

We need to establish a matrix **Q** to transform the increments of the Lagrangian strain vector and the second Piola-Kirchhoff stress vector in the local coordinate system $\overline{r}$, $\overline{s}$, and $\overline{t}$ to those in the global Cartesian coordinate system. Therefore,

$$\Delta \mathbf{e}_{loc}^{6} = \mathbf{Q} \Delta \mathbf{e} \tag{6.3.24a}$$

and

$$\Delta \mathbf{t} = \mathbf{Q}^{T} \Delta \mathbf{t}_{loc}^{6} \tag{6.3.24b}$$

The combining of Eqs. 6.3.22 and 6.3.24 yields the 6 by 6 material tangent stiffness matrix in the global Cartesian coordinate system. The constitutive equation is

$$\Delta \mathbf{t} = \mathbf{D}_{6} \Delta \mathbf{e} \tag{6.3.25}$$

where

$$\Delta \mathbf{t} = \begin{bmatrix} \Delta t_{11} \; \Delta t_{22} \; \Delta t_{33} \; \Delta t_{12} \; \Delta t_{23} \; \Delta t_{31} \end{bmatrix}^{T}$$

$$\Delta \mathbf{e} = \left[\Delta e_{11}\ \Delta e_{22}\ \Delta e_{33}\ 2\Delta e_{12}\ 2\Delta e_{23}\ 2\Delta e_{31}\right]^T$$

and the material tangent stiffness matrix $\mathbf{D}_6$ is 6 by 6 and is obtained by using Eqs. 6.3.23c and 6.3.24 as

$$\mathbf{D}_6 = \mathbf{Q}^T \mathbf{D}_{loc}^6 \mathbf{Q} \tag{6.3.26}$$

The transformation matrix $\mathbf{Q}$ can be computed by means of the second order tensor transformation

$$\begin{aligned} \Delta\varepsilon_{lm} &= L_{li} L_{mj} \Delta\varepsilon_{ij} \\ l,m &= \bar{r},\bar{s},\bar{t} \\ i,j &= 1,2,3 \end{aligned} \tag{6.3.27a}$$

and

$$\begin{aligned} L_{li} &= \mathbf{e}_l \cdot \mathbf{e}_i \\ L_{mj} &= \mathbf{e}_m \cdot \mathbf{e}_j \end{aligned} \tag{6.3.27b}$$

where $\mathbf{e}_i$ $(i = 1, 2, 3)$ are the unit vectors along the global Cartesian coordinate axes x_i in space, whereas $\mathbf{e}_l$ $(l = \bar{r}, \bar{s}, \bar{t})$ are the unit vectors along the local orthogonal coordinate axes $\bar{r}$, $\bar{s}$, and $\bar{t}$. We write $L_{li}L_{mj}$ in the matrix form as

$$\mathbf{Q} = \begin{bmatrix} (L_{\bar{r}1})^2 & (L_{\bar{r}2})^2 & (L_{\bar{r}3})^2 & L_{\bar{r}1}L_{\bar{r}2} & L_{\bar{r}2}L_{\bar{r}3} & L_{\bar{r}3}L_{\bar{r}1} \\ (L_{\bar{s}1})^2 & (L_{\bar{s}2})^2 & (L_{\bar{s}3})^2 & L_{\bar{s}1}L_{\bar{s}2} & L_{\bar{s}2}L_{\bar{s}3} & L_{\bar{s}3}L_{\bar{s}1} \\ (L_{\bar{t}1})^2 & (L_{\bar{t}2})^2 & (L_{\bar{t}3})^2 & L_{\bar{t}1}L_{\bar{t}2} & L_{\bar{t}2}L_{\bar{t}3} & L_{\bar{t}3}L_{\bar{t}1} \\ 2L_{\bar{r}1}L_{\bar{s}1} & 2L_{\bar{r}2}L_{\bar{s}2} & 2L_{\bar{r}3}L_{\bar{s}3} & (L_{\bar{r}1}L_{\bar{s}2} + L_{\bar{r}2}L_{\bar{s}1}) & (L_{\bar{r}2}L_{\bar{s}3} + L_{\bar{r}3}L_{\bar{s}2}) & (L_{\bar{r}3}L_{\bar{s}1} + L_{\bar{r}1}L_{\bar{s}3}) \\ 2L_{\bar{s}1}L_{\bar{t}1} & 2L_{\bar{s}2}L_{\bar{t}2} & 2L_{\bar{s}3}L_{\bar{t}3} & (L_{\bar{s}1}L_{\bar{t}2} + L_{\bar{s}2}L_{\bar{t}1}) & (L_{\bar{s}2}L_{\bar{t}3} + L_{\bar{s}3}L_{\bar{t}2}) & (L_{\bar{s}3}L_{\bar{t}1} + L_{\bar{s}1}L_{\bar{t}3}) \\ 2L_{\bar{t}1}L_{\bar{r}1} & 2L_{\bar{t}2}L_{\bar{r}2} & 2L_{\bar{t}3}L_{\bar{r}3} & (L_{\bar{t}1}L_{\bar{r}2} + L_{\bar{t}2}L_{\bar{r}1}) & (L_{\bar{t}2}L_{\bar{r}3} + L_{\bar{t}3}L_{\bar{r}2}) & (L_{\bar{t}3}L_{\bar{r}1} + L_{\bar{t}1}L_{\bar{r}3}) \end{bmatrix} \tag{6.3.28}$$

Finally, we can establish the element tangent stiffness matrix for a generic element in the global Cartesian coordinate system. From Eq. 6.3.17, the element tangent stiffness matrix for the linear strain increments has the same form as that for the total Lagrangian formulation in Eq. 6.2.15 except integration over the current volume v_m of the element. Therefore,

$$\mathbf{K}_l = \int_{v_m} \mathbf{B}_l^T\, \mathbf{D}_6\, \mathbf{B}_l\, dv \tag{6.3.29}$$

where the matrix $\mathbf{B}_l$ is from Eq. 6.3.21.

Similar to the derivation of Eq. 6.2.28, we express (using Eq. 6.3.14c)

$$\sigma_{ij}\delta\Delta\eta_{ij} = \sum_{l=1}^{3}\delta\Delta u_{l,i}\sigma_{ij}\Delta u_{l,j} \tag{6.3.30a}$$

Inserting Eq. 6.3.21c into Eq. 6.3.30a, we have

$$\sigma_{ij}\delta\Delta\eta_{ij} = \delta\Delta\mathbf{U}^T\mathbf{B}_{nl}^T\boldsymbol{\sigma}_9\mathbf{B}_{nl}\Delta\mathbf{U} \tag{6.3.30b}$$

for the nonlinear term in Eq. 6.3.17 to derive the element tangent stiffness matrix for the nonlinear strain increments as

$$\mathbf{K}_{nl} = \int_v \mathbf{B}_{nl}^T\,\boldsymbol{\sigma}_9\,\mathbf{B}_{nl}\,dv \tag{6.3.31a}$$

where the matrix $\mathbf{B}_{nl}$ is from Eq. 6.3.21d, and the initial stress matrix is

$$\boldsymbol{\sigma}_9 = \begin{bmatrix} \sigma_{11}\mathbf{I}_3 & \sigma_{12}\mathbf{I}_3 & \sigma_{13}\mathbf{I}_3 \\ \sigma_{12}\mathbf{I}_3 & \sigma_{22}\mathbf{I}_3 & \sigma_{23}\mathbf{I}_3 \\ \sigma_{13}\mathbf{I}_3 & \sigma_{23}\mathbf{I}_3 & \sigma_{33}\mathbf{I}_3 \end{bmatrix} \tag{6.3.31b}$$

where $\mathbf{I}_3$ is a 3 by 3 identity matrix.

We insert these two matrices into Eq. 6.3.17, and we have

$$\delta\Delta w_m = \Delta(\mathbf{Fi}_m)^T\,\delta\Delta\mathbf{U}_m = (\mathbf{K}_l\Delta\mathbf{U}_m)^T\,\delta\Delta\mathbf{U}_m + (\mathbf{K}_{nl}\Delta\mathbf{U}_m)^T\,\delta\Delta\mathbf{U}_m$$

Because $\delta\Delta\mathbf{U}_m$ is arbitrary, we have, for a generic element m,

$$\begin{aligned} \Delta(\mathbf{Fi}_m)^T &= (\mathbf{K}_l\Delta\mathbf{U}_m)^T + (\mathbf{K}_{nl}\Delta\mathbf{U}_m)^T \\ \Delta\mathbf{Fi}_m &= \mathbf{K}\Delta\mathbf{U}_m \end{aligned} \tag{6.3.32}$$

Dividing both sides of Eq. 6.3.32 by $\Delta\mathbf{U}_m$ and setting $\Delta\mathbf{U}_m$ approaching zero, we obtain the partial derivative

$$\frac{\partial\mathbf{Fi}_m}{\partial\mathbf{U}_m} = \mathbf{K} \quad (\text{nso } m) \tag{6.3.33}$$

where the tangent stiffness matrix for a generic element m is the sum of the linear and nonlinear parts

$$\mathbf{K} = \mathbf{K}_l + \mathbf{K}_{nl} \tag{6.3.34}$$

The nonlinear tangent stiffness matrix is due to the initial stresses or is called the geometric stiffness matrix. We can assemble all element tangent stiffness matrices of the shell elements in the finite element model to obtain the global tangent stiffness matrix.

The internal nodal forces for a generic element $P_1P_2P_3P_4$ are obtained from the virtual work in Eq. 6.3.12. After obtaining the incremental solution at time t, we update the configuration of the shell at time $t + \Delta t$ and transform the second Piola-Kirchhoff stress vector to the Cauchy stress vector (as will be shown in Section 6.3.9), so that the virtual work done by the internal nodal forces in each element of the shell finite element model is

$$\left(\mathbf{Fi}_m\right)^T \delta\Delta\mathbf{U}_m = \int_{\bar{v}_m} \boldsymbol{\sigma}^T \delta\Delta\mathbf{e}\, dv \quad (\text{nso } m) \tag{6.3.35a}$$

where the Cauchy stress vector $\boldsymbol{\sigma}$ is

$$\boldsymbol{\sigma} = \left[\sigma_{11}\ \sigma_{22}\ \sigma_{33}\ \sigma_{12}\ \sigma_{23}\ \sigma_{31}\right]^T$$

Here, $\bar{v}_m$ is the volume of shell element m at time $t + \Delta t$. Substituting the matrix $\mathbf{B}_l$ in Eq. 6.3.21 (evaluated at $t + \Delta t$) to transform the increment of the displacement vector to the increment of the linear strain vector in Eq. 6.3.35a, we obtain the internal nodal force vector

$$\mathbf{Fi}_m = \int_{\bar{v}_m} \mathbf{B}_l^T\, \boldsymbol{\sigma}\, dv \tag{6.3.35b}$$

We also can assemble all element internal nodal forces of the shell elements in the finite element model to obtain the global internal nodal force vector.

6.3.6 Shape (Interpolation) Functions

We use the bilinear interpolation functions for the shell element. They have the simple form

$$\phi_1 = \frac{1}{4}(1 + r)(1 + s) \tag{6.3.36a}$$

$$\phi_2 = \frac{1}{4}(1 - r)(1 + s) \tag{6.3.36b}$$

$$\phi_3 = \frac{1}{4}(1-r)(1-s) \tag{6.3.36c}$$

$$\phi_4 = \frac{1}{4}(1+r)(1-s) \tag{6.3.36d}$$

Figure 6.6b shows the locations of the nodes on the plane defined by the natural coordinate axes r and s.

After we have the interpolation functions, we summarize the procedure to establish the element stiffness matrix for the quadrilateral shell element, as shown in Table 6.3.

TABLE 6.3
PROCEDURE TO ESTABLISH AN ELEMENT STIFFNESS MATRIX FOR THE QUADRILATERAL SHELL ELEMENT

1. Establish the interpolation functions ϕ_k $(k = 1,2,3,4)$ and the derivatives with respect to the natural coordinates from Eq. 6.3.36.
2. Transform from the natural coordinate system to the global coordinate system by means of the inverse of the Jacobian matrix $\mathbf{J}$ in Eq. 6.3.19.
3. Establish the linear transformation matrix $\mathbf{B}_l$ and the nonlinear transformation matrix $\mathbf{B}_{nl}$ in Eq. 6.3.21.
4. Use Eq. 6.3.23c to establish the material tangent stiffness matrix $\mathbf{D}_{loc}^{6}$ referred to the local orthogonal coordinate system.
5. Transform the material tangent stiffness matrix to the global coordinate system using Eq. 6.3.26.
6. Use the integral in Eq. 6.3.29 to compute the element tangent stiffness matrix for the linear strain increments.
7. Use the integral in Eq. 6.3.31 to compute the element tangent stiffness matrix for the nonlinear strain increments.
8. Finally, the element tangent stiffness matrix is shown in Eq. 6.3.34 as $\mathbf{K} = \mathbf{K}_l + \mathbf{K}_{nl}$.

6.3.7 Numerical Integration for a Quadrilateral Shell Element

Similar to the total Lagrangian method, the interpolation functions are expressed in terms of the natural coordinates r, s, and t'. Therefore, a typical integral has the form

$$I = \int_v f(r,s,t')\,dv \tag{6.3.37}$$

where v is the volume of the element at time t, $dv = Jdrdsdt'$, and J is the determinant of the Jacobian matrix in Eq. 6.3.19c. Because the ranges of r, s, and t' are from −1 to +1 for the quadrilateral element, the integral becomes

$$I = \int_{-1}^{1}\int_{-1}^{1}\int_{-1}^{1} f(r,s,t')Jdrdsdt' \tag{6.3.38}$$

Let

$$F(r,s,t') = f(r,s,t')J \tag{6.3.39}$$

The numerical integration scheme gives

$$I = \sum_{j=1}^{m} w_h^j \sum_{l=1}^{n} w_s^l F\left(r_l, s_l, t'_j\right) \tag{6.3.40}$$

We use Simpson's rule for the numerical integration with respect to t'. Because we need to analyze loading as well as unloading in integration of the constitutive equations, the seven or more points for Simpson's rule for numerical integration with respect to t' are recommended. We equally divide the coordinate t' from −1 to +1 into six divisions. Therefore, there are seven equally spaced integration points ($m = 7$), and the weights w_h^j are $[1\ 4\ 2\ 4\ 2\ 4\ 1]/9$, as show in Figure 6.9.

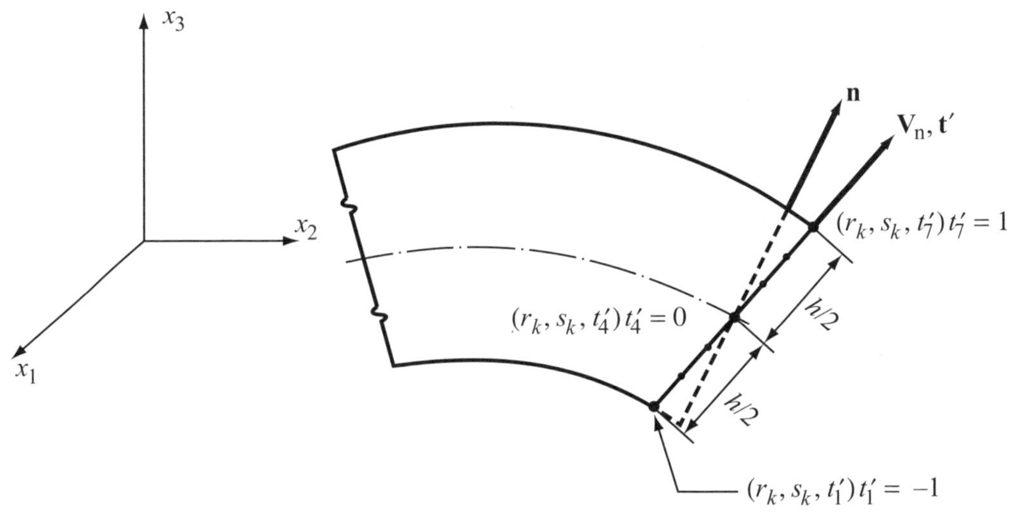

Figure 6.9 Seven integration points through the thickness of a quadrilateral shell element.

Because the interpolation functions are bilinear, a low-order numerical integration with respect to r and s is used. We suggest the 2 by 2 Gaussian numerical integration scheme ($n = 4$), and the location of each integration point is shown in Figure 6.10. We use equal weight of $w_s^l = 1/4$ for these four integration points.

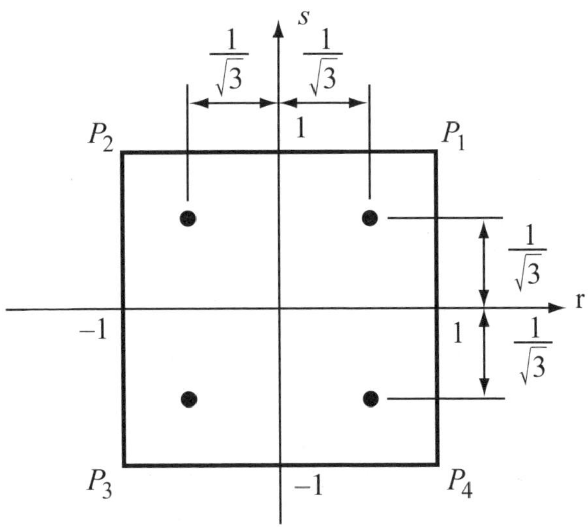

Figure 6.10 Locations of the 2 by 2 Gaussian numerical integration points for a quadrilateral element on the r–s plane.

Note that the variables at a Gaussian integration point such as the thickness of the deformed shell may be computed by Eq. 6.3.11 where the partial derivative $\partial\mathbf{x}/\partial t'$ can be interpolated from those at the four nodes.

6.3.8 Five to Six Degrees of Freedom per Node

A typical node k of the shell element contains an incremental (generalized) displacement vector of five degrees of freedom, Δu_1^k, Δu_2^k, Δu_3^k, α^k, and β^k. The increments of rotation α^k and β^k are in the directions of the vectors $\mathbf{V}_1^k$ and $\mathbf{V}_2^k$, respectively. We transform the increments of the nodal rotational vector $\lfloor \alpha^k \ \beta^k \ 0 \rceil$ to the common global Cartesian coordinate system for assembly. This means that we transform the increments of nodal rotational vector from three components α^k, β^k, and 0 referred to the local coordinate system $\mathbf{V}_1^k$, $\mathbf{V}_2^k$, and $\mathbf{V}_n^k$ to the three components referred to the global Cartesian coordinate system. Therefore, the total number of components is six, but they are not linearly independent. In general, there is no problem because each element connected to the common node is not necessarily C^1 continuous. If the elements connected to a node are C^1 continuous, the global stiffness matrix would be singular. We may introduce a small artificial torsional spring to remove the singularity. To do so might affect wrinkling or buckling properties of the thin shell. The best way to overcome this problem is to use five components for this node in the coordinate system defined by two axes on the tangent plane and the normal about which there is no rotation. We may use this local coordinate system instead of transforming the five components to six in the global coordinate system.

6.3.9 Updating Configurations, Strains, and Stresses

In this section, we first update the generalized displacement vector at a node k. Because displacements and displacement increments are referred to the Cartesian coordinate system, Eq. 6.3.5 gives the coordinates of the updated configuration as

$$\bar{x}_i = x_i + \Delta u_i \tag{6.3.41}$$

For simplicity, we omit the superscript k for the nodal number in the position vector and the incremental displacement vector.

If the rotations α and β (the subscripts k omitted) of the director at a node are sufficiently small (e.g., less than $1/2$ degree), we may update the director at the node k by Eq. 6.3.7b as

$$\bar{V}_{ni} = V_{ni} + \Delta V_{ni} \tag{6.3.42}$$

where $\Delta V_{ni} = -V_{2i}\alpha + V_{1i}\beta$ from Eq. 6.3.8. Note that we also omit the superscript k on these three vectors at node k. To keep the system orthogonal, we may use the updated director in Eq. 6.3.42 and the procedure in Eq. 6.3.2 to redefine V_{1i} and V_{2i}.

For the rotations α and β that are not very small, we can use the exact method to update the director $\mathbf{V}_n$ and two associated orthogonal vectors $\mathbf{V}_1$ and $\mathbf{V}_2$. Following Argyris [1982], a vector $\mathbf{p}$ becomes $\bar{\mathbf{p}}$ after rotating about an axis $\mathbf{e}_R$ by a finite angle ω, as shown in Figure 6.11. Let

$$\boldsymbol{\omega} = \omega\, \mathbf{e}_R \tag{6.3.43a}$$

where $\boldsymbol{\omega}$ is called a finite rotational pseudo vector and has Cartesian components as

$$\boldsymbol{\omega} = \left[\omega_1 \ \omega_2 \ \omega_3\right]^T \tag{6.3.43b}$$

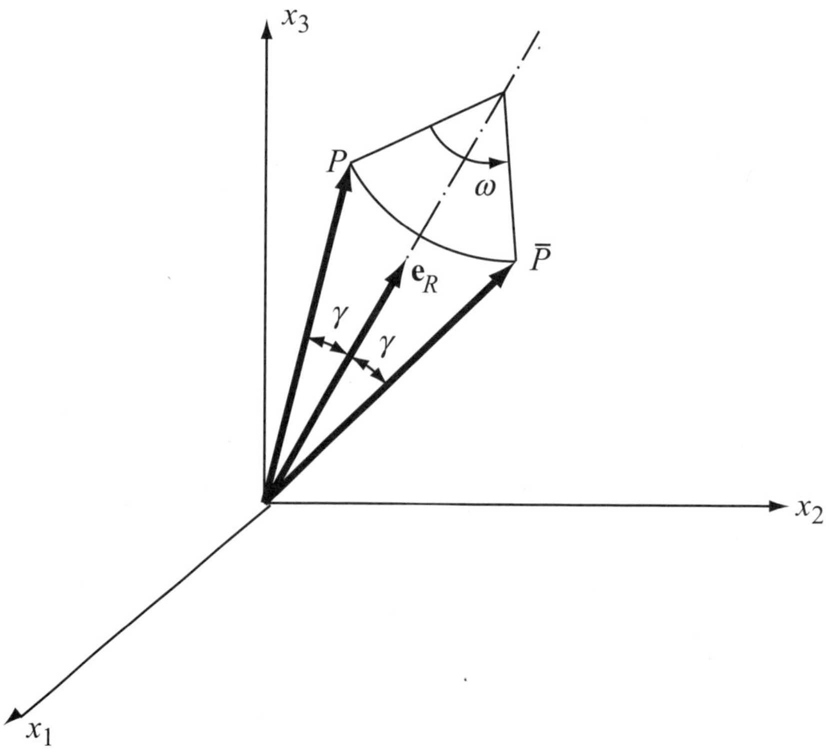

Figure 6.11 Vector $\mathbf{p}$ rotating to $\bar{\mathbf{p}}$ about an axis $\mathbf{e}_R$.

A transformation matrix **T** can be established, such that

$$\bar{\mathbf{p}} = \mathbf{T}(\boldsymbol{\omega})\mathbf{p} \tag{6.3.44a}$$

where

$$\begin{aligned} \mathbf{T}(\boldsymbol{\omega}) &= \mathbf{I} + \frac{\sin\omega}{\omega}\mathbf{S} + \frac{1}{2}\left(\frac{\sin(\omega/2)}{\omega/2}\right)^2 \mathbf{S}^2 \\ \omega &= \|\boldsymbol{\omega}\| \end{aligned} \tag{6.3.44b}$$

and

$$\mathbf{S} = \begin{bmatrix} 0 & -\omega_3 & \omega_2 \\ \omega_3 & 0 & -\omega_1 \\ -\omega_2 & \omega_1 & 0 \end{bmatrix} \tag{6.3.44c}$$

Using Eqs. 6.3.43 and 6.3.44, we update $\mathbf{V}_1$, $\mathbf{V}_2$, and $\mathbf{V}_n$ (defined in Section 6.3.1) by the rigid body rotation. Referring to the coordinate system defined by $\mathbf{V}_1$, $\mathbf{V}_2$, and $\mathbf{V}_n$ as shown in Figure 6.7, we set

$$\boldsymbol{\omega} = [\alpha \ \beta \ 0]^T \tag{6.3.45a}$$

Therefore,

$$\mathbf{S} = \begin{bmatrix} 0 & 0 & \beta \\ 0 & 0 & -\alpha \\ -\beta & \alpha & 0 \end{bmatrix} \tag{6.3.45b}$$

and

$$\mathbf{S}^2 = \begin{bmatrix} -\beta^2 & \alpha\beta & 0 \\ \alpha\beta & -\alpha^2 & 0 \\ 0 & 0 & -(\alpha^2 + \beta^2) \end{bmatrix} \tag{6.3.45c}$$

Finally, the transformation matrix is

$$\mathbf{T} = \begin{bmatrix} 1 - g_2\beta^2 & g_2\alpha\beta & g_1\beta \\ g_2\alpha\beta & 1 - g_2\alpha^2 & -g_1\alpha \\ -g_1\beta & g_1\alpha & 1 - g_2(\alpha^2 + \beta^2) \end{bmatrix} \tag{6.3.46}$$

where

$$g_1 = \frac{\sin\omega}{\omega}$$

$$g_2 = \frac{1}{2}\left(\frac{\sin(\omega/2)}{\omega/2}\right)^2$$

$$\omega^2 = \alpha^2 + \beta^2$$

Using Eq. 6.3.44a to transform $\mathbf{V}_1$, $\mathbf{V}_2$, and $\mathbf{V}_n$, respectively, we obtain

$$\bar{\mathbf{V}}_1 = \mathbf{T}\begin{bmatrix}1\\0\\0\end{bmatrix} = \begin{bmatrix}1 - g_2\beta^2\\ g_2\alpha\beta\\ -g_1\beta\end{bmatrix}$$

$$\bar{\mathbf{V}}_2 = \mathbf{T}\begin{bmatrix}0\\1\\0\end{bmatrix} = \begin{bmatrix}g_2\alpha\beta\\ 1 - g_2\alpha^2\\ g_1\alpha\end{bmatrix}$$

$$\bar{\mathbf{V}}_n = \mathbf{T}\begin{bmatrix}0\\0\\1\end{bmatrix} = \begin{bmatrix}g_1\beta\\ -g_1\alpha\\ 1 - g_2\left(\alpha^2 + \beta^2\right)\end{bmatrix}$$

Arranging in the matrix form gives

$$\begin{bmatrix}\bar{\mathbf{V}}_1\\ \bar{\mathbf{V}}_2\\ \bar{\mathbf{V}}_n\end{bmatrix} = \begin{bmatrix}1 - g_2\beta^2 & g_2\alpha\beta & -g_1\beta\\ g_2\alpha\beta & 1 - g_2\alpha^2 & g_1\alpha\\ g_1\beta & -g_1\alpha & 1 - g_2\left(\alpha^2 + \beta^2\right)\end{bmatrix}\begin{bmatrix}\mathbf{V}_1\\ \mathbf{V}_2\\ \mathbf{V}_n\end{bmatrix} \tag{6.3.47}$$

For sufficiently small α and β, $g_1 \approx 1$ and $g_2 \approx 1/2$, we omit the quadratic terms in Eq. 6.3.47 and simplify Eq. 6.3.47 to

$$\begin{bmatrix}\bar{\mathbf{V}}_1\\ \bar{\mathbf{V}}_2\\ \bar{\mathbf{V}}_n\end{bmatrix} = \begin{bmatrix}1 & 0 & -\beta\\ 0 & 1 & \alpha\\ \beta & -\alpha & 1\end{bmatrix}\begin{bmatrix}\mathbf{V}_1\\ \mathbf{V}_2\\ \mathbf{V}_n\end{bmatrix} \tag{6.3.48}$$

We leave to the readers as an exercise to prove that $\bar{\mathbf{V}}_1$, $\bar{\mathbf{V}}_2$, and $\bar{\mathbf{V}}_n$ in Eq. 6.3.47 are mutually orthonormal, but $\bar{\mathbf{V}}_1$, $\bar{\mathbf{V}}_2$, and $\bar{\mathbf{V}}_n$ in Eq. 6.3.48 are not necessarily mutually orthonormal.

For updating stresses at any point (Gaussian integration point), first we can use simple addition of the increment of the second Piola-Kirchhoff stress tensor (vector) to the Cauchy stress tensor (vector) as in Eq. 6.3.13, both in the deformed configuration at time t. Note that this Cauchy stress tensor is equal to the second Piola-Kirchhoff stress tensor because both are in the same configuration. The sum in Eq. 6.3.13 is the second Piola-Kirchhoff stress tensor at time $t + \Delta t$ but in the configuration at time t. The increment of the second Piola-Kirchhoff stress vector (tensor) $\Delta\mathbf{t}$ can be obtained from multiplying the material tangent stiffness matrix in Eq. 6.3.25a by the increment of the Lagrangian strain vector (tensor) $\Delta\boldsymbol{\varepsilon}$ computed from the increments of the nodal displacement vectors. To do so, the updated Cauchy stress tensor after transforming from the second Piola-Kirchhoff stress tensor does not necessarily satisfy the yield function in Eq. 3.9.15, even if we subdivide the strain increments into 200 sub-intervals. In Chapter 7, we discuss the radial return method to update the stresses that can satisfy the yield function.

After we obtain the second Piola-Kirchhoff stress tensor at time $t + \Delta t$ in the configuration at time t, we need to transform the stress tensor to the Cauchy stress tensor and then to that referred to the local coordinate system both in the same configuration at time $t + \Delta t$. First we update the second Piola-Kirchhoff stress tensor. Following the J_2 flow theory, we assume that the sheet metal is incompressible for plastic deformation. Because most deformation is plastic and because elastic deformation is very small during sheet metal forming processes, we can neglect the change in the sheet volume. Consequently, the Cauchy stress tensor can be obtained from the second Piola-Kirchhoff stress tensor in Eq. 6.3.13 at time $t + \Delta t$ by the transformation (see Eringen [1967]) as

$$\bar{\sigma}_{ij} = \frac{\partial \bar{x}_i}{\partial x_k}\frac{\partial \bar{x}_j}{\partial x_l}\bar{t}_{kl} \qquad (6.3.49)$$

where $\bar{x}_i = {}_0x_i + u_i + \Delta u_i = x_i + \Delta u_i$. Note that this transformation involves two different configurations but in the common global coordinate system. After we have $\bar{\sigma}_{ij}$ (converted to stress vector $\bar{\boldsymbol{\sigma}}$) in the global Cartesian coordinate system, we transform it to $\bar{\sigma}_{lm}$ $(l, m = \bar{r}, \bar{s}, \bar{t})$ (converted to stress vector $\bar{\boldsymbol{\sigma}}_{loc}$) referred to the local coordinate system $l, m = \bar{r}, \bar{s}, \bar{t}$. Note that this transformation refers to the same configuration at time $t + \Delta t$. The matrix in Eq. 6.3.28 gives the transformation

$$\bar{\boldsymbol{\sigma}}_{loc} = \mathbf{Q}\bar{\boldsymbol{\sigma}} \qquad (6.3.50)$$

where

$$\bar{\boldsymbol{\sigma}} = \begin{bmatrix} \bar{\sigma}_{11} & \bar{\sigma}_{22} & \bar{\sigma}_{33} & \bar{\sigma}_{12} & \bar{\sigma}_{23} & \bar{\sigma}_{31} \end{bmatrix}^T$$

$$\bar{\boldsymbol{\sigma}}_{loc} = \begin{bmatrix} \bar{\sigma}_{\bar{r}\bar{r}} & \bar{\sigma}_{\bar{s}\bar{s}} & \bar{\sigma}_{\bar{t}\bar{t}} & \bar{\sigma}_{\bar{r}\bar{s}} & \bar{\sigma}_{\bar{s}\bar{t}} & \bar{\sigma}_{\bar{t}\bar{r}} \end{bmatrix}^T$$

We set the normal stress component $\overline{\sigma}_{\overline{tt}}$ in the normal direction, with small deviation from the director $\bar{\mathbf{a}}_3$ or $\overline{\mathbf{V}}_n$ because of the transverse shear deformation, which is zero for a thin shell element. Using the current Cauchy stress vector $\overline{\boldsymbol{\sigma}}_{loc}$, we can update the material tangent stiffness matrix in Eq. 6.3.23c for computation of the next time step.

For strain updating, we still need to transform the increment of the strain vector from the global to the local coordinate system in the configuration at time t, and then we may compute the thickness at a point (node) in the shell element. The matrix in Eq. 6.3.28 gives the transformation

$$\Delta\boldsymbol{\varepsilon}_{loc} = \mathbf{Q}\Delta\boldsymbol{\varepsilon} \tag{6.3.51}$$

where

$$\Delta\boldsymbol{\varepsilon} = \begin{bmatrix} \Delta\varepsilon_{11} & \Delta\varepsilon_{22} & \Delta\varepsilon_{33} & 2\Delta\varepsilon_{12} & 2\Delta\varepsilon_{23} & 2\Delta\varepsilon_{31} \end{bmatrix}^T$$

$$\Delta\boldsymbol{\varepsilon}_{loc} = \begin{bmatrix} \Delta\varepsilon_{\overline{rr}} & \Delta\varepsilon_{\overline{ss}} & \Delta\varepsilon_{\overline{tt}} & 2\Delta\varepsilon_{\overline{rs}} & 2\Delta\varepsilon_{\overline{st}} & 2\Delta\varepsilon_{\overline{tr}} \end{bmatrix}^T$$

Similar to the stress vector, we ignore the normal component $\Delta\varepsilon_{\overline{tt}}$. From the incremental strain tensor referred to the deformed configuration at time t, we can compute the principal values by solving a quadratic equation for ε as

$$\det\left|\Delta\varepsilon_{\alpha\beta} - \varepsilon\delta_{\alpha\beta}\right| = 0 \quad (\alpha, \beta = \bar{r}, \bar{s}) \tag{6.3.52}$$

Two roots ε_α for ε of this equation are the principal components of the Lagrangian strain referred to the configuration at time t. Tang *et al.* [1988] also gave the stretch ratios with the configuration at time t as the reference

$$\lambda_\alpha = \sqrt{1 + 2\varepsilon_\alpha} \tag{6.3.53}$$

Using the stretch ratios at a Gaussian integration point on the reference surface of the deformed shell element, we can update the thickness at a point (node) of a shell element. Note that the bending strains do not affect significantly the thickness change, as mentioned in Chapter 5. As already explained, most deformation of a shell element during forming processes is plastic that is incompressible (i.e., no volume change). Therefore,

$$\lambda_1\lambda_2\lambda_3 = 1$$

where λ_3 is the stretch ratio in the thickness direction. Thus, the deformed thickness is updated approximately as

$$\bar{h} = \lambda_3 h \tag{6.3.54}$$

where the thicknesses h and $\bar{h}$ are at time t and $t + \Delta t$, respectively. This is one of the methods to update the thickness. The next paragraph shows how to update the total strain.

The easiest way to compute the total Lagrangian strains is by adding all the increments. To do so, all the strain increments must be referred to the undeformed configuration at time $t = 0$, and then we can add them together. First, we transform the global components of the increment of the deformation tensor (from t to $t + \Delta t$) referred to the configuration at time $t + \Delta t$ to the increment of the Lagrangian strain tensor referred to the undeformed configuration (see Hibbit *et al.* [1970]), such as

$$\Delta_0 \varepsilon_{pq} = \frac{\partial \bar{x}_i}{\partial_0 x_p} \frac{\partial \bar{x}_j}{\partial_0 x_q} \Delta e_{ij} \quad (p,q = 1,2,3) \tag{6.3.55}$$

where $\bar{x}_i = {}_0x_i + u_i + \Delta u_i$ (the coordinates of the deformed configuration at time $t + \Delta t$) and the increment of the deformation tensor $\Delta e_{ij} = \left(\Delta u_{i,j} + \Delta u_{j,i}\right)/2$ (see Eq. 3.6.36 for the definition of the rate of a deformation tensor). Now we may transform the increment of the Lagrangian strain tensor referred to the local coordinate system by means of the matrix in Eq. 6.3.28 as

$$\Delta_0 \boldsymbol{\varepsilon}_{loc} = \mathbf{Q} \Delta_0 \boldsymbol{\varepsilon} \tag{6.3.56}$$

where the transformation matrix is established based on the configuration at time 0 and

$$\Delta_0 \boldsymbol{\varepsilon} = \left[\Delta_0 \varepsilon_{11} \quad \Delta_0 \varepsilon_{22} \quad \Delta_0 \varepsilon_{33} \quad 2\Delta_0 \varepsilon_{12} \quad 2\Delta_0 \varepsilon_{13} \quad 2\Delta_0 \varepsilon_{23}\right]^T$$

$$\Delta_0 \boldsymbol{\varepsilon}_{loc} = \left[\Delta_0 \varepsilon_{\overline{rr}} \quad \Delta_0 \varepsilon_{\overline{ss}} \quad \Delta_0 \varepsilon_{\overline{tt}} \quad 2\Delta_0 \varepsilon_{\overline{rs}} \quad 2\Delta_0 \varepsilon_{\overline{rt}} \quad 2\Delta_0 \varepsilon_{\overline{st}}\right]^T$$

Therefore, the total Lagrangian strain vector at a point at time $t + \Delta t$ is

$${}_0\boldsymbol{\varepsilon}_{loc}\left(t + \Delta t\right) = {}_0\boldsymbol{\varepsilon}_{loc}\left(t\right) + \Delta_0 \boldsymbol{\varepsilon}_{loc} \tag{6.3.57}$$

Based on the total Lagrangian strains on the tangent plane of the undeformed reference surface for the interests of stamping engineers, we may use Eqs. 6.3.52 and 6.3.53 to compute the total stretch ratios λ_α $(\alpha = 1,2)$. The natural strains are given by

$$\varepsilon_\alpha^l = \ln \lambda_\alpha \tag{6.3.58a}$$

and the principal engineering strains are

$$e_\alpha = \lambda_\alpha - 1 \tag{6.3.58b}$$

Table 6.4 summarizes the procedure to update the nodal coordinates and rotations, and the strain and stress vectors at time $t + \Delta t$ for the quadrilateral shell element.

TABLE 6.4
PROCEDURE FOR UPDATING VARIABLES OF THE QUADRILATERAL SHELL ELEMENT

1. Update the nodal coordinates by simple addition of the displacement increments to the nodal coordinates at time t, as shown in Eq. 6.3.41.
2. Establish the transformation matrix **T** in Eq. 6.3.46 with the given rotational incremental vector $\boldsymbol{\omega} = [\alpha \;\; \beta \;\; 0]^T$ of the director $\mathbf{V}_n$ about the vectors $\mathbf{V}_1$ and $\mathbf{V}_2$ in Eq. 6.3.45a at a node.
3. Update the orthonormal system $[\mathbf{V}_1 \;\; \mathbf{V}_2 \;\; \mathbf{V}_n]^T$ at time t to the orthonormal system $[\overline{\mathbf{V}}_1 \;\; \overline{\mathbf{V}}_2 \;\; \overline{\mathbf{V}}_n]^T$ at time $t + \Delta t$ using Eq. 6.3.47.
4. Transform the stress vector referred to the global coordinate system to the local coordinate system such that it would be physically meaningful for thin shells.
   a. Update the second Piola-Kirchhoff stress tensor $\bar{t}_{ij} = \sigma_{ij} + \Delta t_{ij}$ from Eq. 6.3.13.
   b. Convert the second Piola-Kirchhoff stress tensor to the Cauchy stress tensor at time $t + \Delta t$ using Eq. 6.3.49.
   c. Use Eq. 6.3.50 to transform the Cauchy stress vector referred to the global coordinate system to the local coordinate system. We may ignore the normal stress component in the thickness direction.
5. Use the updated Cauchy stress vector $\bar{\boldsymbol{\sigma}}_{loc}$ to update the material tangent stiffness matrix in Eq. 6.3.23c to compute the next time step.
6. Transform the strain vector referred to the global coordinate system to the local coordinate system, such that it would be physically meaningful for thin shells.
   a. Transform the global components of the increment of the deformation tensor (from t to $t + \Delta t$) referred to the configuration at time $t + \Delta t$ to the increment of the Lagrangian strain tensor referred to the undeformed configuration (see Hibbit *et al.* [1970]) using Eq. 6.3.55

   $$\Delta_0 \varepsilon_{pq} = \frac{\partial \bar{x}_i}{\partial_0 x_p} \frac{\partial \bar{x}_j}{\partial_0 x_q} \Delta e_{ij} \qquad (p, q = 1,2,3)$$

   where $\bar{x}_i = x_i + \Delta u_i$.
   b. Now transform the increments from the global to the local coordinate system by means of the matrix in Eq. 6.3.28 as in Eq. 6.3.56 $\Delta_0 \boldsymbol{\varepsilon}_{loc} = \mathbf{Q} \Delta_0 \boldsymbol{\varepsilon}$, where the transformation matrix is established based on the configuration at time 0 and

   $$\Delta_0 \boldsymbol{\varepsilon} = \left[\Delta_0 \varepsilon_{11} \;\; \Delta_0 \varepsilon_{22} \;\; \Delta_0 \varepsilon_{33} \;\; 2\Delta_0 \varepsilon_{12} \;\; 2\Delta_0 \varepsilon_{13} \;\; 2\Delta_0 \varepsilon_{23}\right]^T$$

   $$\Delta_0 \boldsymbol{\varepsilon}_{loc} = \left[\Delta_0 \varepsilon_{\overline{rr}} \;\; \Delta_0 \varepsilon_{\overline{ss}} \;\; \Delta_0 \varepsilon_{\overline{tt}} \;\; 2\Delta_0 \varepsilon_{\overline{rs}} \;\; 2\Delta_0 \varepsilon_{\overline{rt}} \;\; 2\Delta_0 \varepsilon_{\overline{st}}\right]^T$$

TABLE 6.4 *(Continued)*

c. The total Lagrangian strain vector at a point at time $t + \Delta t$ is from Eq. 6.3.57 ${}_0\boldsymbol{\varepsilon}_{loc}(t + \Delta t) = {}_0\boldsymbol{\varepsilon}_{loc}(t) + \Delta_0\boldsymbol{\varepsilon}_{loc}$.

d. Based on the total Lagrangian strains on the tangent plane of the undeformed reference surface for the interests of stamping engineers, use Eqs. 6.3.52 and 6.3.53 to compute the total stretch ratios λ_α $(\alpha = 1, 2)$. The natural strains are given by Eq. 6.3.58a $\varepsilon^l_\alpha = \ln \lambda_\alpha$.

e. The principal engineering strains are $e_\alpha = \lambda_\alpha - 1$ from Eq. 6.3.58b.

6.3.10 Shear Lock and Membrane Lock

Because the lower-order functions are used to interpolate separately the in-plane and bending deformation in this quadrilateral shell element, shear lock and membrane lock might occur. Membrane lock results from inability of shell elements to represent inextensional modes of deformation. In a thin shell, the bending stiffness is small, but the membrane stiffness is large. When a shell element cannot bend without stretching, the strain energy is shifted incorrectly to that due to the membrane strains, resulting in under-prediction of deformation. Therefore, the shell element is too stiff because of membrane lock. For the shell elements discussed in this book, no severe membrane lock is observed. Shear lock does not appear in C^1 continuous elements such as the curved triangular element discussed in Section 6.2, because in C^1 continuous elements, the displacement is defined so that a normal remains normal after deformation. There are no transverse shear strains for C^1 continuous elements. In the quadrilateral element, there are significant transverse shear strains for very thin shells. These transverse shear strains cause shear lock. Once there is a lock, the element is extremely stiff. To obtain accurate results, an excessive number of finite elements must be used. Therefore, the computing time is not practical for production use.

The easiest way to overcome shear lock is to apply selective or reduced integration and numerical factors. However, the remedies are not general but problem dependent. They are difficult for production use. A more practical method, called the assumed strain method, can be used such that the transverse shear strains are interpolated differently. The bending and membrane strains are calculated as usual from the displacement interpolation. Because we interpolate the transverse shear strains differently, strains referred to the natural coordinate system r, s, and t' on the deformed reference surface of a shell at time t, as shown in Figure 6.12(a), are used.

Figure 6.12(a) also shows a schematic view of the shell element and the Cartesian and the natural coordinate systems for the element. Following Bathe and Dvorkin [1986], we assume the increments of transverse shear strains as

$$\Delta\varepsilon_{rt'} = \frac{1}{2}(1 + s)\Delta\varepsilon^A_{rt'} + \frac{1}{2}(1 - s)\Delta\varepsilon^C_{rt'} \tag{6.3.59a}$$

$$\Delta\varepsilon_{st'} = \frac{1}{2}(1 + r)\Delta\varepsilon^D_{st'} + \frac{1}{2}(1 - r)\Delta\varepsilon^B_{st'} \tag{6.3.59b}$$

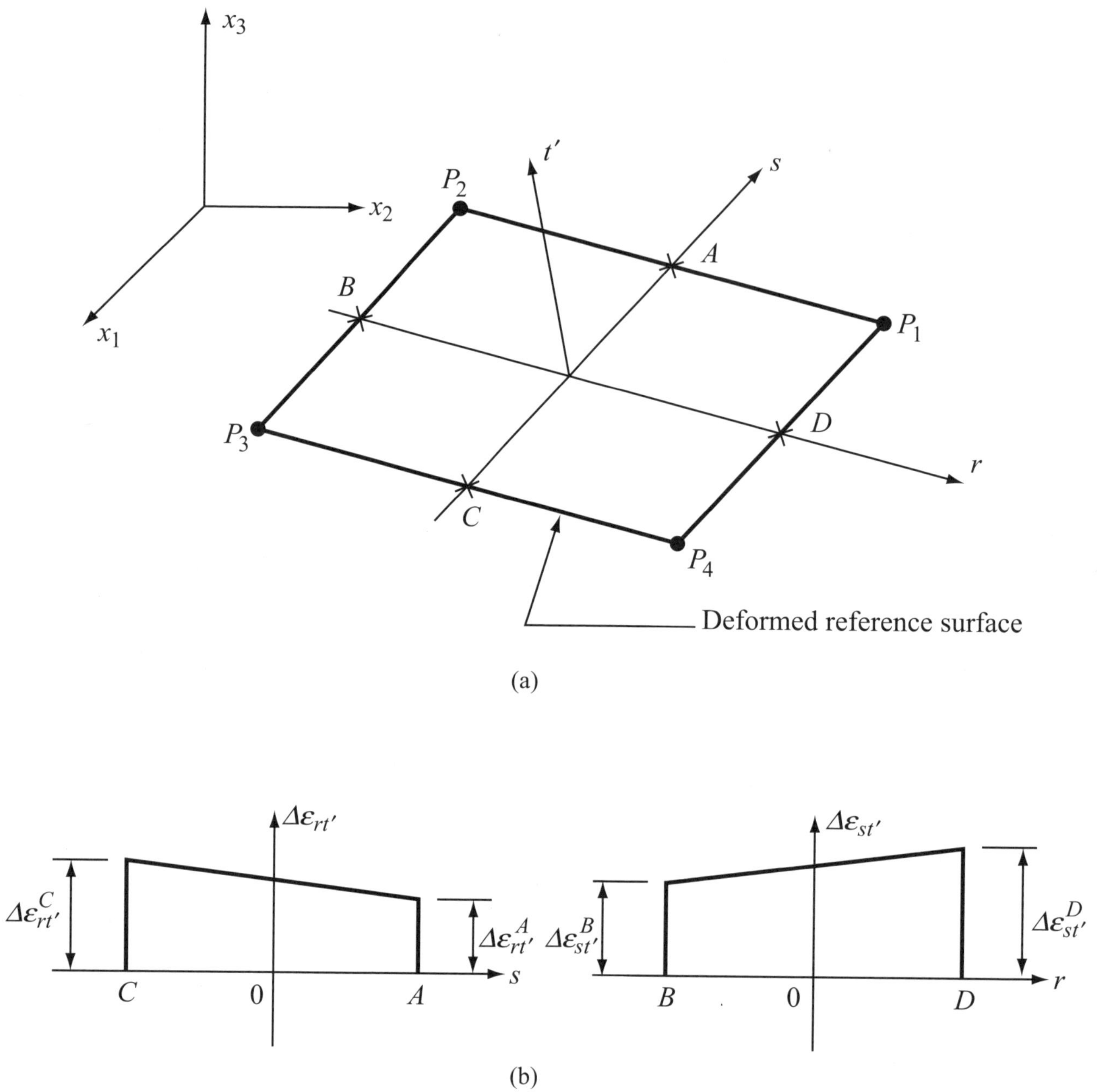

Figure 6.12 Assumed transverse shear strain increments for the four-node quadrilateral element: (a) locations of points A, B, C, and D, and (b) distributions of the transverse shear strain increments.

where $\Delta\varepsilon_{rt'}^A$, $\Delta\varepsilon_{rt'}^C$, $\Delta\varepsilon_{st'}^D$, and $\Delta\varepsilon_{st'}^B$ are the increments of the transverse shear strains at points A, C, D, and B (Figure 6.12(a)) calculated directly from the displacement interpolations. Figure 6.12(b) also shows the linear distributions of the increments of the transverse shears where the magnitudes are assigned arbitrarily.

6.4 Discussion of C^1 and C^0 Continuous Elements

In this chapter, we have discussed two typical thin shell elements. One is C^1 continuous, while the other is only C^0 continuous. Table 6.5 compares these two elements for applying to simulations of sheet metal forming processes.

TABLE 6.5
COMPARISON OF THE TWO SHELL ELEMENTS

	Curved Triangular Thin Shell	Quadrilateral Shell
Conforming	Yes, C^1 continuous	No, only C^0 continuous
Interpolation functions	Complicated	Simple, bilinear
Isoparametric	Yes	Yes
Number of elements for complex shapes	Less	More
Number of strain and stress components	3	5; 6 during computation
Numerical integration	7 points through the thickness, and 7 points on the surface	7 points through the thickness, and 4 points on the surface
Updating procedures	Simple addition	Transformations required
Locking	No shear locking	Shear locking; special treatment required
Node with 5 degrees of freedom	No special treatment	Removing the sixth degree of freedom or adding a small torsional spring
Coordinate systems	Cartesian and convected systems	Cartesian and natural coordinate systems and another coordinate system for rotation; Cartesian coordinate system used in computation, but a system generated from the natural coordinate system used in strain and stress computation; complex transformations required
Contact problems*	Easy to converge for Newton-Raphson iteration	Difficult to converge for Newton-Raphson iteration

* Will be discussed in Chapter 7.

6.5 References

Argyris, J., "An Excursion of Large Rotations," *Computer Method in Applied Mechanics and Engineering*, Vol. 32, 1982, p. 85.

Bathe, K.-J. and Bolourch, S., "A Geometric and Material Nonlinear Plate and Shell Element," *Computer and Structures*, Vol. 11, 1980, p. 23.

Bathe, K.-J. and Dvorkin, E.N., "A Formulation of General Shell Elements—The Use of Mixed Interpolation of Tensorial Components," *International Journal for Numerical Methods in Engineering*, Vol. 22, 1986, p. 697.

Dupuis, G., "Application of Ritz's Method to Thin Elastic Shell Analysis," *Journal of Applied Mechanics*, American Society of Mechanical Engineers, Vol. 38, 1971, p. 987.

Eringen, A.C., *Mechanics of Continua*, John Wiley & Sons, New York, 1967.

Goël, J.-J., "Construction of Basic Functions for Numerical Utilisation of Ritz's Method," *Numer. Math.*, Vol. 12, 1968, p. 435.

Hibbit, H.D, Marcal, P.V., and Rice, J. R., "A Finite Element Formulation for Problems of Large Strain and Large Displacement," *International Journal of Solids and Structures*, Vol. 6, 1970, p. 1069.

Lin, T.C. and Morgan, G.W., "A Study of Axisymmetric Vibrations of Cylindrical Shells as Affected by Rotary Inertia and Transverse Shear," *Journal of Applied Mechanics*, American Society of Mechanical Engineers, Vol. 23, 1956, p. 255.

Mindlin, R.D., "Influence of Rotary Inertia and Shear on Flexural Motion of Isotropic Elastic Plates," *Journal of Applied Mechanics*, American Society of Mechanical Engineers, Vol. 18, 1951, p. 31.

Needleman, A., "Void Growth in an Elastic-Plastic Medium," *Journal of Applied Mechanics*, American Society of Mechanical Engineers, Vol. 39, 1972, p. 964.

Reissner, E., "The Effect of Transverse Shear Deformation on the Bending of Elastic Plates," *Journal of Applied Mechanics*, ASME Transactions, American Society of Mechanical Engineers, Vol. 67, 1945, p. A69.

Tang, S.C., "Analysis of Springback in Sheet Forming Operation," *Advanced Technology of Plasticity*, Proceedings of the Second International Conference on the Technology of Plasticity (2ICTP), Vol. 1, Stuttgart, Germany, 1987, p. 193.

Tang, S.C., Gress, J., and Ling, P., "Sheet Metal Forming Modeling of Automobile Body Panels," Proceedings of 15th Biennial Congress of the International Deep Drawing Research Group (IDDRG), Dearborn, MI, ASM International, Metals Park, OH, 1988, p. 185.

Zienkiewicz, O.C. and Taylor, R.L., *The Finite Element Method*, Fourth Edition, McGraw-Hill, London and New York, 1988.

CHAPTER

7

Methods of Solution and Numerical Examples

7.1 Introduction to Methods for Solving Equations of Motion

In Chapter 4, we derived the equations of motion in the discrete form for an entire sheet metal part. For sheet metal forming processes, the main driving forces come from tool contact. In Chapter 4, formulas are presented to compute the contact forces, including the frictional forces governed by the Coulomb friction law. Because the sheet metal forming process is not a steady-state phenomenon, we need to find the transient solution of the equations of motion. By using the direct integration method to integrate numerically the equations of motion, we use the finite difference method to discretize the time coordinate t in the equations of motion and to solve the equations incrementally step by step to a specified time or until the tools bottom out to the home position. With the properly specified boundary and initial conditions, we may compute the response at any time t with a step size of Δt. For accurate and efficient computations, we may change the step size of Δt during the computations.

7.1.1 Equations of Motion and Constraint Conditions

From Chapter 4, the equations of motion in Eq. 4.3.12 in the discrete form for an entire sheet metal part are

$$\mathbf{M\ddot{U}} + \mathbf{C\dot{U}} + \mathbf{Fi} = \mathbf{Fe} + \mathbf{Fc} \qquad (7.1.1)$$

where $\mathbf{M}$ is the mass matrix, $\mathbf{C}$ is the damping matrix, $\mathbf{\dot{U}}$ is the generalized velocity vector, $\mathbf{\ddot{U}}$ is the generalized acceleration vector, $\mathbf{Fi}$ is the generalized internal nodal force vector including moments, $\mathbf{Fe}$ is the external force vector (including the weight of a sheet), and $\mathbf{Fc}$ is the contact nodal force vector. Because we emphasize using shell elements in our sheet metal forming analysis, the equations of motion are formulated for a finite element model using shell elements. Therefore, $\mathbf{U}$, $\mathbf{\dot{U}}$, and $\mathbf{\ddot{U}}$ are the generalized displacement, velocity, and acceleration vectors in the equations of motion, respectively.

As discussed in Chapter 6, there are nine degrees of freedom at each node for the curved triangular shell element and six degrees of freedom for the four-node shell element at each node after being assembled in the common global Cartesian coordinate system. The conjugate internal and external nodal force vectors at each node contain the bending and twisting moments in addition to the forces in the equations of motion, but the contact nodal force vector at each contact node contains only the forces.

For efficient computation, we use the diagonal mass and damping matrices in the equations of motion instead of the consistent mass and damping matrices in Eqs. 4.3.9 and 4.3.11, respectively. We lump the mass or damping equally at all nodes of a finite element, and then we assemble all elements together for the finite element model to form the diagonal mass or damping matrix. The equations of motion for rotation have different forms from those in Eq. 7.1.1, and they are Euler's equations. For example, the x_1 component of the Euler's equations is

$$I_1\dot{\omega}_1 - \omega_2\omega_3\left(I_2 - I_3\right) = m_1 \tag{7.1.2}$$

where I_1, I_2, and I_3 are the principal moments of inertia for the lumped mass at a node, $\dot{\omega}_1$ is the x_1 component of the angular acceleration, ω_2 and ω_3 are the x_2 and x_3 components of the angular velocity, respectively, and m_1 is the x_1 component of the moment vector. Note that the principal moments of inertia and the product term of the principal moments of inertia and the angular velocities are used in Eq. 7.1.2. To avoid difficulty in numerical integration, we set all principal moments of inertia equal to I (the moment of inertia matrix is isotropic). Therefore, Eq. 7.1.2 is generalized to

$$I\dot{\omega}_i = m_i \tag{7.1.3}$$

of which the form is similar to that of translation. Here the subscript i denotes the x_1 component. Because the moment of inertia matrix is isotropic, the form in Eq. 7.1.3 is valid for any component in any rectangular Cartesian coordinate system.

The lumped mass at a nodal point in a finite element model with thin shell elements is isotopic for translational motion. For rotation, the lumped mass also should be isotropic in order to omit the product term in the Euler's equations of rotation, as mentioned in the previous paragraph. Following Hughes [1987], the rotational mass is obtained by multiplying the translational mass with a scaling factor α, so that a large critical time step for numerical integration of the equations of motion is permitted without loss of stability. Let

$$Mr = \alpha Mt \tag{7.1.4}$$

where Mr is the rotational mass or the moment of inertia I in Eq. 7.1.3 at a node, Mt is the translational mass lumped on a node, and the scaling factor α is

$$\alpha = \max\left[\alpha_1, \alpha_2\right] \tag{7.1.5}$$

where

$$\alpha_1 = \frac{h_0^2}{12} \tag{7.1.6a}$$

$$\alpha_2 = \frac{V_0}{8h_0} \tag{7.1.6b}$$

Here, h_0 and V_0 are the thickness and the volume, respectively, for an undeformed shell element. Although a combination of the implicit method of solution and the consistent mass matrix would minimize error, complications exist for the consistent mass matrix in Euler's equations of motion for large rotations, as mentioned in the previous paragraph. We recommend that the diagonal mass and the damping matrices be used, even for the implicit method. In fact, dynamic effect does not play an important role in simulations of sheet metal forming processes.

For surface contact in sheet metal forming processes, the constraint conditions also are presented in Chapter 4. Because we assume the tool is rigid, the sheet cannot penetrate into the tool. Therefore, the constraint condition is that the gap function $g_l(\mathbf{x})$ at a node l should satisfy

$$g_l(\mathbf{x}) = (\mathbf{x} - \mathbf{x}_t) \cdot \mathbf{n} \geq 0 \tag{7.1.7}$$

where $\mathbf{x}$ is the position vector of a node on the reference surface of a deformed shell, $\mathbf{x}_t$ is the position vector of a point on the tool surface such that $\|\mathbf{x} - \mathbf{x}_t\|$ is minimum, and $\mathbf{n}$ is the normal of the tool surface at the point $\mathbf{x}_t$. Note that when $g_l > 0$, there is no contact. When $g_l = 0$, there is a contact. From Eq. 4.5.2 in Chapter 4, the position x_3 of the tool surface, which is moving downward, is

$$x_3 = S(x_\alpha) - z(t) \tag{7.1.8a}$$

where $z(t) = \int_0^t \dot{U}_t dt$. Here, $\dot{U}_t$ is the tool speed (a scalar). For a stationary tool surface, we set $z(t) = 0$ ($\dot{U}_t = 0$), and the position x_3 of the tool surface is

$$x_3 = S(x_\alpha) \tag{7.1.8b}$$

Note that for a rigid tool surface, $\mathbf{x}_t$ does not depend on any deformation. Taking the material derivative of the gap function in Eq. 7.1.7 and using $\|\mathbf{x} - \mathbf{x}_t\| = 0$ at the contact node l, we have

$$\frac{D}{Dt} g_l(\mathbf{x}, t) = \left(\dot{\mathbf{U}}^l + \dot{U}_t \mathbf{e}_3\right) \cdot \mathbf{n} = 0 \tag{7.1.9a}$$

where $\dot{\mathbf{U}}^l$ is the velocity vector of the contact node l, and the normal $\mathbf{n}$ is along the direction $S_{,\alpha} = \partial S/\partial x_\alpha, -1$. For a moving tool surface,

$$S_{,\alpha}\dot{U}_\alpha^l - \left(\dot{U}_3^l + \dot{U}_t\right) = 0 \tag{7.1.9b}$$

For a stationary tool surface,

$$S_{,\alpha}\dot{U}_\alpha^l - \dot{U}_3^l = 0 \tag{7.1.9c}$$

The physical meaning of Eq. 7.1.9a is that the normal component of the relative velocity of a contact node vanishes, or the node cannot penetrate into the tool.

The Coulomb friction law states that the frictional nodal force vector $\mathbf{T}^l$ is

$$\mathbf{T}^l \approx -\mu P^l \frac{\dot{\mathbf{U}}_r^l}{\left\|\dot{\mathbf{U}}_r^l\right\|}, \qquad \left\|\dot{\mathbf{U}}_r^l\right\| \neq 0 \tag{7.1.10}$$

where P^l is the magnitude of the normal component of the contact nodal force vector, and $\dot{\mathbf{U}}_r^l$ is the relative velocity vector at the contact node l on the reference surface of a deformed shell. For sliding contact, $\left\|\dot{\mathbf{U}}_r^l\right\| \neq 0$, and Eq. 7.1.10 is valid. For sticking contact, $\left\|\dot{\mathbf{U}}_r^l\right\| = 0$, the frictional force vector cannot be determined by Eq. 7.1.10, and its magnitude must satisfy

$$\left\|\mathbf{T}^l\right\| - \mu P^l < 0 \tag{7.1.11}$$

7.1.2 Boundary and Initial Conditions

Along the periphery of a sheet, we may specify either a displacement or a force boundary condition. For a clamped edge, the displacement and rotation components are zero. For a free edge, the force and moment components are zero. Also, we must consider draw-bead restraint forces along the draw-bead lines, as discussed in Chapter 4.

We must specify the initial conditions in order to have a unique solution. For the initial step, we assume the sheet is quiescent before time $t = 0$ and set the initial conditions at $n = 0$ for the initial step as

$$\mathbf{U}_0 = 0 \tag{7.1.12a}$$

and

$$\dot{\mathbf{U}}_0 = 0 \tag{7.1.12b}$$

When the tool, a punch, or a binder ring starts to move downward as specified in Eq. 7.1.8a to press the sheet, we may compute the initial acceleration vector from the equations of motion in Eq. 7.1.1 using the initial conditions and then the transient response for the subsequent time intervals.

7.1.3 Explicit and Implicit Integration

Two general methods, the dynamic and the quasi-static ones, are discussed here for direct integration of the equations of motion for the time variable t. The dynamic explicit (time) integration method is the most popular one used for simulations of sheet metal forming processes. In this time integration method, the central difference with the increment Δt of the time coordinate is used to discretize accelerations. A diagonal mass matrix often is used. The computation then becomes simple because no simultaneous equations are solved. The features for strain rate effects and adaptive meshing (see Section 7.7) can easily be implemented into the software based on the dynamic explicit integration method. The disadvantage of this method is its numerical instability problem. For simulations of sheet metal forming processes, the time interval for numerical integration should be taken as small as 10^{-7} second for a thin shell element model and even smaller for a three-dimensional solid element model. Researchers have tried to artificially increase the forming speed or the mass density ρ. In this way, the total computing time is reduced to become acceptable.

For the dynamic implicit integration method, a consistent mass matrix should be used for accurate results. However, as mentioned in the previous paragraph, the equations of motion for rotations involve quadratic terms of angular velocities. We still use the lumped mass approach, which results in an isotropic mass matrix for translation as well as rotation at each node in a finite element model. A much larger time step may be taken than that for the explicit integration. Oscillations due to vibration or ringing cannot be avoided. When a large change of response occurs, there might be convergent problems in the iteration. However, this large change can be taken care of by the inertia term and the use of smaller time increments in the numerical integration. Not much research has been done to apply the dynamic implicit integration method to simulations of sheet metal forming processes.

7.1.4 Quasi-Static Equations

Because the tool speed to form an automotive panel is relatively low, the inertia term can be neglected. The equations of motion in Eq. 7.1.1 become the equations of equilibrium as

$$\mathbf{Fi} = \mathbf{Fe} + \mathbf{Fc} \tag{7.1.13}$$

Therefore, a quasi-static problem can be formulated for an incremental solution. Another advantage of neglecting the inertia term is that undesired oscillations of the response can be avoided. In this analysis, the total tool travel distance is divided into a few hundred steps. Each increment is about $1/1000$ of the largest linear dimension of a panel.

7.2 Explicit Integration of Equations of Motion with Constraint Conditions

In an explicit integration, we write the equations of motion as

$$\mathbf{M}\ddot{\mathbf{U}}_n + \mathbf{C}\dot{\mathbf{U}}_n = \mathbf{Fe}_n + \mathbf{Fc}_n - \mathbf{Fi}_n \tag{7.2.1}$$

subject to the constraint conditions and the Coulomb friction law in Eqs. 7.1.9 through 7.1.11 for the nth time step, where the subscript n denotes the vector or the matrix for the step n or at time t. With given $\mathbf{U}_n$, $\dot{\mathbf{U}}_n$, $\mathbf{Fe}_n$, and $\mathbf{Fi}_n$, we compute these variables for the next step $n + 1$ at time $t + \Delta t$. Note that the contact nodal force vector $\mathbf{Fc}_n$ must satisfy the Coulomb friction law.

7.2.1 Discretization and Solutions

We use the central difference method to discretize the velocity and acceleration vectors in terms of the displacement vectors

$$\dot{\mathbf{U}}_n = \frac{1}{2\Delta t}\left(\mathbf{U}_{n+1} - \mathbf{U}_{n-1}\right) \tag{7.2.2}$$

$$\ddot{\mathbf{U}}_n = \frac{1}{\Delta t^2}\left(\mathbf{U}_{n+1} - 2\mathbf{U}_n + \mathbf{U}_{n-1}\right) \tag{7.2.3}$$

where Δt is the time interval, and the subscripts $n - 1$, n, and $n + 1$ denote the vectors at time $t - \Delta t$, t, and $t + \Delta t$, respectively. Because of potential large rotation in sheet metal forming analyses, we cannot compute the total rotation by simple addition of the increments (see Section 6.3.9 in Chapter 6 on computing the total rotation). Therefore, we use only the incremental (generalized) displacement vector to compute the incremental strain vector. Let the incremental displacement vector at step n be defined as

$$\Delta\mathbf{U}_n = \mathbf{U}_{n+1} - \mathbf{U}_n \tag{7.2.4}$$

The velocity and acceleration vectors in Eqs. 7.2.2 and 7.2.3 can be expressed in terms of the incremental displacement vectors as

$$\dot{\mathbf{U}}_n = \frac{1}{2\Delta t}\left(\Delta\mathbf{U}_n + \Delta\mathbf{U}_{n-1}\right) \tag{7.2.5}$$

$$\ddot{\mathbf{U}}_n = \frac{1}{\Delta t^2}\left(\Delta\mathbf{U}_n - \Delta\mathbf{U}_{n-1}\right) \tag{7.2.6}$$

Substituting Eqs. 7.2.5 and 7.2.6 into Eq. 7.2.1 yields

$$\left(\frac{1}{\Delta t^2}\mathbf{M} + \frac{1}{2\Delta t}\mathbf{C}\right)\Delta\mathbf{U}_n = \left(\mathbf{Fe}_n + \mathbf{Fc}_n - \mathbf{Fi}_n\right) + \frac{1}{\Delta t^2}\mathbf{M}\Delta\mathbf{U}_{n-1} - \frac{1}{2\Delta t}\mathbf{C}\Delta\mathbf{U}_{n-1} \quad (7.2.7)$$

The computation of the contact nodal force vector $\mathbf{Fc}_n$ by various methods will be discussed in Section 7.2.3.

For the first step, we assume the sheet is quiescent before time $t = 0$ and set the initial conditions at $n = 0$ $(t = 0)$ as

$$\mathbf{U}_0 = 0 \quad (7.2.8a)$$

and

$$\dot{\mathbf{U}}_0 = 0 \quad (7.2.8b)$$

All terms on the right side of Eq. 7.2.7 vanish, except the external and contact nodal force vectors $\mathbf{Fe}_0$ and $\mathbf{Fc}_0$. Therefore, we can solve for $\Delta\mathbf{U}_0$ by inverting the matrix on the left side

$$\Delta\mathbf{U}_0 = \left(\frac{1}{\Delta t^2}\mathbf{M} + \frac{1}{2\Delta t}\mathbf{C}\right)^{-1}\left(\mathbf{Fe}_0 + \mathbf{Fc}_0\right) \quad (7.2.9)$$

when $\mathbf{Fe}_0$ is specified and $\mathbf{Fc}_0$ is set in advance.

For a general step $n > 0$ $(t > 0)$, we assume the mass and damping matrices do not change. Because $\Delta\mathbf{U}_{n-1}$ is known, all terms on the right side of Eq. 7.2.7 can be computed. The internal nodal force vector is updated by the principle of virtual work in Eq. 6.2.36 (see Chapter 6) for the total Lagrangian formulation and in Eq. 6.3.35 for the updated Lagrangian formulation. Note that Eqs. 6.2.36 and 6.3.35 are for the internal nodal force vector of an element. The total internal nodal force vector is obtained by assembling all element internal nodal force vectors. After all terms on the right side are computed, we can solve Eq. 7.2.7 for the incremental displacement vector as

$$\Delta\mathbf{U}_n = \left(\frac{1}{\Delta t^2}\mathbf{M} + \frac{1}{2\Delta t}\mathbf{C}\right)^{-1}\left[\left(\mathbf{Fe}_n + \mathbf{Fc}_n - \mathbf{Fi}_n\right) + \frac{1}{\Delta t^2}\mathbf{M}\Delta\mathbf{U}_{n-1} - \frac{1}{2\Delta t}\mathbf{C}\Delta\mathbf{U}_{n-1}\right] \quad (7.2.10)$$

when $\mathbf{Fe}_n$ is specified and $\mathbf{Fc}_n$ is known at the step n. We update all responses such as the configuration of the sheet, the strains, and the stresses. We repeat this process until the final step is reached.

The diagonal mass and damping matrices often are used. The computation then becomes simple because no solving of simultaneous equations is involved. Instead, only a set of uncoupled single equations must be solved. One equation for each degree of freedom is with only one unknown. Therefore, this explicit time integration method is very efficient for a detailed finite element model with as many as millions of degrees of freedom in simulations of sheet metal forming processes. The disadvantage of this method is its numerical instability problems because the internal, external, and contact nodal force vectors on the right side of Eq. 7.2.10 are evaluated at step n instead of step $n + 1$.

7.2.2 Numerical Instability

The time increment is restricted by the Courant-Fredrich-Lewy condition of numerical stability. For a simple linear truss element, the condition is expressed by the time increment

$$\Delta t < l/c \tag{7.2.11}$$

where l is the length of a truss element, and c represents the speed of the sound wave in the truss. Physically, it means that the time incrçment should be smaller than the time for the sound wave traveling across the element. The wave speed c can be estimated as

$$c = \sqrt{\frac{E}{\rho}} \tag{7.2.12}$$

where E represents Young's modulus, and ρ is the material mass density of the truss element. For a finite element model of nonlinear shell elements, the time increment Δt may be smaller than that shown in Eq. 7.2.12.

According to the criteria in the theoretical manual of LS-DYNA (Hallquist [1998]), the time step for a thin shell element is restricted as

$$\Delta t < L_s/c_p \tag{7.2.13}$$

where L_s is the characteristic length for the smallest finite element in a thin shell finite model, and c_p is the speed of the sound wave in a thin plate (shell), which is expressed by

$$c_p = \sqrt{\frac{E}{\rho\left(1 - v^2\right)}} \tag{7.2.14}$$

where E is Young's modulus, v is Poisson's ratio, and ρ is the mass density of the shell material. The characteristic length L_s for the smallest shell element is

$$L_s = \frac{(1 + \beta) A_s}{\max\left[L_1, L_2, L_3, (1 - \beta) L_4\right]} \tag{7.2.15}$$

where $\beta = 0$ for quadrilateral shell elements and 1 for triangular shell elements; A_s is the area of a shell element; and L_1, L_2, L_3, and L_4 are the lengths of the sides of the shell element. A more conservative value of the characteristic length for a quadrilateral shell element is defined as

$$L_s = \frac{(1+\beta)A_s}{\max(D_1, D_2)} \tag{7.2.16}$$

where D_1 and D_2 are the lengths of the diagonals of a quadrilateral shell element.

To form an automobile panel, the tool speed is relatively low compared with the speed of a sound wave, and a forming process usually takes about a second. For a steel sheet, if an element with its smallest side of 2 mm is used to model the contact with the tool corner of a tight radius, then $\Delta t < 0.4 \times 10^{-6}$ sec should be used. For the duration of 1 sec, the total number of steps is greater than 2.5 million. When using three-dimensional solid elements to model a curved area with a tight radius, four elements through the thickness of a sheet of 0.8 mm are used. The smallest side of an element will be 0.2 mm. The time increment should be smaller than $\Delta t < 0.4 \times 10^{-7}$ sec, and the total number of steps should be greater than 25 million. This would take too much computing time. Researchers have tried to artificially adjust the forming speed by using a factor between 10 to 1000 times faster than the actual speed or by artificially increasing the mass density ρ. Hence, the number of time steps to simulate a stretch/draw forming process is reduced to several thousand, which would be practical for the simulation of the forming process. However, unnecessary dynamic effects may be generated to give inaccurate responses. Unfortunately, most commercial codes for forming analyses are based on the dynamic explicit integration method originally developed for high-speed impact analyses.

7.2.3 Computing Contact Nodal Forces

When the contact nodal forces by the direct and Lagrangian multiplier methods (see Section 4.5 in Chapter 4) are computed, the constraint equations are coupled with the equations of motion. A set of simultaneous equations for all contact nodes must be solved. The main advantage of using the explicit method to solve the equations of motion in order to avoid solving the simultaneous equations is lost. Therefore, it is not recommended to use these formulations without appropriate modification to compute the contact nodal force vector $\mathbf{Fc}_n$ to be used in Eq. 7.2.7 and then to solve for the incremental displacement vector $\Delta\mathbf{U}_n$.

For the penalty method, the contact nodal force vector $\mathbf{Fc}_n$ can be computed by the use of Eqs. 4.5.28 to 4.5.31 from Chapter 4 for step n. Inserting the contact nodal force vector into the right side of the equations of motion, we can solve for the incremental displacement vector separately for each degree of freedom. The main advantage of the explicit integration is kept. We use only the penalty method to compute the contact nodal force vector in the subsequent discussion on the explicit integration.

7.2.4 Updating Variables for Dynamic Explicit Integration

Once the incremental (generalized) displacement vector $\Delta\mathbf{U}_n$ is solved from Eq. 7.2.7, we can update all terms on the right side of Eq. 7.2.7 for establishing the equations of motion for the next time step. Meanwhile, we may update the deformation and stresses by accumulating the increments of the variables (not the rotation). For the shell element with the updated Lagrangian formulation, we should update the reference configuration and its associated coordinate system.

7.2.5 Summary of the Dynamic Explicit Integration Method with Contact Nodal Forces Computed by the Penalty Method

Table 7.1 summarizes the procedures to compute the deformation and stresses for a sheet under a forming process using the dynamic explicit integration method with contact nodal forces computed by the penalty method.

7.2.6 Application of the Dynamic Explicit Integration Method to Sheet Metal Forming Analysis

As mentioned, researchers have tried to artificially adjust the forming speed by using a factor between 10 to 1000 times faster than the actual speed or by artificially increasing the mass density ρ. Hence, the number of time steps to simulate a stretch/draw forming process is reduced to several thousand. Using this scaling technique, the numerical results for deformation often are good for simulating a stretch/draw forming process with reasonable computing time, because most areas of a sheet contact the tool surfaces during the process. However, the computed stresses are questionable.

In analyzing a binder wrap, severe vibrations occur because a wide area on a sheet is not supported by the tools. Therefore, a final static solution is hard to obtain from the dynamic explicit integration method. The same difficulty would occur in applying the explicit method to the analysis of springback after a part is formed. A few dynamic explicit integration codes have been switched to the quasi-static method for the final analysis of springback. Because the forming stresses from dynamic explicit integration analyses are questionable, the results of springback from the dynamic analyses would not be reliable. The features for strain rate effects and adaptive meshing can easily be implemented into the software based on the dynamic explicit integration method.

7.3 Implicit Integration of Equations of Motion with Constraint Conditions

In an implicit integration, we write the equations of motion (Eq. 7.1.1) in the form of

$$\mathbf{M}\ddot{\mathbf{U}}_{n+1} + \mathbf{C}\dot{\mathbf{U}}_{n+1} + \mathbf{Fi}_{n+1} = \mathbf{Fe}_{n+1} + \mathbf{Fc}_{n+1} \tag{7.3.1}$$

subject to the constraint conditions and the Coulomb friction law in Eqs. 7.1.9 through 7.1.11 for the $(n+1)$th time step, where the subscript $n+1$ denotes the vectors for the time step $n+1$. With given $\mathbf{U}_n$, $\dot{\mathbf{U}}_n$, $\mathbf{Fi}_n$, and $\mathbf{Fe}_{n+1}$, we compute these other than the external nodal force vector, which is specified, at the time step $n+1$.

TABLE 7.1
DYNAMIC EXPLICIT INTEGRATION METHOD WITH CONTACT NODAL FORCES COMPUTED BY THE PENALTY METHOD

1. Start at time $t = 0$ (step $n = 0$).
   a. Initialize the nodal displacement and velocity vectors.
   b. Initialize the element strain and stress vectors.
   c. Assign the penalty parameters α_N and α_T.
   d. Set the tool positions.
   e. Establish the mass and damping matrices **M** and **C**.
2. Choose a time increment based on the criterion for numerical stability, and determine the increment of the tool travel Δz.
3. Compute the contact nodal force $P_n = \alpha_N \langle -g_l(\mathbf{x}) \rangle$ from Eq. 4.5.26a in Chapter 4 at time t, where $g_l(\mathbf{x}) = (\mathbf{x} - \mathbf{x}_t) \cdot \mathbf{n}$. Here, $\mathbf{x}$ is the position vector of a contact node at time t, and $\mathbf{x}_t$ is the contact point on the tool surface at the tool travel with the increment Δz.
4. Compute the frictional force vector $\mathbf{T}_n$ at time t (set $\mathbf{T}_0 = 0$) from Eqs. 4.5.28 to 4.5.31 in Chapter 4.
5. Solve for the displacement increment vector $\Delta \mathrm{U}_n$ in Eq. 7.2.10 for each node in the finite element model.
6. Update the following:
   a. The nodal displacement and velocity vectors.
   b. The element strain and stress vectors following Section 6.2.8 or 6.3.9 in Chapter 6.
7. $t + 1$ becomes t, and $n + 1$ becomes n.
8. Check $t > t_{max}$ (the last time step) or if the tools reach the home position.
   a. If yes, then the simulation is done.
   b. If no, then go to Step 2.
9. Print the final results for the deformed position, strains, and stresses in the formed sheet metal part.

7.3.1 Newmark's Integration Scheme

We want to express the velocity and acceleration vectors in Eq. 7.1.1 in terms of the unknown incremental displacement vector $\Delta\mathbf{U}_n = \mathbf{U}_{n+1} - \mathbf{U}_n$. To do so, we need a numerical integration scheme. We apply the integration scheme by Newmark [1959], with the assumptions of

$$\mathbf{U}_{n+1} = \mathbf{U}_n + \dot{\mathbf{U}}_n\Delta t + \left[\left(\frac{1}{2} - \alpha\right)\ddot{\mathbf{U}}_n + \alpha\ddot{\mathbf{U}}_{n+1}\right]\Delta t^2 \tag{7.3.2}$$

$$\dot{\mathbf{U}}_{n+1} = \dot{\mathbf{U}}_n + \left[(1-\delta)\ddot{\mathbf{U}}_n + \delta\ddot{\mathbf{U}}_{n+1}\right]\Delta t \tag{7.3.3}$$

where α and δ are the parameters that can be assigned to obtain the accuracy and stability of the numerical integration. We use $\alpha = 1/4$ and $\delta = 1/2$ in Eqs. 7.3.2 and 7.3.3, which are called the trapezoidal rules in numerical integration. For linear problems, this numerical scheme for time integration is unconditionally stable. However, the user should be careful because it is not guaranteed for numerical stability for highly nonlinear problems such as analyses of sheet metal forming processes. Because of the complexity of the nonlinearity, no numerical stability criterion has been established in the literature. The practical way to achieve stability is by varying the time step until a satisfactory result is obtained.

Inserting $\alpha = 1/4$, $\delta = 1/2$, and $\Delta\mathbf{U}_n = \mathbf{U}_{n+1} - \mathbf{U}_n$ into Eqs. 7.3.2 and 7.3.3, we obtain

$$\ddot{\mathbf{U}}_{n+1} = \frac{4}{\Delta t^2}\Delta\mathbf{U}_n - \frac{4}{\Delta t}\dot{\mathbf{U}}_n - \ddot{\mathbf{U}}_n \tag{7.3.4a}$$

$$\dot{\mathbf{U}}_{n+1} = \frac{2}{\Delta t}\Delta\mathbf{U}_n - \dot{\mathbf{U}}_n \tag{7.3.4b}$$

7.3.2 Newton-Raphson Iteration

Because the equations of motion, the constraint conditions, and the Coulomb friction law are highly nonlinear, we may use the Newton-Raphson iteration to solve the incremental displacement vector and the contact nodal force vector. We set up an iteration scheme for the equations of motion (Eq. 7.3.1) as

$$\mathbf{M}\ddot{\mathbf{U}}_{n+1}^{(k)} + \mathbf{C}\dot{\mathbf{U}}_{n+1}^{(k)} + \mathbf{Fi}_{n+1}^{(k)} = \mathbf{Fe}_{n+1} + \mathbf{Fc}_{n+1}^{(k)} \tag{7.3.5}$$

for the kth iteration of the load step $n+1$. For the iteration, Eqs. 7.3.4a and 7.3.4b become, respectively,

$$\ddot{\mathbf{U}}_{n+1}^{(k)} = \frac{4}{\Delta t^2}\left[\sum_{j=1}^{k-1} \Delta\mathbf{U}^{(j)} + \Delta\mathbf{U}^{(k)}\right] - \frac{4}{\Delta t}\dot{\mathbf{U}}_n - \ddot{\mathbf{U}}_n \qquad (7.3.6)$$

$$\dot{\mathbf{U}}_{n+1}^{(k)} = \frac{2}{\Delta t}\left[\sum_{j=1}^{k-1} \Delta\mathbf{U}^{(j)} + \Delta\mathbf{U}^{(k)}\right] - \dot{\mathbf{U}}_n \qquad (7.3.7)$$

We assume the external nodal force vector **Fe** is a specified function of time, and Taylor's expansion gives

$$\begin{aligned} \mathbf{Fi}_{n+1}^{(k)} &\approx \mathbf{Fi}_{n+1}^{(k-1)} + \mathbf{K}_{n+1}^{(k-1)}\Delta\mathbf{U}^{(k)} \\ \Delta\mathbf{U}^{(k)} &\equiv \mathbf{U}_{n+1}^{(k)} - \mathbf{U}_{n+1}^{(k-1)} \end{aligned} \qquad (7.3.8)$$

and

$$\mathbf{Fc}_{n+1}^{(k)} \approx \mathbf{Fc}_{n+1}^{(k-1)} + \mathbf{Kc1}_{n+1}^{(k-1)}\Delta\mathbf{U}^{(k)} + \mathbf{Kc2}_{n+1}^{(k-1)}\Delta\mathbf{P}^{(k)} \qquad (7.3.9)$$

where $\Delta\mathbf{P}^{(k)}$ is the increment of the magnitude of the normal contact nodal force vector **P** with a component P^l at contact node l in a finite element model. The superscript means for the $(k-1)$th iteration of load step $n+1$, $\mathbf{K}_{n+1}^{(k-1)} = \partial\mathbf{Fi}/\partial\mathbf{U}_{at\,(k-1)}$ is the conventional global tangent stiffness matrix evaluated for the $(k-1)$th iteration, and $\mathbf{Kc1}_{n+1}^{(k-1)}$ and $\mathbf{Kc2}_{n+1}^{(k-1)}$ (to be derived in Eq. 7.3.10) are the global tangent stiffness matrices due to the contact nodal forces evaluated for the $(k-1)$th iteration. For the initial iteration, we set the force vectors and the stiffness matrices equal to those at the last iteration of the load step n.

We still try to use the three methods mentioned in Chapter 4 to establish the contact nodal force vector, the stiffness matrices due to the contact and constraint conditions.

7.3.3 Computing the Contact Nodal Force Vector by the Direct Method

Following Wang and Budiansky [1978] to compute the contact nodal force vector, we derive the contribution to the global tangent stiffness matrix for the kth iteration of load step $n+1$ as

$$\begin{aligned} \mathbf{Kc1}_{n+1}^{(k-1)} &= \frac{\partial\mathbf{Fc}}{\partial\mathbf{U}}_{\text{at}\,(k-1)} \\ \mathbf{Kc2}_{n+1}^{(k-1)} &= \frac{\partial\mathbf{Fc}}{\partial\mathbf{P}}_{\text{at}\,(k-1)} \end{aligned} \qquad (7.3.10)$$

To simplify the presentation, we use only the contact nodal force vector at a node l, and we may often omit the superscript or subscript l in the following derivations. For a sliding contact node, the contact nodal force vector from Eqs. 4.5.6 and 4.5.7 from Chapter 4 is

$$\begin{aligned}\mathbf{Fc} &= P\mathbf{n} - \mu P \frac{\dot{\mathbf{U}}_r}{\left\|\dot{\mathbf{U}}_r\right\|} \\ &= P\left[\left(S,_\alpha N^{-1} - \mu \dot{U}_\alpha V^{-1}\right)\mathbf{e}_\alpha - \left(N^{-1} + \mu S,_\alpha \dot{U}_\alpha V^{-1}\right)\right]\mathbf{e}_3\end{aligned} \tag{7.3.11a}$$

where S represents the tool surface in Eq. 7.1.8, $\mathbf{e}_i$ (i = 1, 2, 3) are unit vectors parallel to the global Cartesian coordinate axes, $\dot{\mathbf{U}}_r$ is the relative velocity vector, and $\mathbf{n}$ is the normal to the tool surface at the contact point. In Eq. 7.3.11a,

$$\dot{\mathbf{U}}_r = \dot{U}_\alpha \mathbf{e}_\alpha + S,_\alpha \dot{U}_\alpha \mathbf{e}_3 \tag{7.3.11b}$$

$$\mathbf{n} = \left(S,_\alpha \mathbf{e}_\alpha - \mathbf{e}_3\right)/N \tag{7.3.11c}$$

$$N = \left(1 + S,_\alpha S,_\alpha\right)^{1/2} \tag{7.3.11d}$$

$$V = \left\|\dot{\mathbf{U}}_r\right\| = \left[\dot{U}_\alpha \dot{U}_\alpha + \left(S,_\alpha \dot{U}_\alpha\right)^2\right]^{1/2} \neq 0 \tag{7.3.11e}$$

We note that the constraint condition $\dot{U}_3 + \dot{U}_t = S,_\alpha \dot{U}_\alpha$ in Eq. 7.1.9b has been invoked. Before taking the partial differentiation with respect to the displacement components U_α $(\alpha = 1, 2)$, we would like to show the complexity involved in establishing the tangent stiffness matrix. Let us differentiate the terms PN^{-1} and $\mu PS,_\alpha \dot{U}_\alpha V^{-1}$ in Eq. 7.3.11a to demonstrate

$$\frac{\partial PN^{-1}}{\partial U_\gamma} = -PS,_\alpha S,_{\alpha\gamma} N^{-3}, \quad \gamma = 1, 2 \tag{7.3.12}$$

and

$$\frac{\partial \mu PS,_\alpha \dot{U}_\alpha V^{-1}}{\partial U_\gamma} = \mu P\left[\left(\dot{U}_\alpha S,_\gamma - \dot{U}_\gamma S,_\alpha\right)\dot{U}_\alpha / \Delta t + S,_{\alpha\gamma} \dot{U}_\alpha \dot{U}_\beta \dot{U}_\beta\right]V^{-3} \tag{7.3.13}$$

Here, $\partial \dot{U}_\alpha / \partial U_\gamma = \delta_{\alpha\gamma} / \Delta t$ is based on the backward Euler integration. The displacement components U_α and the velocity components $\dot{U}_\alpha$ are for the contact node l. The partial derivatives include the second derivative $S,_{\alpha\gamma}$ and the nonlinear terms $\dot{U}_\alpha \dot{U}_\gamma$

and $\dot{U}_\alpha \dot{U}_\beta \dot{U}_\beta$. For a tool surface with a complex geometry to form automotive panels, the second derivatives of a surface are hard to obtain. Although we may model a tool surface by a set of flat elements and consequently $S_{,\alpha\gamma} = 0$ as mentioned in Chapter 4, nonlinear terms of unknown velocities still exist in the global tangent stiffness matrix. Therefore, they may cause problems that must be solved. Later in this book, we will use only a set of flat elements to model the tool surfaces.

Because there is another unknown magnitude P of the normal component of the contact nodal force vector at each contact node, we must have another equation to have a sufficient number of equations to solve the incremental global displacement vector $\Delta\mathbf{U}$ and the incremental contact nodal force vector $\Delta\mathbf{P}$ (for all contact nodes) in the normal direction. The additional equation is the constraint condition for the contact node l by taking the equal sign in Eq. 7.1.7,

$$Fct_l = g_l(\mathbf{x}) = (\mathbf{x} - \mathbf{x}_t) \cdot \mathbf{n} = 0 \tag{7.3.14}$$

where $\mathbf{x} = {}_0\mathbf{x} + \mathbf{u}$. Here, $\mathbf{u}$ is the displacement vector at the contact node. In Eq. 7.3.14, $\mathbf{x}_t$ is the tool position from Eq. 4.5.2a in Chapter 4, and $\mathbf{x}_t = x_\alpha \mathbf{e}_\alpha + \left[S(x_\alpha) - z(t)\right]\mathbf{e}_3$. Here, x_α are independent of time t because of the assumption of the rigid tool surface. In the following derivations, we will use the partial derivatives $\partial Fct_l / \partial U_i^l = n_i$ (with the assumption of contacting a flat element with the normal n_i) and $\partial Fct_l / \partial t = n_3 dz/dt$ (with $dz/dt = \dot{U}_t$) for the initial iteration and a traveling tool. Otherwise, $\partial Fct_l / \partial t = 0$.

Taylor's expansion of Eq. 7.3.14 for all contact nodes for the $(k-1)$th iteration of the step $n+1$ (subscript $n+1$ omitted for simplicity) gives

$$\mathbf{Fct}^{(k)} = \mathbf{Fct}^{(k-1)} + \left.\frac{\partial \mathbf{Fct}}{\partial \mathbf{U}}\right|_{\text{at }(k-1)} \Delta\mathbf{U} + \left.\frac{\partial \mathbf{Fct}}{\partial t}\right|_{\text{at }(k-1)} \Delta t = 0 \tag{7.3.15}$$

Now we extend the derivation to all nodes in the finite element model. Therefore, the vector $\mathbf{U}$ contains the generalized displacement for all nodes. For a stationary tool or any iteration other than the initial one, the term $\partial\mathbf{Fct}/\partial t = \mathbf{0}$. The constraint conditions contribute to the global tangent stiffness

$$\begin{aligned} \mathbf{Kct1}^{(k-1)} &= \frac{\partial \mathbf{Fct}}{\partial \mathbf{U}_{\text{at }(k-1)}} \\ \mathbf{Kct2}^{(k-1)} &= \frac{\partial \mathbf{Fct}}{\partial \mathbf{P}} = 0 \end{aligned} \tag{7.3.16}$$

Finally, the global tangent stiffness matrix has the form

$$\begin{bmatrix} \tilde{\mathbf{K}}^{(k-1)} - \mathbf{Kc1}^{(k-1)} & -\mathbf{Kc2}^{(k-1)} \\ -\mathbf{Kct1}^{(k-1)} & \mathbf{0} \end{bmatrix} \begin{bmatrix} \Delta\mathbf{U}^{(k)} \\ \Delta\mathbf{P}^{(k)} \end{bmatrix} = \begin{bmatrix} \mathbf{Fe} + \mathbf{Fc}^{(k-1)} - \mathbf{Fi}^{(k-1)} + \mathbf{R}^{(k-1)} \\ \mathbf{Fct}^{(k-1)} + \left.\dfrac{\partial \mathbf{Fct}}{\partial t}\right|_{\text{at }(k-1)} \Delta t \end{bmatrix} \tag{7.3.17}$$

where

$$\tilde{\mathbf{K}}^{(k-1)} = \frac{4}{\Delta t^2}\mathbf{M} + \frac{2}{\Delta t}\mathbf{C} + \mathbf{K}^{(k-1)} \tag{7.3.18}$$

$$\mathbf{R}^{(k-1)} = \mathbf{M}\left[-\frac{4}{\Delta t^2}\sum_{j=1}^{k-1}\Delta\dot{\mathbf{U}}^{(j)} + \frac{4}{\Delta t}\dot{\mathbf{U}}_n + \ddot{\mathbf{U}}_n\right] + \mathbf{C}\left[-\frac{2}{\Delta t}\sum_{j=1}^{k-1}\Delta\dot{\mathbf{U}}^{(j)} + \dot{\mathbf{U}}_n\right] \tag{7.3.19}$$

and $\mathbf{Kc1}^{(k-1)}$ and $\mathbf{Kc2}^{(k-1)}$ are from Eq. 7.3.10. The dynamic tangent stiffness matrix $\tilde{\mathbf{K}}^{(k-1)}$ contains the mass and damping matrices.

Although a combination of the implicit method of solution and the consistent mass matrix would minimize errors in computation, complications exist for the consistent mass matrix in Euler's equations of motion for large rotations, as mentioned in Section 7.1.1. We recommend that the diagonal mass matrix and the damping matrix be used, even for the implicit method. In fact, dynamic effect does not play an import role in simulations of sheet metal forming processes because of the low speed of the tool travel.

For sticking contact, the relative velocity vector $\dot{\mathbf{U}}_r = \dot{U}_\alpha \mathbf{e}_\alpha + \left(\dot{U}_3 + \dot{U}_t\right)\mathbf{e}_3 = 0$. Therefore, $\Delta U_\alpha = 0$ and $\Delta U_3 = -\dot{U}_t \Delta t$. Inserting these specified displacement increments into Eq. 7.3.17, we can solve for the remaining components of the incremental global displacement vector $\Delta\mathbf{U}$. This completes the direct method of solution of the contact problems by the direct method.

For the initial iteration $k = 1$, we set the unknown variables equal to those for the last iteration of the step n such as $\mathbf{x}_{n+1}^{(0)} = \mathbf{x}_n$ and $\mathbf{U}_{n+1}^{(0)} = \mathbf{U}_n$. We also assume each contact node contacts the same flat element as that in the previous time step for the initial iteration, and we will adjust the contact element (in the term $\mathbf{Fct}^{(k-1)}$ in Eq. 7.3.17) on the tool surface for the remaining iterations.

7.3.4 Computing the Contact Nodal Force Vector by the Lagrangian Multiplier Method

For the Lagrangian multiplier method, the contact nodal force vector from Eq. 4.5.22 in Chapter 4 at a contact point l is

$$\mathbf{Fc}^l(\mathbf{U}) = \lambda_l\left(\mathbf{n} - \mu\tau_l\mathbf{s}\right) \tag{7.3.20}$$

Its contribution to the global tangent stiffness matrix contains two parts:

$$\begin{aligned}\mathbf{Kc1}_{n+1}^{(k-1)} &= \frac{\partial \mathbf{Fc}}{\partial \mathbf{U}}_{\text{at }(k-1)} \\ \mathbf{Kc2}_{n+1}^{(k-1)} &= \frac{\partial \mathbf{Fc}}{\partial \mathbf{\Lambda}}_{\text{at }(k-1)}\end{aligned} \tag{7.3.21}$$

where **Fc** is the global contact nodal force vector, and the vector $\mathbf{\Lambda}$ contains λ and τ for all nodes (including those at node l). The first equation in Eq. 7.3.21 is identical to that in Eq. 7.3.10 from the direct method. It involves the second derivatives $S_{,\alpha\gamma}$ and the nonlinear terms such as $\dot{U}_\alpha \dot{U}_\gamma$ and $\dot{U}_\alpha \dot{U}_\beta \dot{U}_\beta$. If we use flat elements to model a tool surface, $S_{,\alpha\gamma}$ vanish; however, the nonlinear terms $\dot{U}_\alpha \dot{U}_\gamma$ and $\dot{U}_\alpha \dot{U}_\beta \dot{U}_\beta$ cannot be eliminated.

The constraint conditions also contribute to the global tangent stiffness matrix. The additional contribution by the constraint conditions expressed by $\mathbf{Fct}^l = \left[w(g_l, \lambda_l)\, v(\dot{u}_l, \tau_l)\right]^T$ from Eq. 4.5.23 in Chapter 4 at contact node l can be computed by

$$\begin{aligned}\mathbf{Kct1}^{(k-1)} &= \frac{\partial \mathbf{Fct}}{\partial \mathbf{U}}_{\text{at }(k-1)} \\ \mathbf{Kct2}^{(k-1)} &= \frac{\partial \mathbf{Fct}}{\partial \mathbf{\Lambda}}_{\text{at }(k-1)}\end{aligned} \tag{7.3.22}$$

where the subscript $n+1$ is omitted for simplicity, **Fct** is the global constraint condition vector, and the vector $\mathbf{\Lambda}$ contains λ and τ for all nodes (including those at node l). Approximate differentiable functions in Eqs. 4.5.20a and 4.5.20b in Chapter 4 are highly recommended to be used in Eq. 4.5.23 to compute the contribution to the global tangent stiffness matrix in Eq. 7.3.22. Another advantage to using the approximate equations for $v(\dot{u}, \tau)$ in Eq. 4.5.20b is that we may unify to treat sticking and sliding contacts. Taylor's expansion gives

$$\mathbf{Kct}^{(k)} = \mathbf{Kct}^{(k-1)} + \mathbf{Kct1}^{(k-1)}\Delta\mathbf{U} + \mathbf{Kct2}^{(k-1)}\Delta\mathbf{\Lambda} + \frac{\partial \mathbf{Fct}}{\partial t}\Delta t = 0 \tag{7.3.23}$$

We add the contributions from all nodes to the global tangent stiffness matrix, which has the final form for the kth iteration of the $(n+1)$th step

$$\begin{bmatrix} \tilde{\mathbf{K}}^{(k-1)} - \mathbf{Kc1}^{(k-1)} & -\mathbf{Kc2}_{n+1}^{(k-1)} \\ -\mathbf{Kct1}^{(k-1)} & -\mathbf{Kct2}^{(k-1)} \end{bmatrix} \begin{bmatrix} \Delta\mathbf{U}^{(k)} \\ \Delta\mathbf{\Lambda}^{(k)} \end{bmatrix} = \begin{bmatrix} \mathbf{Fe} + \mathbf{Fc}^{(k-1)} - \mathbf{Fi}^{(k-1)} + \mathbf{R}^{(k-1)} \\ \mathbf{Fct}^{(k-1)} + \dfrac{\partial \mathbf{Fct}}{\partial t}_{\text{at }(k-1)} \Delta t \end{bmatrix} \tag{7.3.24}$$

For a stationary tool or any iteration other than the initial one, the term $\partial \mathbf{Fct}/\partial t = \mathbf{0}$. The dynamic global tangent stiffness matrix and the force vectors are identical to those in

Eqs. 7.3.17 through 7.3.19. For the initial iteration $k = 1$, we set the unknown variables equal to those for the last iteration of the step n such as $\mathbf{x}_{n+1}^{(0)} = \mathbf{x}_n$ and $\mathbf{U}_{n+1}^{(0)} = \mathbf{U}_n$.

As mentioned, we may avoid the second derivatives of the tool surfaces in establishing the global tangent stiffness matrix. The nonlinear terms of the nodal velocities (the products of the velocity components) such as those in Eq. 7.3.13 must be included, and they would cause convergent problems during the Newton-Raphson iteration. The solution vector in Eq. 7.3.24 involves an additional vector $\Delta\mathbf{\Lambda}$ that contains $\Delta\lambda$ and $\Delta\tau$ for each contact node. In addition, the matrix is not symmetrical, and the diagonal terms of the matrix in Eq. 7.3.24 may contain zeros. Therefore, there may be difficulties for some linear simultaneous equation solvers. For these reasons, we do not think this is an appropriate method for solving contact problems in simulations of automotive sheet metal forming processes.

7.3.5 Computing the Contact Nodal Force Vector by the Penalty Method

For the penalty method, we compute the contact nodal force vector at a node l (we may omit the superscript or subscript l for simplicity)

$$\mathbf{Fc} = P\mathbf{n} + \mathbf{T} \tag{7.3.25}$$

For the kth iteration of the $(n+1)$th step, we rewrite Eq. 4.5.28 from Chapter 4 as

$$P_{n+1}^{(k)} \approx \alpha_N \left\langle -g\left(\mathbf{x}_{n+1}^{(k-1)}\right)\right\rangle \tag{7.3.26}$$

Later, the subscript $n + 1$ will be omitted in most equations of this section for simplicity. The gap function at a node is defined in Eq. 7.1.7 as

$$g(\mathbf{x}) = (\mathbf{x} - \mathbf{x}_t) \cdot \mathbf{n}$$

and the Macauley bracket is defined in Eq. 4.5.26b as

$$\langle -y \rangle = \frac{1}{2}\left[-y + |y|\right]$$

Using the state of sticking contact to start, we establish the trial frictional force vector as

$$\overline{\mathbf{T}}_{n+1}^{(k)} = \mathbf{T}_n + \alpha_T\left(\mathbf{U}_{r_{n+1}}^{(k-1)} - \mathbf{U}_{r_n}\right) \tag{7.3.27}$$

$$\overline{\Phi}_{n+1}^{(k)} = \left\|\overline{\mathbf{T}}_{n+1}^{(k)}\right\| - \mu P_{n+1}^{(k)} \tag{7.3.28}$$

where $\mathbf{U}_r$ is the relative displacement vector with respect to the tool position $\mathbf{x}_t$. Here, $\mathbf{U}_{r_{n+1}}^{(k-1)} - \mathbf{U}_{r_n} \approx \dot{\mathbf{U}}_r^{(k-1)}\Delta t$ where $\dot{\mathbf{U}}_r^{(k-1)}$ is the relative velocity from Eq. 7.3.11b. From the return mapping strategy discussed in Chapter 4, we rewrite Eq. 4.5.31a as

$$\mathbf{T}_{n+1}^{(k)} = \overline{\mathbf{T}}_{n+1}^{(k)} - \Delta\xi \frac{\overline{\mathbf{T}}_{n+1}^{(k)}}{\left\|\overline{\mathbf{T}}_{n+1}^{(k)}\right\|} \tag{7.3.29a}$$

where

$$\Delta\xi = \begin{Bmatrix} 0 & \text{if } \overline{\Phi}_{n+1} \le 0 \\ \dfrac{\overline{\Phi}_{n+1}}{\alpha_T} & \text{if } \overline{\Phi}_{n+1} > 0 \end{Bmatrix} \tag{7.3.29b}$$

For the initial iteration $k = 1$, we set the unknown variables equal to those for the last iteration of step n such as $\mathbf{x}_{n+1} = \mathbf{x}_n$ and $\mathbf{U}_{n+1}^{(0)} = \mathbf{U}_n$. Similar to the previous method to compute contact nodal forces, we also assume each contact node contacts the same flat element as that for the previous time step for the initial iteration, and we will make adjustment of the contact element on the tool surface for the remaining iterations.

For a tool surface with complex geometry to form automotive panels, the second derivatives of a surface are hard to obtain. We may model a tool surface by a set of flat elements as mentioned in Chapter 4 and consequently the second derivatives with respect to the coordinates, $S,_{\alpha\beta} = 0$, for a flat element. The Taylor expansion of the normal component of the contact nodal force at node l in the finite element model using Eq. 7.3.26 gives

$$p^{(k)}\mathbf{n}^l \approx \alpha_N \left[\left\langle -g^{(k-1)} \right\rangle + \frac{\partial\langle -g \rangle}{\partial \mathbf{U}}\bigg|_{\text{at }(k-1)} \Delta\mathbf{U} + \frac{\partial\langle -g \rangle}{\partial t}\bigg|_{\text{at }(k-1)} \Delta t \right]_{\text{for } l} \mathbf{n}^l \tag{7.3.30}$$

where the gap function g is defined at the node l in the finite element model, and $\mathbf{n}^l$ is the normal to a flat element on the tool surface model. (Note that $\mathbf{n}^l$ does not change for a flat element.) The terms of the partial derivatives with respect to $\mathbf{U}$ in Eq. 7.3.30 contribute to the global tangent stiffness matrix, and the final form is

$$\left[\tilde{\mathbf{K}} + \alpha_N \mathbf{G}\right]^{(k-1)} \left[\Delta\mathbf{U}^{(k)}\right] = \mathbf{Fe} + \mathbf{N}^{(k-1)} + \mathbf{T}^{(k)} - \mathbf{Fi}^{(k-1)} + \mathbf{R}^{(k-1)} \tag{7.3.31a}$$

where

$$\tilde{\mathbf{K}}^{(k-1)} = \frac{4}{\Delta t^2}\mathbf{M} + \frac{2}{\Delta t}\mathbf{C} + \mathbf{K}^{(k-1)}$$

$$\mathbf{R}^{(k-1)} = \mathbf{M}\left[-\frac{4}{\Delta t^2}\sum_{j=1}^{k-1}\Delta\dot{\mathbf{U}}^{(j)} + \frac{4}{\Delta t}\dot{\mathbf{U}}_n + \ddot{\mathbf{U}}_n\right] + \mathbf{C}\left[-\frac{2}{\Delta t}\sum_{j=1}^{k-1}\Delta\dot{\mathbf{U}}^{(j)} + \dot{\mathbf{U}}_n\right]$$

In Eq. 7.3.31a, $\mathbf{N}^{(k-1)}$ contains the normal component of the contact nodal force vector for all nodes in the finite element model, and $\mathbf{N}^{(k-1)}$ contains the terms

$$\alpha_N\left[\left\langle -g^{(k-1)}\right\rangle + \frac{\partial\langle -g\rangle}{\partial t}_{\text{at }(k-1)}\Delta t\right]_{\text{for } l}\mathbf{n}^l$$

which are the first and last terms on the right side of Eq. 7.3.30. Here, $\mathbf{T}^{(k)}$, computed by Eq. 7.3.29, contains the tangential component of the contact nodal force vector for all nodes. For a stationary tool or any iteration other than the initial one, the term $\partial\langle -g\rangle/\partial t$ in Eq. 7.3.30 equals 0. The global dynamic tangent stiffness matrix is evaluated by means of the values for the previous iteration. In Eq. 7.3.31a, $\mathbf{G}$ contains the normal component from the middle term on the right side of Eq. 7.3.30 of the contact nodal force vector. $\mathbf{G}$ is a diagonal matrix assembled from $\mathbf{G}^l$ (the superscript denotes the node number in the finite element mode) of

$$\mathbf{G}^l = \begin{bmatrix} \mathbf{n}^l\left(\mathbf{n}^l\right)^T & \mathbf{0}_1 \\ \mathbf{0}_2 & \mathbf{0}_3 \end{bmatrix}, \quad \langle -g\rangle \neq 0$$
$$\mathbf{G}^l = \mathbf{0}, \quad \langle -g\rangle = 0 \tag{7.3.31b}$$

where $\mathbf{0}_1$ is a 3 by $(m-3)$ zero matrix, $\mathbf{0}_2 = \left(\mathbf{0}_1\right)^T$, $\mathbf{0}_3$ is an $(m-3)$ by $(m-3)$ zero matrix, and m is the number of degrees of freedom for each finite element node. Note that the first three elements of the incremental displacement vector are for translations along the global coordinate axes. For any other arrangement of the elements in the incremental displacement vector, we may reshuffle the rows and columns in the matrix $\mathbf{G}^l$.

As mentioned in Chapter 4, there is no additional vector $\Delta\mathbf{P}$ or $\Delta\boldsymbol{\Lambda}$ in the global stiffness equations to be solved. Also, the global tangent stiffness matrix is symmetrical. However, we should choose the optimum penalty parameters α_N and α_T such that the solution would be accurate without any numerical problems.

7.3.6 Solving a Large Number of Simultaneous Equations

Most computation time that is consumed in the implicit method is in inverting the large dynamic tangent stiffness matrices that are not symmetrical for the contact problems formulated by the direct or Lagrangian multiplier method. For the penalty method, the large global tangent stiffness matrix can be made symmetrical. Therefore, it can be decomposed into

$$\mathbf{K} = \mathbf{LDL}^T \tag{7.3.32}$$

where **L** is a lower triangular matrix, and **D** is a diagonal matrix. After decomposition, it is easy to solve for the incremental displacement vector such as in Eq. 7.3.31a. Because a thin shell element is stiff in stretching but soft in bending, iterative methods for solution of the incremental displacement vector do not easily converge. The classic Gaussian elimination to decompose the matrix is possibly the best method. For the current generation of supercomputers with vector and parallel processors, a system with a million unknowns is still practical, even with one thousand total iterations for the simulation of a stretching/drawing forming a complex panel. On a massive parallel workstation with shared memory, the decomposition still could be done (see Gupta and Kumar [2001]).

7.3.7 Convergence of the Newton-Raphson Iteration

By inverting the global tangent stiffness matrix in Eq. 7.3.31a, for example, we can solve for the displacement increment $\Delta\mathbf{U}^{(k)}$ for the kth iteration of the $(n+1)$th load step. We update the internal, external, and contact nodal force vectors and check the convergence of the iteration by the following criteria:

$$\left\|\Delta\mathbf{U}^{(k)}\right\| < \varepsilon_1 \left\|\sum_{j=1}^{k} \Delta\mathbf{U}^{(j)}\right\| \tag{7.3.33a}$$

and

$$\left\|\mathbf{Rs}^{(k)}\right\| < \varepsilon_2 \text{Max}\left(\left\|\mathbf{Fe}^{(k)}\right\|, \left\|\mathbf{Fc}^{(k)}\right\|, \left\|\mathbf{Fi}^{(k)}\right\|, \left\|\mathbf{M}\ddot{\mathbf{U}}^{(k)}\right\|, \left\|\mathbf{C}\dot{\mathbf{U}}^{(k)}\right\|\right) \tag{7.3.33b}$$

where

$$\mathbf{Fc}^{(k)} = \mathbf{N}^{(k)} + \mathbf{T}^{(k)}$$

and ε_1 and ε_2 are some assigned small numbers with an order of 10^{-3} or smaller. Equation 7.3.31a gives

$$\left\|\mathbf{Rs}^{(k)}\right\| = \left\|\mathbf{Fe}_{n+1} + \mathbf{N}^{(k)} + \mathbf{T}^{(k)} - \mathbf{Fi}^{(k)} + \mathbf{R}^{(k)}\right\|$$

and

$$\mathbf{R}^{(k)} = \mathbf{M}\left[-\frac{4}{\Delta t^2}\sum_{j=1}^{k}\Delta\dot{\mathbf{U}}^{(j)} + \frac{4}{\Delta t}\dot{\mathbf{U}}_n + \ddot{\mathbf{U}}_n\right] + \mathbf{C}\left[-\frac{2}{\Delta t}\sum_{j=1}^{k}\Delta\dot{\mathbf{U}}^{(j)} + \dot{\mathbf{U}}_n\right]$$

In addition, we must check the constraint conditions that should be satisfied at each contact point. First, we compute how close a contact point should be to the tool surface. Secondly, we check any violation of the Coulomb friction law at each contact point. If any of the criteria with some small tolerance is violated, we go to the next iteration $k + 1$ until they are completely satisfied. When all the criteria and constraint conditions are satisfied, the Newton-Raphson iteration for this load step is convergent. After the convergence, we repeat the Newton-Raphson iteration for the next load step until the desired response is reached. A much larger time step than that for the explicit integration may be taken. Oscillations due to vibration or ringing cannot be avoided. When a large change of response occurs, there might be convergent problems in the iteration. However, this large change can be taken care of by the inertia term and the use of a smaller time increment in the numerical integration.

7.3.8 Updating Variables for Dynamic Implicit Integration

After the Newton-Raphson iteration to solve the equations of motion with contact constraint conditions converges for the mth iteration, we obtain the incremental displacement vector from the nth to the $(n+1)$th load step as

$$\Delta \mathbf{U}_n = \sum_{j=1}^{m} \Delta \mathbf{U}^{(j)} \tag{7.3.34}$$

Then we can update the displacement, velocity, and acceleration vectors

$$\mathbf{U}_{n+1} = \mathbf{U}_n + \Delta \mathbf{U}_n$$

$$\dot{\mathbf{U}}_{n+1} = \frac{2}{\Delta t} \Delta \mathbf{U}_n - \dot{\mathbf{U}}_n$$

$$\ddot{\mathbf{U}}_{n+1} = \frac{4}{\Delta t^2} \Delta \mathbf{U}_n - \frac{4}{\Delta t} \dot{\mathbf{U}}_n - \ddot{\mathbf{U}}_n$$

We may directly obtain the contact nodal force vector for the $(n+1)$th load step by using Eqs. 7.3.25 through 7.3.29 after updating the nodal displacement and velocity vectors. Using $\Delta \mathbf{U}_n$, we also can compute the incremental strains and the incremental stresses (Section 7.5 will discuss a better method) and can update the strains and stresses for all elements in the finite element model at the $(n+1)$th load step. Using the updated variables, we can continue to compute the solution for the next load step until the final load step is reached.

7.3.9 Summary of the Implicit Integration Method with Contact Nodal Forces Computed by the Penalty Method

Table 7.2 summarizes the procedures to compute the deformation and stresses in a sheet under a forming process using the dynamic implicit integration method with contact nodal forces computed by the penalty method.

TABLE 7.2
DYNAMIC IMPLICIT INTEGRATION METHOD WITH CONTACT NODAL FORCES COMPUTED BY THE PENALTY METHOD

1. Start at time $t = 0$ (step $n = 0$).
   a. Initialize the nodal displacement and velocity vectors.
   b. Initialize the element strain and stress vectors.
   c. Assign the penalty parameters α_N and α_T.
   d. Set the tool positions.
   e. Establish the mass and damping matrices $\mathbf{M}$ and $\mathbf{C}$.
2. Choose a time increment, and determine the increment of the tool travel Δz.
3. Start the iteration for the time step $n + 1$, and initialize the iteration number $k = 1$.
4. Set and compute the variables.
   a. Set $\mathbf{U}_{n+1}^{(k)} = \mathbf{U}_{n+1}^{(k-1)}$ and $\mathbf{x}_{n+1}^{(k)} = \mathbf{x}_{n+1}^{(k-1)}$ for $k > 1$. Otherwise, $\mathbf{U}_{n+1}^{(1)} = \mathbf{U}_n$ and $\mathbf{x}_{n+1}^{(1)} = \mathbf{x}_n$.
   b. Compute $P_{n+1}^{(k)} \approx \alpha_N \left\langle -g\left(\mathbf{x}_{n+1}^{(k-1)}\right)\right\rangle$ from Eq. 7.3.26. Note that $P_{n+1}^{(1)} = \alpha_N \left\langle -g\left(\mathbf{x}_n\right)\right\rangle$.
   c. Compute $\mathbf{T}_{n+1}^{(k)}$ from Eqs. 7.3.27 to 7.3.29. Note that $\mathbf{T}_{n+1}^{(1)} = \mathbf{T}_n$ and $\mathbf{T}_0 = \mathbf{0}$.
5. Establish the element tangent stiffness equations (see Chapter 6) using the consistent material tangent stiffness matrix in Eq. 6.2.31 or Eq. 6.3.16.
6. Assemble the element tangent stiffness equations to form the global tangent stiffness in Eq. 7.3.31.
7. Solve for the incremental nodal displacement vector $\Delta\mathbf{U}^{(k)}$ for the iteration k.

TABLE 7.2 ***(Continued)***

8. Update the following:
   a. The nodal displacement and velocity vectors.
   b. The element strain and stress vectors following Section 6.2.8 or Section 6.3.9 from Chapter 6.
   c. The contact nodal force vector using Eqs. 7.3.25 through 7.3.29 and the updated nodal displacement and velocity vectors.
9. Check for iteration convergence using the criteria defined in Eq. 7.3.33.
   a. Convergent: $n + 1$ become n, or $t + \Delta t$ become t. Use a more accurate method in Section 7.5 to update the stresses for the current time step, and go to Step 10.
   b. Not convergent: Set $k = k + 1$, and go to Step 4.
10. Check $t > t_{max}$ (the last time step) or if the tools are in the home position.
    a. If yes, then the simulation is done.
    b. If no, then go to Step 2.
11. Print the final results for the deformed position, strains, and stresses in the formed sheet metal part.

7.3.10 Application of Dynamic Implicit Integration to Sheet Metal Forming Analysis

The dynamic implicit integration method has the potential to overcome the difficulty of the convergence of the Newton-Raphson iteration. The difficulty is due to rigid body motion, rapid change of displacements, and/or the buckling or wrinkling that occur in the corresponding quasi-static solution. Although the time step used in solution can be much larger than that in the explicit integration, a smaller time step is still required for the convergence of the Newton-Raphson iteration if there is an abrupt change of response or an occurrence of buckling or wrinkling. Not many engineers use this method to simulate sheet metal forming processes for practical design. Researchers should further develop this method so that practical engineers can use the method for more accurate and reliable results.

7.4 Quasi-Static Solutions

Because the tool speed to form an automotive panel is relatively low, the inertia term can be neglected. The equations of motion become the equations of equilibrium. Therefore, a quasi-static problem can be formulated for a load step. Another advantage of neglecting the inertia term is that undesired oscillations of the response can be avoided. It also is justified by comparing the plastic work and the kinetic energy per unit volume of a sheet during a forming process. For example, we assume that the sheet material is steel and that the material follows a rigid perfect plastic model. For the yield stress $\sigma = 150$ MPa and a plastic strain of $\varepsilon = 0.1$, the energy dissipated by the plastic work for a volume of 1 mm^3 is

$$W_p = \sigma\varepsilon = 0.150 \times 10^{-1}\ \mathrm{N \cdot m}$$

For a sheet with a speed of 200 mm/sec (equal to the typical tool speed), the kinetic energy for a volume of 1 mm^3 is

$$E_k = \frac{1}{2} m V^2 = 0.158 \times 10^{-6}\ \mathrm{N \cdot m}$$

where $\rho = 7.87 \times 10^3 \mathrm{Kg/m^3}$ for the density of steel used. The kinetic energy is five orders of magnitude smaller than the energy dissipated from the plastic work. For this reason, the inertia effect usually can be neglected.

7.4.1 Equations of Equilibrium and Constraint Conditions

When we neglect the dynamic terms in Eq. 7.1.1, we obtain the equations of equilibrium in Eq. 7.1.13. Equation 7.1.13 is written here again

$$\mathbf{Fi} = \mathbf{Fe} + \mathbf{Fc} \tag{7.4.1}$$

with the contact constraint conditions in Eq. 7.1.7 and the frictional forces in Eq. 7.1.10. Because the equations of equilibrium are highly nonlinear for analyses of sheet metal forming processes, we must linearize the equations in an incremental form and use small load increments to obtain the desired solution by adding the incremental field quantities of each increment. We divide the load or the tool travel into several hundred steps. Similar to the dynamic analysis, we use the virtual time t equal to the tool travel distance. The equation of the tool surface at the virtual time t (with the dimension of length) becomes

$$x_3 = S\left(x_\alpha\right) - t \tag{7.4.2}$$

For a stationary tool surface, $t = 0$ in Eq. 7.4.2.

Using a virtual time increment or step size Δt and the properly specified boundary and initial conditions, we may compute the response at any virtual time $n\Delta t$ ($n = 1, 2, 3...$) until the desired solution is reached. During the computation, we may change the step size Δt for an accurate and efficient computation.

For each step, the equilibrium equations must be satisfied. For the $\left(n + 1\right)$th load step, the equilibrium equation can be written as

$$\mathbf{Fi}_{n+1} = \mathbf{Fe}_{n+1} + \mathbf{Fc}_{n+1}$$

For the global tangent stiffness matrix, we drop the dynamic terms of the mass and damping matrices in Eq. 7.3.31.

7.4.2 Boundary and Initial Conditions for Quasi-Static Analysis

Because no inertia forces are involved in the equations of equilibrium, a sheet must be properly supported to eliminate completely the rigid body motion. Therefore, sufficient boundary conditions must be specified. For the initial conditions, we assume the sheet is without any deformation until the punch or binder ring moves to press the sheet at the virtual time $t = 0^+$. The effect of the weight of a sheet is taken care of during the binder wrap stage.

7.4.3 Quasi-Static Solutions Without an Equilibrium Check

Because the global tangent stiffness matrix computed by the direct or Lagrangian multiplier method involves nonlinear terms of the velocity vector, the method of solution without iteration cannot be applied. Although we may use the velocity vector of the previous step, accuracy of the solution would be sacrificed.

We may try to establish the global tangent stiffness equations by the penalty method. We omit the dynamic terms in the global tangent stiffness matrix and drop the iteration number k in Eq. 7.3.31, so that

$$\tilde{\mathbf{K}} = \mathbf{K}$$

and

$$\mathbf{R} = 0$$

Eq. 7.3.31 is reduced to

$$\left[\mathbf{K} + \alpha_N \mathbf{G}\right]_n \left[\Delta \mathbf{U}_n\right] = \mathbf{Fe}_{n+1} + \mathbf{N}_n + \mathbf{T}_n - \mathbf{Fi}_n \tag{7.4.3}$$

We can solve for $\Delta \mathbf{U}_n$ when the global tangent stiffness matrix in Eq. 7.4.3 is not singular. If it is singular, the sheet may not be properly supported, or there may be a structural instability of buckling or wrinkling. We will discuss the treatment of structural stability in Chapter 8. Once $\Delta \mathbf{U}_n$ is found, we can use the procedure discussed in Section 7.2.4 to update all variables. Subsequently, we can continue to find the solution for the next step until the final solution is computed.

Because the contact nodal force vectors $\mathbf{N}_n$ and $\mathbf{T}_n$ on the right side of Eq. 7.4.3 are evaluated for the previous step, these equations in Eq. 7.4.3 are approximate in nature. Therefore, the incremental solution without equilibrium check is not a good approximation, and it is not recommended to apply to simulations of sheet metal forming processes.

7.4.4 Quasi-Static Solutions with an Equilibrium Check

Because the equilibrium equations in Eq. 7.4.1 are highly nonlinear, we cannot solve them directly. We use the Newton-Raphson iteration to compute the incremental solution from the nth load step (where all field quantities are known) to the $(n+1)$th load step. If the iteration is convergent, the equilibrium equations should be satisfied for the $(n+1)$th load step.

The iterative scheme is identical to those for the solutions of the dynamic tangent stiffness equations when we set the dynamic terms to zero. Taking the stiffness equations in Eq. 7.3.31 for the penalty method to compute the contact nodal force vector as an example, we set

$$\tilde{\mathbf{K}} = \mathbf{K}$$

and

$$\mathbf{R}^{(k-1)} = 0$$

Equation 7.3.31 is reduced to

$$\left[\mathbf{K} + \alpha_N \mathbf{G}\right]^{(k-1)} \left[\Delta\mathbf{U}^{(k)}\right] = \mathbf{Fe} + \mathbf{N}^{(k-1)} + \mathbf{T}^{(k)} - \mathbf{Fi}^{(k-1)} \tag{7.4.4}$$

for the kth iteration for the $(n+1)$th load step. Note that $\mathbf{T}^{(k)}$ can be expressed by the values for the $(k-1)$th iteration.

For simplicity, we may establish the tangent stiffness equations referred to a local coordinate system at contact node l, that is defined by the normal $\mathbf{n}^l$ of a flat element on the tool surface and two tangent vectors $\mathbf{t}_a$ (we may set $\mathbf{t}_1 = \mathbf{e}_2 \times \mathbf{n}^l$ if $\mathbf{n}^l$ is not parallel to $\mathbf{e}_2$ and $\mathbf{t}_2 = \mathbf{n}^l \times \mathbf{t}_1$). We transform the incremental displacement vector $\Delta\mathbf{U}$ at node l (referred to the global coordinate system) to $\Delta\bar{\mathbf{U}}$ (referred to the local coordinate system at node l) with the transformation matrix

$$\mathbf{Tr}_l = \begin{bmatrix} \mathbf{e}_1 \\ \mathbf{e}_2 \\ \mathbf{e}_3 \end{bmatrix} \cdot \begin{bmatrix} \mathbf{n}^l & \mathbf{t}_1 & \mathbf{t}_2 \end{bmatrix} \tag{7.4.5a}$$

such that

$$\Delta\mathbf{U} = \mathbf{Tr}_l \Delta\bar{\mathbf{U}} \tag{7.4.5b}$$

Applying this transformation to Eq. 7.4.4 for every contact node in the finite element model, we obtain the modified tangent stiffness equation at contact node l for the increments of only the three displacement components (not rotation or gradient), for example, as

$$\left[\bar{\mathbf{K}} + \alpha_N \bar{\mathbf{G}}\right]^{(k-1)} \left[\Delta \bar{\mathbf{U}}^{(k)}\right] = \bar{\mathbf{F}}\mathbf{e} + \bar{\mathbf{N}}^{(k-1)} + \bar{\mathbf{T}}^{(k)} - \bar{\mathbf{F}}\mathbf{i}^{(k-1)} \tag{7.4.6}$$

where (using Eq. 7.3.31b)

$$\bar{\mathbf{K}} = \left(\mathbf{Tr}_l\right)^T \mathbf{K}\mathbf{Tr}_l \tag{7.4.7a}$$

and

$$\bar{\mathbf{G}} = \begin{bmatrix} \bar{\mathbf{G}}_t & \mathbf{0}_1 \\ \mathbf{0}_2 & \mathbf{0}_3 \end{bmatrix}$$

$$\bar{\mathbf{G}}_t = \left(\mathbf{Tr}_l\right)^T \mathbf{n}^l \left(\mathbf{n}^l\right)^T \mathbf{Tr}_l' = \begin{bmatrix} 1 & 0 & 0 \\ 0 & 0 & 0 \\ 0 & 0 & 0 \end{bmatrix} \tag{7.4.7b}$$

$$\bar{\mathbf{F}}\mathbf{e} = \begin{bmatrix} Fe_{\mathbf{n}} \\ Fe_{\mathbf{t}_1} \\ Fe_{\mathbf{t}_2} \end{bmatrix} \tag{7.4.7c}$$

$$\bar{\mathbf{N}}^{(k-1)} = \begin{bmatrix} N^{(k-1)} \\ 0 \\ 0 \end{bmatrix} \tag{7.4.7d}$$

$$\bar{\mathbf{T}}^{(k)} = \begin{bmatrix} 0 \\ T_{\mathbf{t}_1}^{(k)} \\ T_{\mathbf{t}_2}^{(k)} \end{bmatrix} \tag{7.4.7e}$$

$$\bar{\mathbf{F}}\mathbf{i}^{(k-1)} = \begin{bmatrix} Fi_{\mathbf{n}}^{(k-1)} \\ Fi_{\mathbf{t}_1}^{(k-1)} \\ Fi_{\mathbf{t}_2}^{(k-1)} \end{bmatrix} \tag{7.4.7f}$$

Here, the subscripts $\mathbf{n}$, $\mathbf{t}_1$, and $\mathbf{t}_2$ represent the components of the vector in the direction of $\mathbf{n}$, $\mathbf{t}_1$, and $\mathbf{t}_2$, respectively. Note that

$$N^{(k-1)} = \alpha_N \left[\left\langle -g^{(k-1)} \right\rangle + \frac{\partial \langle -g \rangle}{\partial t}_{\text{at }(k-1)} \Delta t \right]$$

Therefore, for each contact node, only the first equation (the tangent stiffness in the normal direction) in Eq. 7.4.6 involves α_N. The other two equations are the tangent stiffnesses in the tangential directions $\mathbf{t}_1$ and $\mathbf{t}_2$, respectively. Note that all matrices and vectors are at a generic contact node l. All other tangent stiffness equations involve the vector $\Delta\mathbf{U}^l$ that must be transformed to $\Delta\bar{\mathbf{U}}$.

For the initial iteration of each load step, we set the frictional force vector as

$$\begin{aligned} \mathbf{T}_{n+1}^{(0)} &= \mathbf{T}_n \\ \mathbf{T}_0 &= \mathbf{0} \end{aligned} \tag{7.4.8}$$

and then update it by means of Eqs. 7.3.27 to 7.3.29 for the subsequent iterations. In the normal direction at contact node l, the terms with the penalty parameter α_N on both sides of the tangent stiffness equations dominate, as long as we pick a large number for α_N. This is equivalent to a problem with the normal component of the displacement specified at the contact node and solved by the penalty method in a book by Bathe [1996]. The specified displacement increment computed from $N^{(k-1)}$ in Eq. 7.4.7d is

$$\Delta\mathbf{U}^l \cdot \mathbf{n} = \left\langle -g^{(k-1)} \right\rangle + \frac{\partial \langle -g \rangle}{\partial t}_{\text{at }(k-1)} \Delta t \tag{7.4.9}$$

where $\mathbf{n}$ is the normal to the tool surface at contact node l, $\left\langle -g^{(k-1)} \right\rangle$ is the term to correct the contact node drifting away from the tool surface, and the term $\left(\partial\langle -g \rangle/\partial t\right)_{at\,(k-1)} \Delta t = -\dot{U}_t \Delta t\, \mathbf{e}_3 \cdot \mathbf{n}$ (using Eq. 7.1.7) is from the increment of the tool movement (downward) in a time interval Δt. For a stationary tool or any iteration other than the initial, this term $\left(\partial\langle -g \rangle/\partial t\right)_{at\,(k-1)} \Delta t$ vanishes.

This method is equivalent to a mixed boundary value problem with the displacement specified in the normal direction and the two force components specified along the two tangential directions. The advantage of this modified penalty method to solve a surface contact problem is that the global tangent stiffness matrix can be kept symmetrical. There are no additional variables involved in the solution vector, and the solution is almost exact for very large α_N that would not generate numerical problems in a linear simultaneous equation solver. This method has been used in MTLFRM developed by Ford Motor Company for simulations of sheet metal forming processes of complex automotive panels.

The boundary and initial conditions are discussed in Section 7.4.2. As mentioned in Section 7.4.3, if we drop the dynamic terms in the global tangent stiffness matrix, the matrix may become singular because of the rigid body motion (not properly supported) or a structural instability. Therefore, we fail to compute $\Delta\mathbf{U}$ for that load step. If the sheet is properly supported, the rigid body motion can be eliminated. In the next chapter, we will discuss how to deal with structural instability.

The convergence criteria for the iteration and the updating procedure also are identical to those in the dynamic implicit integration solution. After the convergence of iterations, the equilibrium in Eq. 7.4.1 is satisfied for the $(n+1)$th load step. Therefore, this solution should be much more reliable and accurate than that without iterations in the previous section.

7.4.5 Summary of the Quasi-Static Method with the Contact Nodal Force Vector Computed by the Penalty Method

Table 7.3 summarizes the procedures to compute the deformation and stresses in a sheet under a forming process using the quasi-static method with an equilibrium check and contact nodal forces computed by the penalty method.

TABLE 7.3
QUASI-STATIC METHOD WITH AN EQUILIBRIUM CHECK AND CONTACT NODAL FORCES COMPUTED BY THE PENALTY METHOD

1. Start at virtual time $t = 0$ (step $n = 0$).
   a. Initialize the nodal displacement vector.
   b. Initialize the element strain and stress vectors.
   c. Assign the penalty parameters α_N and α_T.
   d. Set the tool positions.
2. Choose a virtual time increment, and determine the increment of the tool travel Δz.
3. Start iteration for the time step $n + 1$, and initialize the iteration with the iteration number $k = 1$.
4. Set variables $\mathbf{U}_{n+1}^{(k)} = \mathbf{U}_{n+1}^{(k-1)}$ and $\mathbf{x}_{n+1}^{(k)} = \mathbf{x}_{n+1}^{(k-1)}$ for $k > 1$.
   Otherwise, $\mathbf{U}_{n+1}^{(1)} = \mathbf{U}_n$ and $\mathbf{x}_{n+1}^{(1)} = \mathbf{x}_n$. Establish the element tangent stiffness equations using the consistent material tangent stiffness matrices in Eq. 6.2.31 or Eq. 6.3.26.
5. Assemble the element tangent stiffness equations to form the global tangent stiffness in Eq. 7.4.4.

TABLE 7.3 *(Continued)*

6. Establish the local coordinate system defined by the normal of the tool surface $\mathbf{n}^l$ and two tangent vectors $\mathbf{t}_1$ and $\mathbf{t}_2$ (we may set $\mathbf{t}_1 = \mathbf{e}_2 \times \mathbf{n}^l$, if $\mathbf{n}^l$ is not parallel to $\mathbf{e}_2$ and $\mathbf{t}_2 = \mathbf{n}^l \times \mathbf{t}_1$) for each contact node. Transform the tangent stiffness equations to the local coordinate system for each contact node. Note that it is equivalent to a mixed boundary value problem with the displacement increment specified in the normal direction and the forces specified in the two tangent directions. The forces are due to friction. The frictional force vector is approximated by $\mathbf{T}_{n+1}^{(k)}$ from Eq. 7.3.27 to 7.3.29. Note that $\mathbf{T}_{n+1}^{(1)} = \mathbf{T}_n$ and $\mathbf{T}_0 = \mathbf{0}$ for the initial iteration.
7. Solve for the transformed nodal displacement increment $\Delta\bar{\mathbf{U}}_n$ in Eq. 7.4.6. Transform $\Delta\bar{\mathbf{U}}_n$ back to $\Delta\mathbf{U}_n$ using Eq. 7.4.5.
8. Update the following:
   a. The nodal displacement and velocity vectors.
   b. The element strain and stress vectors following the discussion in Sections 6.2.8 and 6.3.9 from Chapter 6.
   c. The contact nodal force vector at every contact node.
9. Check for iteration convergence using the criteria defined in Eq. 7.3.33.
   a. Convergent: $n + 1$ becomes n or $t + \Delta t$ becomes t. Use a more accurate method in Section 7.5 to update the stresses for the current time step, and go to Step 10.
   b. Not convergent: Set $k = k + 1$, and go to Step 4.
10. Check $t > t_{\max}$ (the last time step) or if the tools are in the home position.
    a. If yes, then the simulation is done.
    b. If no, then go to Step 2.
11. Print the final results for deformed position, strains, and stresses in the formed sheet metal part.

7.4.6 Application of the Quasi-Static Method to Sheet Metal Forming Analysis

In this section, we discuss the quasi-static method with an equilibrium check applied to sheet metal forming analyses. It is suitable to compute the shape of a sheet during the binder wrap stage. Because the solution can avoid the artificial inertia effect and undesired oscillations, the forming stresses obtained from the quasi-static method with equilibrium check should be reliable. The stresses can be used for computing springback after a sheet is released from the forming tools.

One advantage of the method is that there are no artificial parameters to adjust. Another advantage is that the solution is more reliable if it can be obtained. The disadvantages include the numerical problems of convergence in the equilibrium check and the excessive computing time for a large finite element model of a sheet. In this method, a large linear system of equations from the global tangent stiffness matrix must be solved for each iteration. For a typical automotive panel, we use 150 load increments for the quasi-static solution. Each load step needs about five Newton-Raphson iterations for solution convergence. We are required to solve the large linear system of equations 750 times. For present-generation computers with vector and parallel processors, the computing time for solving this large system 750 times still is practical.

When the Newton-Raphson iteration is used to solve a surface contact problem in a sheet metal forming analysis, severe discontinuities can occur, such as nodes sliding out of contact, abrupt changes in the direction of the frictional force, stress unloading, and changes in the draw-bead force from tension to compression in the draw-bead model. Therefore, it sometimes is difficult for the Newton-Raphson iteration to converge. Tang and Carnes [1995] proposed some methods to circumvent the convergence difficulty in the iteration by smoothing out these discontinuities. During forming processes, instability due to wrinkling and buckling also can occur, despite the fact that most areas on the sheet are being stretched. At the onset of instability, the tangent stiffness matrix is singular. Therefore, the incremental displacement vector cannot be computed by the iteration. In Chapter 8, we will discuss some methods to compute solutions at the onset of buckling or wrinkling and continue to post-buckling or post-wrinkling regions.

7.5 Integration of Constitutive Equations

In Sections 3.6 through 3.9 of Chapter 3, we formulate the constitutive equations applied to simulations of sheet metal forming processes. In this section, we discuss methods to solve these equations by numerical integration. After we find the increment of the displacement vector and then the increment of the strain vector by solving the equations of motion or equilibrium for a load increment or a tool travel from time t to $t + \Delta t$ (from the load step n to $n + 1$), we compute the increment of the stress vector for the load step by numerical integration and then update the stress vector at time $t + \Delta t$.

The simplest way to compute the increment of the stress vector for a given increment of strain vector is by subdividing the increment into many sub-intervals. The total increment of the stress vector is the summation of the increment for each sub-interval. This is Euler's method for numerical integration of a system of ordinary differential equations. To reduce drift from the true solution, the sub-interval should be sufficiently fine. From our experience, because the increment of the strain vector for even a small tool movement can be large in the quasi-static formulation with the implicit method for simulations of sheet metal forming processes, we should take as many as 200 sub-intervals for accurate computation of the increment of the stress vector for a load step (time interval). In addition, there is no guarantee that the yield function at time $t + \Delta t$ is satisfied for the updated stress vector, even when we set a very small sub-interval. We may use the sub-interval method to estimate the increment of the stress vector during the Newton-Raphson iteration. After the convergence of iterations, we should use a better method to update the stress vector using the increment of the strain vector from time t to $t + \Delta t$ computed by the Newton-Raphson iteration. We may use a much smaller number of sub-intervals (e.g., no more than 10, based on the magnitude of the strain increment) to estimate the increment of the stress vector in each Newton-Raphson iteration.

Using the radial return method in a book by Belytschko *et al.* [2000] to update the stress vector should satisfy the yield function at time $t + \Delta t$. Although a nonlinear equation with a single variable must be solved with a small number of the Newton-Raphson iterations, the computation time for the radial return method is much less than that of the sub-interval method.

We briefly describe the radial return method in this paragraph. The radial return method to update the stress vector with a given incremental strain vector is better explained with an isotropic material under both elastic and plastic deformation in the three-dimensional stress state. For the three-dimensional stress problems for plastically isotropic materials, we recall the yield function in terms of the stress deviatoric vector $\mathbf{s}$ as

$$f = \frac{3}{2}\mathbf{s}^T\mathbf{P}_6\mathbf{s} - X^2 = 0 \tag{7.5.1}$$

where

$$\mathbf{s} = \begin{bmatrix} s_{11} & s_{22} & s_{33} & s_{12} & s_{23} & s_{31} \end{bmatrix}^T$$

$$\mathbf{P}_6 = \begin{bmatrix} \mathbf{I}_3 & \mathbf{0} \\ \mathbf{0} & 2\mathbf{I}_3 \end{bmatrix}$$

Here, $\mathbf{I}_3$ is a 3 by 3 identity matrix.

The stress vector at time $t + \Delta t$ is

$$\boldsymbol{\sigma} = \boldsymbol{\sigma}_n + \mathbf{D}_e\left(\Delta \quad - \Delta\lambda\frac{\partial f}{\partial \sigma}\right) = \boldsymbol{\sigma}^{tr} - 3\Delta\lambda\mathbf{D}_e\mathbf{P}_6\mathbf{s} \tag{7.5.2}$$

where $\boldsymbol{\sigma}_n$ is the stress vector at time t (step n), $\Delta\boldsymbol{\varepsilon}$ is the incremental strain vector (including both elastic and plastic contributions) from time t to $t + \Delta t$, and $\mathbf{D}_e$ is the elastic material tangent stiffness matrix (elasticity matrix), which is obtained from Eq. 3.1.28 as

$$\mathbf{D}_e = \frac{E}{1+\nu}\begin{bmatrix} \left(1+\frac{\nu}{1-2\nu}\right) & \frac{\nu}{1-2\nu} & \frac{\nu}{1-2\nu} & 0 & 0 & 0 \\ & \left(1+\frac{\nu}{1-2\nu}\right) & \frac{\nu}{1-2\nu} & 0 & 0 & 0 \\ & & \left(1+\frac{\nu}{1-2\nu}\right) & 0 & 0 & 0 \\ & sym. & & \frac{1}{2} & 0 & 0 \\ & & & & \frac{1}{2} & 0 \\ & & & & & \frac{1}{2} \end{bmatrix}$$

Here, $\Delta\lambda = \Delta\bar{\varepsilon}^p / 2X$ from Eq. 5.5.47b (take the finite difference form) to solve for $\Delta\varepsilon_p$ and X. Also, the trial elastic stress vector $\boldsymbol{\sigma}^{tr} = \boldsymbol{\sigma}_n + \mathbf{D}_e \Delta\boldsymbol{\varepsilon}$. Using the elastic tangent stiffness matrix and $s_{ii} = 0$, it can be shown that

$$\mathbf{D}_e \mathbf{P}_6 \mathbf{s} = \frac{E}{1+\nu}\mathbf{s}$$

Therefore, Eq. 7.5.2 becomes

$$\boldsymbol{\sigma} = \boldsymbol{\sigma}^{tr} - \frac{3E\Delta\lambda}{1+\nu}\mathbf{s} \tag{7.5.3}$$

Converting the stress vector in Eq. 7.5.3 to the deviatoric stress vector and using $s_{ii} = 0$, we obtain

$$\mathbf{s} = \mathbf{s}^{tr} - \frac{3E\Delta\lambda}{1+\nu}\mathbf{s} \tag{7.5.4a}$$

or

$$\left(1 + \frac{3E\Delta\lambda}{1+\nu}\right)\mathbf{s} = \mathbf{s}^{tr} \tag{7.5.4b}$$

Equation 7.5.4b shows that these two deviatoric stress vectors are parallel. Let **n** be the normal vector that coincides with the position vector $\mathbf{s}^{tr}$ in the deivatoric stress space, as shown in Figure 7.1. Therefore, we may set

$$\mathbf{n} = \frac{\mathbf{s}^{tr}}{\left\|\mathbf{s}^{tr}\right\|} = \frac{\mathbf{s}}{\left\|\mathbf{s}\right\|} \tag{7.5.5}$$

because the position vector **s** to be solved for is parallel to $\mathbf{s}^{tr}$, and **s** is in the normal direction from the flow rule. Equation 7.5.5 gives the unknown position vector

$$\mathbf{s} = \left\|\mathbf{s}\right\|\mathbf{n} \tag{7.5.6}$$

where $\left\|\mathbf{s}\right\|$ is the radius of the yield surface in the deviatoric stress space at time $t + \Delta t$, and it is $\sqrt{2/3}X$. Therefore, Eq. 7.5.4a becomes (using $\Delta\lambda = \Delta\bar{\varepsilon}^p / 2X$)

$$\mathbf{s} = \mathbf{s}^{tr} - \frac{E}{1+\nu}\sqrt{\frac{3}{2}}\Delta\varepsilon_p \mathbf{n} \tag{7.5.7}$$

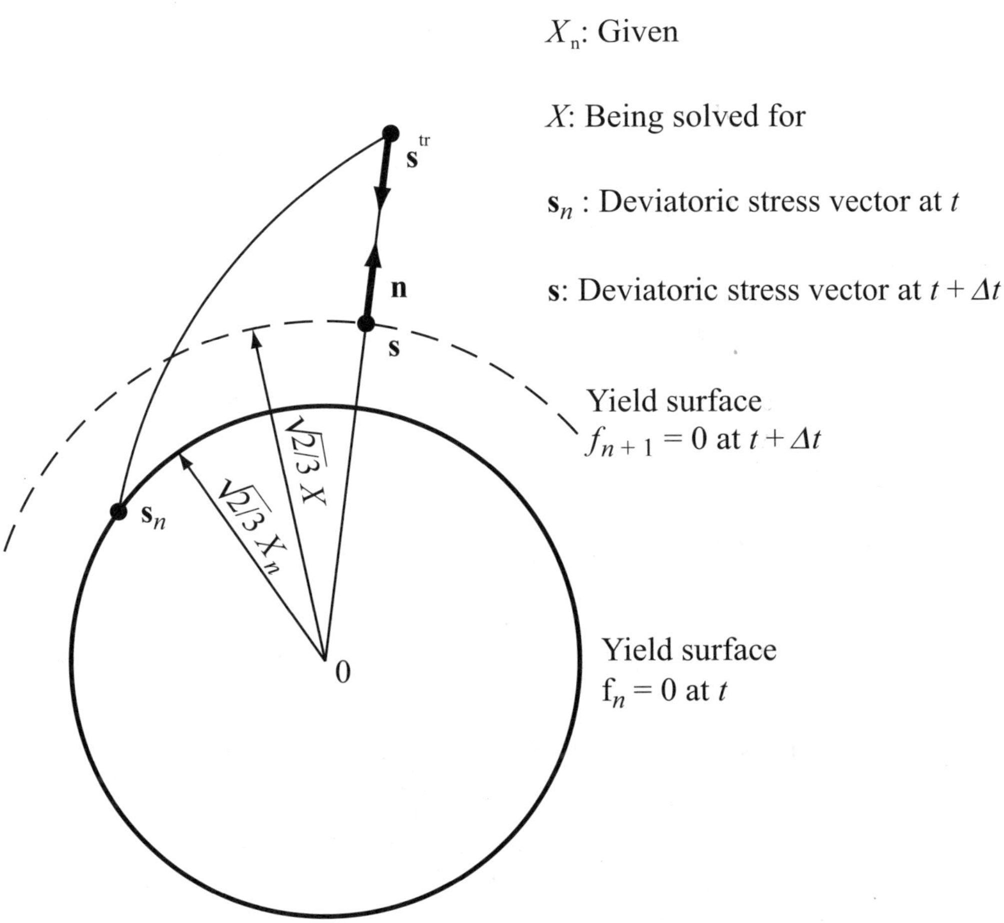

Figure 7.1 Radial return in the deviatoric stress space.

Because the correction vector $-E\sqrt{3/2}\Delta\bar{\varepsilon}^p\mathbf{n}/(1+\nu)$ is in the radial direction, the method thus is named "the radial return method." Although the direction of the correction vector is known, its magnitude in proportion to $\Delta\bar{\varepsilon}^p$ is still unknown and can be computed through solving a nonlinear equation that is obtained by inserting Eq. 7.5.7 into the yield function at time $t + \Delta t$ or for step $n + 1$.

Because we include the transversely anisotropic property of a sheet, there is no advantage to using the deviatoric stress components in the yield criterion. As mentioned, this formulation is for simulating sheet metal forming processes. Therefore, the large (finite) strain must be included. A metal sheet generally can be modeled by thin shell elements under plane stress conditions. Our analysis uses the Cauchy stress tensor referred to a co-rotational rectangular Cartesian coordinate system as shown in Figure 7.2 at a point in a shell element. For the curved triangular thin shell element as described in Chapter 5, two base vectors $\mathbf{g}_1$ and $\mathbf{g}_2$ are defined at an integration point on a parallel surface at a distance of $h\theta^3$ from the deformed reference surface. We establish a local Cartesian coordinate system with two orthonormal tangent vectors $\bar{\mathbf{e}}_1$ and $\bar{\mathbf{e}}_2$ to the surface and the surface normal $\mathbf{n}$ (also normal to the deformed reference surface) as shown in Figure 7.2, where $\varphi = \pi/4 - \alpha/2$ if $\alpha < \pi/2$, and $\varphi = \alpha/2 - \pi/4$ if $\alpha > \pi/2$. This Cartesian coordinate system is approximately the co-rotational coordinate system.

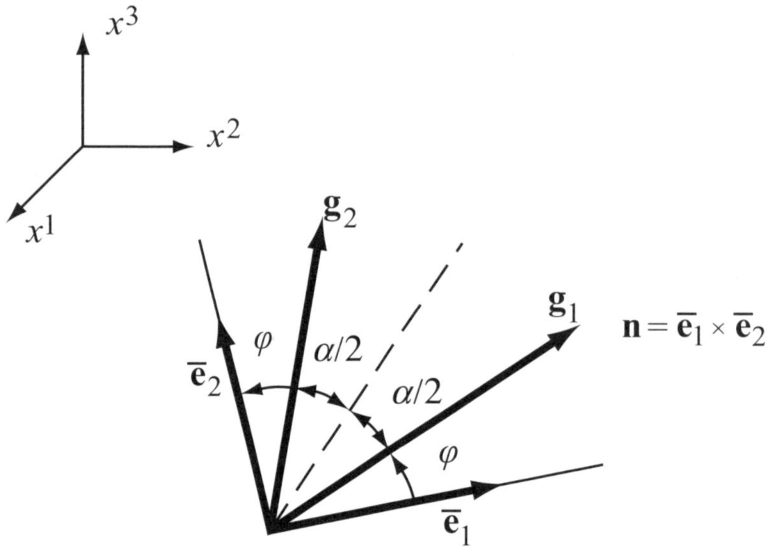

Figure 7.2 A co-rotational Cartesian coordinate system, where $\bar{\mathbf{e}}_1$ and $\bar{\mathbf{e}}_2$ are orthonormal.

7.5.1 Integration of Rate-Insensitive Plane Stress Constitutive Equations with Isotropic Hardening

In reference to the co-rotational coordinate system defined by the vectors $\bar{\mathbf{e}}_1$, $\bar{\mathbf{e}}_2$, and **n**, Eq. 3.9.1 gives the yield function for transversely anisotropic plastic materials with the anisotropy parameter R (not necessarily equal to 1) as

$$f(\boldsymbol{\sigma}) = \sigma_{11}^2 + \sigma_{22}^2 - \frac{2R}{1+R}\sigma_{11}\sigma_{22} + \frac{2(1+2R)}{1+R}\sigma_{12}^2 - X^2 = 0 \tag{7.5.8}$$

where X is the effective flow stress and the Cauchy stress vector $\boldsymbol{\sigma}$ $(\sigma_{11}, \sigma_{22}, \sigma_{12})$ referred to the co-rotational coordinate system. The yield function in the matrix form is

$$f(\boldsymbol{\sigma}) = \boldsymbol{\sigma}^T \mathbf{P} \boldsymbol{\sigma} - X^2 = 0 \tag{7.5.9a}$$

where the matrix **P** is

$$\mathbf{P} = \begin{bmatrix} 1 & -\dfrac{R}{1+R} & 0 \\ -\dfrac{R}{1+R} & 1 & 0 \\ 0 & 0 & \dfrac{2(1+2R)}{1+R} \end{bmatrix} \tag{7.5.9b}$$

The effective stress $\bar{\sigma}$ (or X) is defined as

$$\bar{\sigma} = \left(\boldsymbol{\sigma}^T \mathbf{P}\boldsymbol{\sigma}\right)^{1/2} \tag{7.5.9c}$$

If the incremental strain vector $\Delta\boldsymbol{\varepsilon}$ $(\Delta\varepsilon_{11}, \Delta\varepsilon_{22}, 2\Delta\varepsilon_{12})$ is given, the stress vector can be computed as

$$\boldsymbol{\sigma} = \boldsymbol{\sigma}_n + \mathbf{D}_e\left(\Delta\boldsymbol{\varepsilon} - \Delta\lambda\frac{\partial f}{\partial \boldsymbol{\sigma}}\right) \tag{7.5.10}$$

where the elasticity matrix for the plane stress state is

$$\mathbf{D}_e = \frac{E}{1-\nu^2}\begin{bmatrix} 1 & \nu & 0 \\ \nu & 1 & 0 \\ 0 & 0 & \dfrac{1-\nu}{2} \end{bmatrix}$$

Here, $\Delta\lambda = \Delta\bar{\varepsilon}^p/2X$ is from Eq. 5.5.47b in Chapter 5, and $\Delta\bar{\varepsilon}^p$ is the increment of the effective plastic from the state of stress $\boldsymbol{\sigma}_n$ to $\boldsymbol{\sigma}$. Let the trial stress vector $\boldsymbol{\sigma}^{tr}$ be

$$\boldsymbol{\sigma}^{tr} = \boldsymbol{\sigma}_n + \mathbf{D}_e\Delta\boldsymbol{\varepsilon} \tag{7.5.11}$$

We check this trial stress vector to see whether $f\left(\boldsymbol{\sigma}^{tr}\right) \le 0$ in Eq. 7.5.8. If it is true, this is the solution, and the stress state is elastic. Otherwise, we combine Eqs. 7.5.10 and 7.5.11 to give

$$\boldsymbol{\sigma} = \boldsymbol{\sigma}^{tr} - \Delta\lambda\mathbf{D}_e\frac{\partial f}{\partial \boldsymbol{\sigma}} = \boldsymbol{\sigma}^{tr} - \frac{\Delta\bar{\varepsilon}^p}{2X}\mathbf{D}_e\frac{\partial f}{\partial \boldsymbol{\sigma}} \tag{7.5.12}$$

Partial differentiation of Eq. 7.5.9a with respect to the stress vector gives

$$\frac{\partial f}{\partial \boldsymbol{\sigma}} = 2\mathbf{P}\boldsymbol{\sigma} \tag{7.5.13}$$

Combining Eqs. 7.5.12 and 7.5.13, we obtain

$$\boldsymbol{\sigma} = \boldsymbol{\sigma}^{tr} - \frac{\Delta\bar{\varepsilon}^p}{X}\mathbf{D}_e\mathbf{P}\boldsymbol{\sigma} \tag{7.5.14}$$

and solve for $\boldsymbol{\sigma}$ in the following equation as

$$\left(\mathbf{I} + \frac{\Delta\bar{\varepsilon}^p}{X}\mathbf{D}_e\mathbf{P}\right)\boldsymbol{\sigma} = \boldsymbol{\sigma}^{tr} \tag{7.5.15}$$

Let

$$\boldsymbol{\sigma} = \bar{\mathbf{M}}\boldsymbol{\sigma}^{tr} \tag{7.5.16a}$$

where

$$\bar{\mathbf{M}} = \left(\mathbf{I} + \frac{\Delta\bar{\varepsilon}^p}{X}\mathbf{D}_e\mathbf{P}\right)^{-1} \tag{7.5.16b}$$

Inserting Eq. 7.5.16 into Eq. 7.5.9a yields

$$f\left(\Delta\bar{\varepsilon}^p\right) = \left(\bar{\mathbf{M}}\boldsymbol{\sigma}^{tr}\right)^T \mathbf{P}\left(\bar{\mathbf{M}}\boldsymbol{\sigma}^{tr}\right) - X^2 = 0 \tag{7.5.17}$$

where X is evaluated at $\bar{\varepsilon}^p + \Delta\bar{\varepsilon}^p$. This is a nonlinear equation for $\Delta\bar{\varepsilon}^p$, which is the only unknown because the total effective plastic strain $\bar{\varepsilon}^p$ is given at the end of the previous step n. We may use the Newton-Raphson method to solve for the unknown. Using the Newton-Raphson method to solve this nonlinear equation requires the derivative $df/d\Delta\bar{\varepsilon}^p$ that can be computed as

$$\frac{df}{d\Delta\bar{\varepsilon}^p} = \left(\frac{\partial f}{\partial\boldsymbol{\sigma}}\right)^T \frac{d\boldsymbol{\sigma}}{d\Delta\bar{\varepsilon}^p} - 2XX' \tag{7.5.18}$$

where $X' = df/d\Delta\bar{\varepsilon}^p$. Differentiation of Eq. 7.5.15 and using Eq. 7.5.16b gives

$$\frac{d\boldsymbol{\sigma}}{d\Delta\bar{\varepsilon}^p} = -\frac{\left(1 - \frac{X'}{X}\Delta\bar{\varepsilon}^p\right)}{X}\bar{\mathbf{M}}\mathbf{D}_e\mathbf{P}\bar{\mathbf{M}}\boldsymbol{\sigma}^{tr} \tag{7.5.19}$$

Substituting Eqs. 7.5.13, 7.5.16a, and 7.5.19 into Eq. 7.5.18 yields

$$\frac{df}{d\Delta\bar{\varepsilon}^p} = -2\frac{(1 - \frac{X'}{X}\Delta\bar{\varepsilon}^p)}{X}(\mathbf{P}\bar{\mathbf{M}}\boldsymbol{\sigma}^{tr})^T\bar{\mathbf{M}}\mathbf{D}_e\mathbf{P}\bar{\mathbf{M}}\boldsymbol{\sigma}^{tr} - 2XX' \tag{7.5.20}$$

The Newton-Raphson iteration is set for the kth iteration

$$\Delta\bar{\varepsilon}^{p(k)} = \Delta\bar{\varepsilon}^{p(k-1)} + \delta\bar{\varepsilon}^{p(k)} \tag{7.5.21a}$$

where

$$\delta\bar{\varepsilon}^{p(k)} = -\frac{f\left(\bar{\varepsilon}^{p} + \Delta\bar{\varepsilon}^{p(k-1)}\right)}{\left(\dfrac{df}{d\Delta\bar{\varepsilon}^{p}}\right)_{\text{at }(k-1)}} \tag{7.5.21b}$$

if $\left(df/d\Delta\bar{\varepsilon}^{p}\right)_{at\ (k-1)} \neq 0$. Following a suggestion by Tang *et al.* [2001], we may set a better initial guess of the effective plastic strain $\Delta\bar{\varepsilon}^{p(0)}$. Table 7.4 shows the procedure to estimate $\Delta\bar{\varepsilon}^{p(0)}$.

TABLE 7.4
ESTIMATION OF $\Delta\bar{\varepsilon}^{p(0)}$ FOR FAST CONVERGENCE OF THE NEWTON-RAPHSON ITERATION

1. Given the effective plastic strain $\bar{\varepsilon}^{p}$, the effective stress $\bar{\sigma}$, and the incremental strain (elastic + plastic) vector $\Delta\varepsilon_{\alpha\beta}$

2. Take the finite difference form of Eq. 3.9.13 from Chapter 3 to compute "the effective strain increment"

$$\Delta\bar{\varepsilon}_{tr} = \frac{1+R}{(1+2R)^{1/2}}\left[(\Delta\varepsilon_{11})^2 + (\Delta\varepsilon_{22})^2 + \frac{2R}{1+R}\Delta\varepsilon_{11}\Delta\varepsilon_{22}\frac{2}{1+R}(\Delta\varepsilon_{12})^2\right]^{1/2} \tag{T7.4.1}$$

3. Estimate the updated effective strain

$$\bar{\varepsilon}_{tr} = \left(\frac{\bar{\sigma}}{E} + \bar{\varepsilon}^{p}\right) + \Delta\bar{\varepsilon}_{tr} \tag{T7.4.2}$$

4. Update the estimated effective stress $\bar{\sigma}_{tr}$ from the uniaxial stress-strain equation, for example in Eq. 3.6.29 from Chapter 3 as

$$\bar{\sigma}_{tr} = K\bar{\varepsilon}_{tr}^{n} \tag{T7.4.3}$$

5. Estimate the updated elastic strain

$$\bar{\varepsilon}_{tr}^{e} = \frac{\bar{\sigma}_{tr}}{E} \tag{T7.4.4}$$

6. Update the estimated plastic strain

$$\bar{\varepsilon}_{tr}^{p} = \bar{\varepsilon}_{tr} - \bar{\varepsilon}_{tr}^{e} \tag{T7.4.5}$$

7. Take the difference

$$\Delta\bar{\varepsilon}_{tr}^{p} = \bar{\varepsilon}_{tr}^{p} - \bar{\varepsilon}^{p} \tag{T7.4.6}$$

which is a better initial guess $\Delta\bar{\varepsilon}^{p(0)}$

The iteration process normally takes fewer than five cycles for convergence. After we compute a better value of the effective plastic strain increment, we can compute the stress vector from Eq. 7.5.16a. Finally, we can update the stresses and plastic strains for each integration point in the finite element model and are ready to go to the next step. Note that we use the values at time t as the initial values for the first iteration for computing the effective plastic strain increment at time $t + \Delta t$.

7.5.2 Integration of Rate-Insensitive Plane Stress Constitutive Equations with Anisotropic Hardening

We have chosen one of the anisotropic hardening rules to illustrate the numerical integration to obtain the stress vector with a given incremental strain vector and the strain history. This anisotropic hardening rule may not be the best rule to predict the stresses. However, the hardening rule is general enough to demonstrate the proposed algorithm in the following discussion. The proposed algorithm then can be used for other anisotropic hardening rules.

In Chapter 3, we briefly introduce the hardening rule proposed by Mroz [1967] and generalized by Chu [1984]. With consideration of Chu's generalization, Hill's quadratic orthotropic yield criterion [1950] under plane stress conditions is written as

$$f(\mathbf{r}) \equiv g(\mathbf{r}) - X^2 = 0 \tag{7.5.22a}$$

where $\mathbf{r} = \boldsymbol{\sigma} - \boldsymbol{\alpha}$, and $\boldsymbol{\alpha}$ is the position vector of the center of the active yield surface in the stress space or is called the back stress vector. Here, the function g is written in terms of the components of $\mathbf{r}$ as

$$g\left(r_{\alpha\beta}\right) = r_{11}^2 + r_{22}^2 - \frac{2R}{1+R} r_{11} r_{22} + \frac{2(1+2R)}{1+R} r_{12}^2 \tag{7.5.22b}$$

The partial derivative of the yield function with respect to the Cauchy stress vector takes the form

$$\frac{\partial f}{\partial \boldsymbol{\sigma}} = 2\mathbf{P}\mathbf{r} \tag{7.5.23a}$$

and

$$\frac{\partial f}{\partial \mathbf{r}} = 2\mathbf{P}\mathbf{r} \tag{7.5.23b}$$

Therefore, Eq. 7.5.12 becomes

$$\boldsymbol{\sigma} = \boldsymbol{\sigma}^{tr} - \frac{\Delta \bar{\varepsilon}^p}{X} \mathbf{D}_e \mathbf{P}\mathbf{r} \tag{7.5.24}$$

or

$$\mathbf{r} = \mathbf{r}^{tr} - \frac{\Delta\bar{\varepsilon}^p}{X}\mathbf{D}_e\mathbf{P}\mathbf{r} \tag{7.5.25}$$

where $\boldsymbol{\sigma}^{tr}$ is computed in Eq. 7.5.11, $\mathbf{r}^{tr} = \boldsymbol{\sigma}^{tr} - \boldsymbol{\alpha}$, and $\boldsymbol{\alpha}$ is not a constant vector. Using $\boldsymbol{\alpha}$ from the previous step to obtain $\mathbf{r}^{tr}$ to start a Newton-Raphson iteration, we check this trial vector to see whether $f\left(\mathbf{r}^{tr}\right) \geq 0$ and $\left(\dot{\boldsymbol{\varepsilon}}^p\right)^T \Delta\mathbf{r}^{tr} \geq 0$ ($\Delta\mathbf{r}^{tr} = \mathbf{D}_e\Delta\boldsymbol{\varepsilon}$) or $\left(\mathbf{P}\mathbf{r}\right)^T \Delta\mathbf{r}^{tr} \geq 0$ (from the flow rule and Eq. 7.5.23a, $\dot{\varepsilon}^p = 2\dot{\lambda}\mathbf{P}\mathbf{r}$ where $2\dot{\lambda}$ is positive). If it is true, the stress state is loading in the plastic state.

For the initial loading or after breaking all smaller yield surfaces (see Section 3.7 in Chapter 3), the center of the active yield surface is at the origin $\boldsymbol{\alpha} = \mathbf{0}$. This is identical to that for the isotropic rule discussed in Section 7.5.1, where we compute $\Delta\bar{\varepsilon}^p$ and then update $\bar{\varepsilon}^p$ by adding $\Delta\bar{\varepsilon}^p$ and the yield strength X, which is a function of the updated $\bar{\varepsilon}^p$.

For monotonically increasing loading or reloading after unloading at least once, $\boldsymbol{\alpha} \neq 0$, we assume the center moves along the current vector $\boldsymbol{\beta}$ within the increment of the strain vector. Solving for $\mathbf{r}$ from Eq. 7.5.25, we have

$$\mathbf{r} = \bar{\mathbf{M}}\mathbf{r}^{tr} \tag{7.5.26a}$$

where

$$\bar{\mathbf{M}} = \left(\mathbf{I} + \frac{\Delta\bar{\varepsilon}^p}{X}\mathbf{D}_e\mathbf{P}\right)^{-1} \tag{7.5.26b}$$

Inserting Eq. 7.5.26 into Eq. 7.5.22 gives

$$f\left(\Delta\bar{\varepsilon}^p\right) = \left(\bar{\mathbf{M}}\mathbf{r}^{tr}\right)^T \mathbf{P}\left(\bar{\mathbf{M}}\mathbf{r}^{tr}\right) - X^2 = 0 \tag{7.5.27}$$

where X is evaluated at $\bar{\varepsilon}^p + \Delta\bar{\varepsilon}^p$.

As is the procedure in the isotropic hardening, we solve the effective plastic strain increment in Eq. 7.5.27 by the Newton-Raphson iteration method that requires an additional derivative of $d\boldsymbol{\alpha}/d\Delta\bar{\varepsilon}^p$ because the unknown vector $\boldsymbol{\alpha}$ is contained in Eq. 7.5.27. Modifying Eq. 3.7.3, the differentials of $\boldsymbol{\alpha}$ and X are expressed by

$$d\boldsymbol{\alpha} = A\, dX\, \boldsymbol{\beta} \tag{7.5.28}$$

In Eq. 7.5.28, $\boldsymbol{\beta}$ is a unit vector along which the center of the active yield surface moves and the coefficient A is no longer a constant $\sqrt{2/3}$ as that in the deviatoric stress space for plastically isotropic materials discussed in Chapter 3. The coefficient A can be determined by taking a special case, $\mathbf{r} = \boldsymbol{\sigma} - \boldsymbol{\alpha}$ parallel to $\boldsymbol{\beta}$, and using Eq. 7.5.22. Therefore,

$$A = [g(\boldsymbol{\beta})]^{-1/2} \tag{7.5.29}$$

Here, the function g is given in Eq. 7.5.22b. The derivative of Eq. 7.5.28 with respect to $\Delta\bar{\varepsilon}^p$ is

$$\frac{d\boldsymbol{\alpha}}{d\Delta\bar{\varepsilon}^p} = AX'\boldsymbol{\beta} \tag{7.5.30}$$

because $\boldsymbol{\beta}$ is assumed fixed within the interval. Using this derivative, we can compute

$$\frac{df}{d\Delta\bar{\varepsilon}^p} = \left(\frac{\partial f}{\partial \mathbf{r}}\right)^T \frac{d\mathbf{r}}{d\Delta\bar{\varepsilon}^p} - 2XX' \tag{7.5.31}$$

Differentiating Eq. 7.5.26a and using Eq. 7.5.26b gives

$$\frac{d\mathbf{r}}{d\Delta\bar{\varepsilon}^p} = -\frac{\left(1 - \frac{X'}{X}\Delta\bar{\varepsilon}^p\right)}{X}\bar{\mathbf{M}}\mathbf{D}_e\mathbf{P}\bar{\mathbf{M}}\mathbf{r}^{tr} - AX'\bar{\mathbf{M}}\boldsymbol{\beta} \tag{7.5.32}$$

We note that in Eq. 7.5.23, the last term is due to that $\mathbf{r}^{tr} = \boldsymbol{\sigma}^{tr} - \boldsymbol{\alpha}$, and $\boldsymbol{\alpha}$ is not a constant vector. Inserting Eqs. 7.5.23b and 7.5.32 into Eq. 7.5.31, we obtain the derivative of the yield function with respect to the increment of the effective plastic strain as

$$\frac{df}{d\Delta\bar{\varepsilon}^p} = -2\frac{\left(1 - \frac{X'}{X}\Delta\bar{\varepsilon}^p\right)}{X}\left(\mathbf{P}\bar{\mathbf{M}}\mathbf{r}^{tr}\right)^T \bar{\mathbf{M}}\mathbf{D}_e\mathbf{P}\bar{\mathbf{M}}\mathbf{r}^{tr} - 2AX'\left(\mathbf{P}\bar{\mathbf{M}}\mathbf{r}^{tr}\right)^T \bar{\mathbf{M}}\boldsymbol{\beta} - 2XX' \tag{7.5.33}$$

Now we can solve for $\Delta\bar{\varepsilon}^p$ by the Newton-Raphson method.

After $\Delta\bar{\varepsilon}^p$ is computed in the iteration, we can update $X = X(\bar{\varepsilon}^p + \Delta\bar{\varepsilon}^p)$ and $\Delta X = X\left(\bar{\varepsilon}^p + \Delta\bar{\varepsilon}^p\right) - X\left(\bar{\varepsilon}^p\right)$. Inserting ΔX into the finite difference form of Eq. 7.5.28, we can update the position vector of the center of the active yield surface by $\boldsymbol{\alpha}^{(k)} = \boldsymbol{\alpha}^{(k-1)} + \Delta\boldsymbol{\alpha}$ for the iteration. After the convergence of the Newton-Raphson iteration, we update all the variables $\bar{\varepsilon}^p$, X, $\boldsymbol{\alpha}$, $\mathbf{r}$, and $\boldsymbol{\sigma}$.

After the initial loading, unloading, and reloading for a few cycles, there are a few yield surfaces with different sizes (with different X values) and unit vector $\boldsymbol{\beta}$s stored in the history file. For monotonically increasing loading after a few such cycles, if the unit vector $\boldsymbol{\beta}$ changes its direction (referred to in Chapter 3 as a "break"), we can determine a factor γ such that after the strain increment $\gamma\Delta\boldsymbol{\varepsilon}$ $(0 < \gamma < 1)$, $\boldsymbol{\beta}$ starts to change. We set

$$\boldsymbol{\sigma}^{tr} = \boldsymbol{\sigma}_n + \gamma \mathbf{D}_e \Delta\boldsymbol{\varepsilon} \tag{7.5.34}$$

and

$$\mathbf{r}^{tr} = \mathbf{r}_n + \gamma \mathbf{D}_e \Delta\boldsymbol{\varepsilon} \tag{7.5.35}$$

where $\mathbf{r}^{tr} = \boldsymbol{\sigma}^{tr} - \boldsymbol{\alpha}$ and $\mathbf{r}_n = \boldsymbol{\sigma}_n - \boldsymbol{\alpha}$. The position vector $\boldsymbol{\alpha}$ of the center of the yield surface is along $OO_2^{(1)}$, and the center moves toward $O_1^{(i)}$, but it cannot go beyond $O_1^{(i)}$ during reloading, as shown in Figure 3.22 of Chapter 3. For continuous reloading, the center will move along OA_0. The unit vector $\boldsymbol{\beta}$ changes its direction from $O_1^{(i)}O_2^{(1)}$ to OA_0 (called "break"). Inserting Eq. 7.5.35 into Eq. 7.5.27, we obtain a quadratic equation with the unknown γ

$$f(\gamma) = \left(\bar{\mathbf{M}}\mathbf{r}^{tr}\right)^T \mathbf{P}\left(\bar{\mathbf{M}}\mathbf{r}^{tr}\right) - X^2 = 0 \tag{7.5.36}$$

where X is the size of the yield surface of which the center has the specified position vector $\boldsymbol{\alpha}$. Therefore, we are able to solve for γ, which is the smaller positive root. Note that X (retrieved from the history file) is the size of the active yield surface before unloading occurs in the previous loading cycle. If the smaller root γ is greater than one, there is no change of $\boldsymbol{\beta}$, and the previously mentioned method to update $\boldsymbol{\sigma}$ can be applied. If γ is between 0 and 1, there is a change of $\boldsymbol{\beta}$ within the strain increment vector $\Delta\boldsymbol{\varepsilon}$. With use of the present active yield surface, the strain increment $\gamma\Delta\boldsymbol{\varepsilon}$, and the effective plastic strain increment $\Delta\bar{\varepsilon}^p$ computed from the difference between the sizes of the two consecutive active yield surfaces, we can apply Eqs. 7.5.35 and 7.5.26a to update $\mathbf{r}$ and $\boldsymbol{\sigma}$ without any iteration. For the remaining strain increment $(1-\gamma)\Delta\boldsymbol{\varepsilon}$, we repeat this process and update $\boldsymbol{\sigma}$ until the largest yield surface of which its center is at the origin $\boldsymbol{\alpha} = 0$ becomes active, as mentioned in the previous paragraph. This is the case in which the initial state is restored, and the history of loading and unloading is erased. In this process, if γ is not a real number or is less than zero, there is an error.

We also check for unloading. When $f\left(\mathbf{r}^{tr}\right) < 0$ or $(\mathbf{Pr})^T \Delta\mathbf{r}^{tr} < 0$, the smallest yield surface with the size X_0 becomes active. We reset the effective plastic strain $\bar{\varepsilon}^p = 0$. Its center $\boldsymbol{\alpha}$ is located along the line joining the loading point immediately before unloading and the center of the yield surface with the size X (see Section 3.7 in Chapter 3). The unit vector $\boldsymbol{\beta}$ is along the line. We store $\boldsymbol{\alpha}$, $\boldsymbol{\beta}$, and X in the history file to be retrieved when there is reloading. If $(\mathbf{Pr})^T \Delta\mathbf{r}^{tr} < 0$ but $g\left(\mathbf{r}^{tr}\right) > X_0^2$, reloading or even a "break" may occur with this strain increment. Following the procedure in the previous paragraph, we update $\boldsymbol{\alpha}$, $\boldsymbol{\beta}$, γ, X, $\mathbf{r}$, and $\boldsymbol{\sigma}$. The elastic stress state can happen only when $g\left(\mathbf{r}^{tr}\right) < X_0^2$. Therefore, $\mathbf{r} = \mathbf{r}^{tr}$ and $\boldsymbol{\sigma} = \boldsymbol{\sigma}^{tr}$ is the solution we are seeking.

Because of loading and unloading for a few cycles in simulations of sheet metal forming processes, the computational procedure for Mroz's rule is not as simple as that of the isotropic hardening rule. We summarize the procedures to compute the incremental stress vector with a given incremental strain vector at a point based on Mroz's hardening rule using the radial return method. Initially, we assign the number of data sets $NH = 0$ (zero means no data set) on the history file, $X = X_0$, $\bar{\varepsilon}^p = 0$, and $\boldsymbol{\alpha} = 0$. After the convergence for the first load step, we compute the stress vector with the given incremental strain vector following the procedures in Table 7.5.

7.5.3 Integration of Rate-Insensitive Constitutive Equations with Transverse Shear Strains and Anisotropic Hardening

The stress and strain vectors referred to the local coordinate system $\bar{r}$, $\bar{s}$, and $\bar{t}$ have five distinct components. From Eq. 5.6.9, we can write the yield function fo r the five distinct stress components as

$$\begin{aligned} f &= \sigma_{11}^2 + \sigma_{22}^2 - \frac{2R}{1+R}\sigma_{11}\sigma_{22} + \frac{2(1+2R)}{1+R}\sigma_{12}^2 \\ &+ \frac{6}{1+R}\left(\sigma_{23}^2 + \sigma_{31}^2\right) - X^2 = 0 \end{aligned} \tag{7.5.37}$$

Note that in Eq. 7.5.37, the indices are $1 = \bar{r}$, $2 = \bar{s}$, and $3 = \bar{t}$.

Because the transverse anisotropy parameter R is involved, it is quite complex to transfer to the global Cartesian coordinate system. For keeping the yield function simple and physically meaningful, we suggest the use of the co-rotational coordinate system, such as the one in the plane stress problems. Figure 6.8 in Chapter 6 shows the local coordinate system in the configuration at time t. It rigidly rotates to the configuration at time $t + \Delta t$. We assume the angle between $\bar{\mathbf{r}}$ and $\partial(\mathbf{x} + \Delta\mathbf{u})/\partial r$ is equal to that between $\bar{\mathbf{s}}$ and $\partial(\mathbf{x} + \Delta\mathbf{u})/\partial s$, as shown in Figure 6.8. Therefore, the method used to update the stress vector discussed in the previous section can be applied, except we modify the matrices (5 by 5) as

$$\mathbf{P} = \begin{bmatrix} 1 & -\dfrac{R}{1+R} & 0 & 0 & 0 \\ -\dfrac{R}{1+R} & 1 & 0 & 0 & 0 \\ 0 & 0 & \dfrac{2(1+2R)}{1+R} & 0 & 0 \\ 0 & 0 & 0 & \dfrac{6}{1+R} & 0 \\ 0 & 0 & 0 & 0 & \dfrac{6}{1+R} \end{bmatrix} \tag{7.5.38}$$

TABLE 7.5
PROCEDURES FOR UPDATING THE STRESS VECTOR FROM A GIVEN INCREMENTAL STRAIN VECTOR AT A POINT AFTER THE INITIAL STEP OF SOLUTION BASED ON MROZ'S HARDENING RULE USING THE RADIAL RETURN METHOD

1. Test for any set of data stored on the history file, $NH = 0$.
   a. If yes, go to Step 2.
   b. If no, read X, $\bar{\varepsilon}^p$, $\boldsymbol{\alpha}$, and $\boldsymbol{\beta}$ from the history file.
2. Compute $\mathbf{r}^{tr}$, and check $g\left(\mathbf{r}^{tr}\right) < X_0^2$ (where g is defined in Eq. 7.5.22).
   a. If yes, $\mathbf{r} = \mathbf{r}^{tr}$, and $\boldsymbol{\sigma} = \boldsymbol{\sigma}^{tr}$ is the elastic solution. Exit.
   b. If no, check for unloading: $g\left(\mathbf{r}^{tr}\right) < X^2$ or $(\mathbf{Pr})^T \Delta\mathbf{r}^{tr} < 0$.
      i. If yes, go to Step 3.
      ii. If no, go to Step 4 (monotonically increasing loading).
3. Unloading.
   a. Set $NH = NH + 1$ and $\bar{\varepsilon}^p = 0$.
   b. Establish $\boldsymbol{\beta}$ and then $\boldsymbol{\alpha}$ for the smallest yield surface with the size X_0.
   c. Compute $\mathbf{r}^{tr}$, and check $g\left(\mathbf{r}^{tr}\right) < X_0^2$.
      i. If yes, $\mathbf{r} = \mathbf{r}^{tr}$, and $\boldsymbol{\sigma} = \boldsymbol{\sigma}^{tr}$ is the elastic solution. Exit.
      ii. If no, reload and go to Step 4.
4. Loading or reloading. Check $NH = 0$.
   a. If yes, compute $\Delta\bar{\varepsilon}^p$ following the procedure for isotropic hardening, and update $\bar{\varepsilon}^p$, X, and $\boldsymbol{\sigma}$. Exit.
   b. If no, follow the method in the paragraph to compute the roots γ from Eq. 7.5.36. If the roots are negative or are not real, there is error return. Otherwise, pick the smallest positive root, and test $\gamma > 1$.
      i. If yes, there is no "break." Compute $\Delta\bar{\varepsilon}^p$ using Eq. 7.5.27, and then update $\bar{\varepsilon}^p$, X, $\boldsymbol{\alpha}$, $\mathbf{r}$, and $\boldsymbol{\sigma}$. Exit.
      ii. If no, there is a "break." Use the incremental strain vector $\gamma\Delta\varepsilon$ to compute $\bar{\varepsilon}^p$, X, $\boldsymbol{\alpha}$, $\mathbf{r}$, and $\boldsymbol{\sigma}$ from Eqs. 7.5.35 and 7.5.26.
      iii. Set $(1-\gamma)\Delta\varepsilon$ equal to $\Delta\varepsilon$.
      iv. Set $NH = NH - 1$, and go to Step 4.

and

$$\mathbf{D}_e = \frac{E}{1-\nu^2}\begin{bmatrix} 1 & \nu & 0 & 0 & 0 \\ \nu & 1 & 0 & 0 & 0 \\ 0 & 0 & \frac{1-\nu}{2} & 0 & 0 \\ 0 & 0 & 0 & \frac{k(1-\nu)}{2} & 0 \\ 0 & 0 & 0 & 0 & \frac{k(1-\nu)}{2} \end{bmatrix} \tag{7.5.39}$$

where matrix **P** is established based on the yield function in Eq. 7.5.37, and the elasticity matrix is based on Eq. 5.6.6b from Chapter 5. If the incremental strain vector $\Delta\boldsymbol{\varepsilon}$ $(\Delta\varepsilon_{\overline{rr}}, \Delta\varepsilon_{\overline{ss}}, 2\Delta\varepsilon_{\overline{rs}}, 2\Delta\varepsilon_{\overline{st}}, 2\Delta\varepsilon_{\overline{tr}})$ is given, we may follow the method in Section 7.5.2 to update the stress vector.

For Mroz's hardening rule, we replace the stress vector $\boldsymbol{\sigma}$ by $\mathbf{r} = \boldsymbol{\sigma} - \boldsymbol{\alpha}$ in the yield function in Eq. 7.5.37. The rest exactly follow those for the plane stress derivation in the previous section, except the vectors now contain five instead of three components. Therefore, we do not repeat the derivation here.

7.6 Computing Springback

In sheet metal forming operations, a part that is being formed conforms closely to the die shape while inside the forming tools. After the part is taken out of the forming tools, the part changes the shape. This change of shape is known as springback. Although most of the deformation of the part is plastic, some elastic deformation in the part exists. The elastic deformation is restored, and this restored elastic deformation is known as springback after the constraints by forming tools are removed. Springback is especially severe for aluminum, where Young's modulus is about one-third of that of steel, and for high-strength steel, where the flow stress is higher than that of the mild steel, as shown in Figure 7.3. To compensate for the distortion in shape, the first step is to predict the amount of springback of a part for a given design of forming tools.

To predict springback, which is controlled by the stresses at the end of a stretch/draw die closure, the focus of the prediction accuracy must be shifted to the stress distribution across the entire stamping process. In fact, predicting springback presents many more challenges than those required for predicting fracture and wrinkling or buckling. A quasi-static analysis with implicit time integration is the most appropriate method for accurately predicting the overall strain and stress distributions of an entire stamped part. In addition, because springback involves unloading caused by the release of the part from stamping tools, it requires a material rule that can be used to model accurately the cyclic stress-strain response. A simple method to compute springback proposed by Tang [1987], now extended for use in a three-dimensional analysis, is discussed in the following paragraphs.

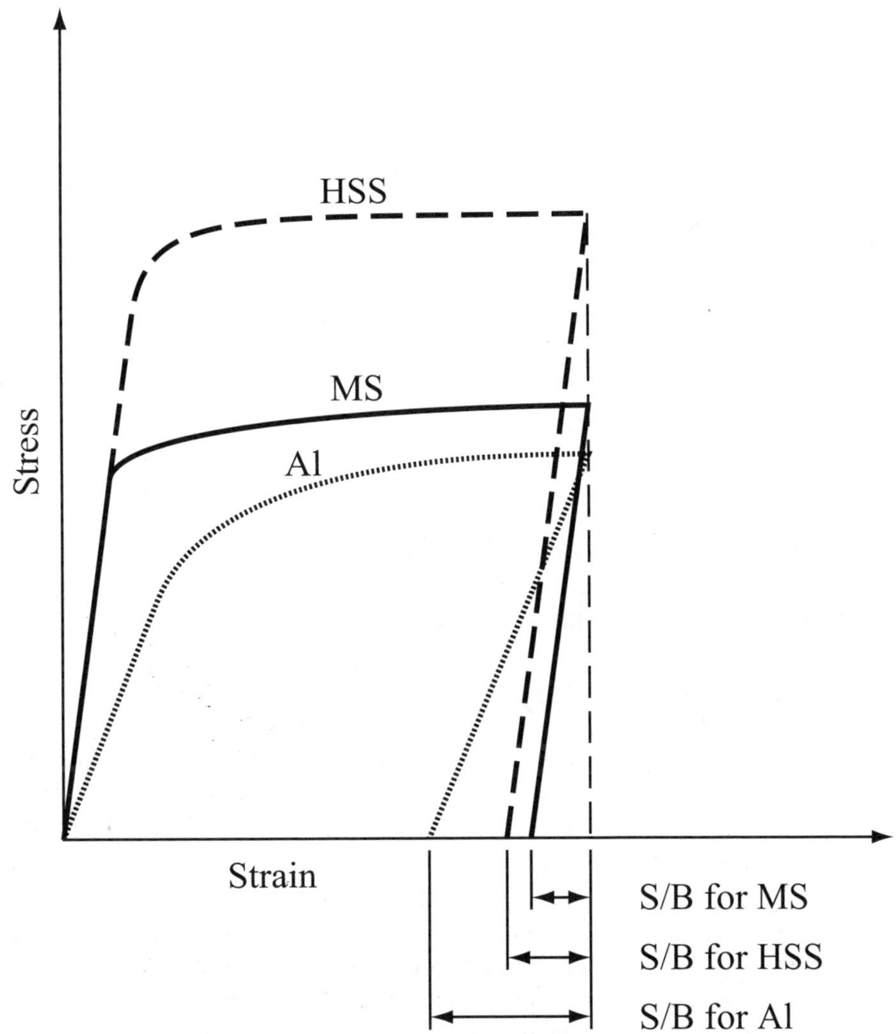

Figure 7.3 Springback (S/B) for mild steel (MS), high-strength steel (HSS), and aluminum (Al) sheets.

7.6.1 Approximate Method for Computing Springback

As in the forming analysis, we must solve a surface contact problem with friction to find the final shape of a part after release from the forming tools. When using the implicit method, the formulation for a surface contact problem involves the material derivatives of a contact nodal force vector in terms of the curvatures of the tool surface. Although some practical methods (see Section 7.3.5) do not require curvatures, the computation is still complex.

An approximate method not requiring the solution of a surface contact problem was proposed by Tang [1987] for a two-dimensional springback analysis. From the forming analysis after die closure, we can compute the tool pressure and the frictional force acting at a contact node on a sheet metal part. Springback can be computed from the deformed shape of the part by releasing the tool pressure and the frictional force at all contacting nodes. Applying these forces with equal magnitudes but opposite signs to all contact nodes of the deformed part with the appropriate support to eliminate any rigid body motion, we can compute the additional deformation of the part due to springback. Figure 7.4(a) shows the tool and the finite element model for a plane strain draw. Figure 7.4(b) shows the deformed sheet after forming. Figure 7.4(c) shows the forces imposed at all contact nodes of the formed sheet to release the tool forces (pressure and frictional forces) for springback computation. This is a two-dimensional plane problem, so we need to fix only the sheet at the center to constrain two translations and one rotation, as shown in Figure 7.4(c).

In this book, we extend this model to a three-dimensional problem. A stamped part must be supported to eliminate all three rigid body translations and rotations for a statically determinate structure. In addition, we do not constrain six degrees of freedom at one point. Instead, to distribute the computational error, we spread the constraints at three or more points that are far apart.

Tang [1987] demonstrated the residual stress distribution in a section of another plane strain draw. Figure 7.5(a) shows the shapes after the punch travels 76 mm and the springback shape after it is released from the forming tools. The sheet is made of mild steel that is 1 mm thick, and the punch has a bottom radius $R = 229$ mm. Figure 7.5(b) shows the normal forming stress distribution after the punch travels 76 mm, and the residual stress distribution at a section passing a material point A on the wall. Figure 7.5(a) shows the deformed location of the material point A. As shown in Figure 7.5(b), the normal forming stresses and especially the residual stresses are not distributed smoothly through the thickness of the sheet. As mentioned in Chapter 6, this is why we require at least seven points of numerical integration to compute the element nodal force vector and the tangent stiffness matrix for an accurate result. In addition, the residual stress is completely reversed from compression to tension at the bottom of the section of the sheet, as shown in Figure 7.5(b). For accurate prediction of the amount of springback, a more sophisticated hardening rule other than the isotropic one discussed earlier is needed. An anisotropic hardening rule in the mathematical theory of plasticity is a good candidate for use in predicting both forming and residual stresses. To implement this theory, the constitutive equation is suggested to be modified in the following subsection.

7.6.2 Constitutive Equations for Springback Analysis

The material response in a springback analysis is mostly in the elastic range. For a part under severe bending in forming, the material may go to reversed plastic loading because of the Bauschinger effect, which lowers the yield stress during springback. Therefore, a theoretically sound but computationally simple anisotropic hardening rule such as that proposed by Mroz [1967] and modified by Chu [1984] and Tang [1990] should be used in a springback analysis. Tang [1990] applied both the isotropic and anisotropic hardening rules to the two-dimensional plane strain draw shown in Figure 7.4.

Figure 7.6(a) shows the principal stresses σ_1 and σ_2 on the top surface of element 15, as shown in Figure 7.4, versus the punch travel based on the isotropic hardening rule. Figure 7.6(b) shows the principal stresses σ_1 and σ_2 on the top surface of the same element based on Mroz's anisotropic hardening rule. The figures also display the principal engineering strains e_1 and e_2. Note that $e_2 = 0$ because of the plane strain conditions. Element 15 was bent over the die corner until the punch traveled by 14 mm. After that instant, this element was under unbending. Therefore, unloading occurred in that element until the punch traveled to 18 mm, based on the isotropic hardening rule, as shown in Figure 7.6(a). After that, there was reloading in element 15. As shown in Figure 7.6(b), reloading occurred earlier at the punch travel of 16 mm, based on the

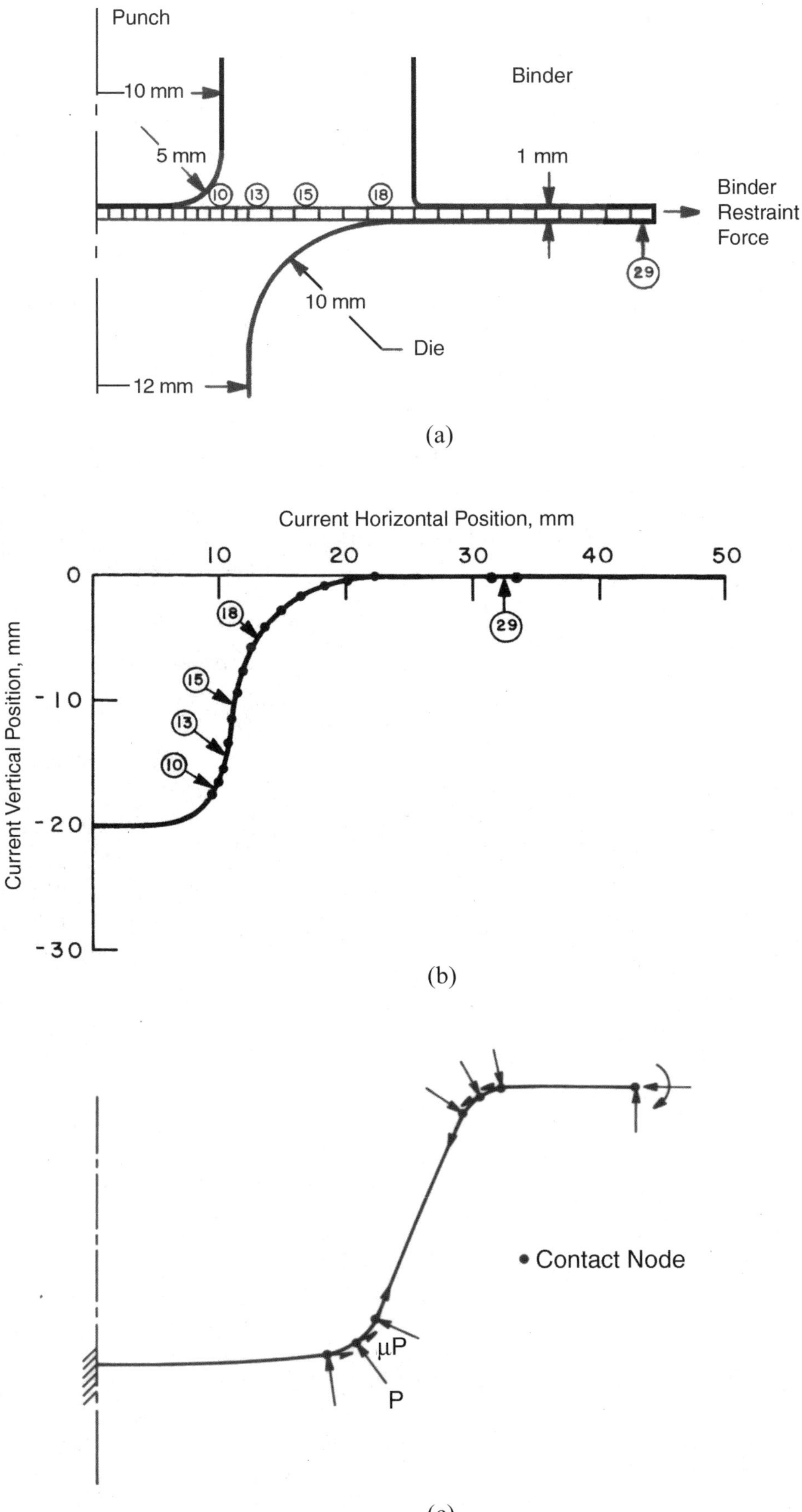

Figure 7.4 A springback model for a plane strain draw with a flat bottom punch: (a) tools and sheet finite element models, (b) deformed sheet after forming, and (c) forces imposed on the formed sheet to release the tool forces for springback computation.

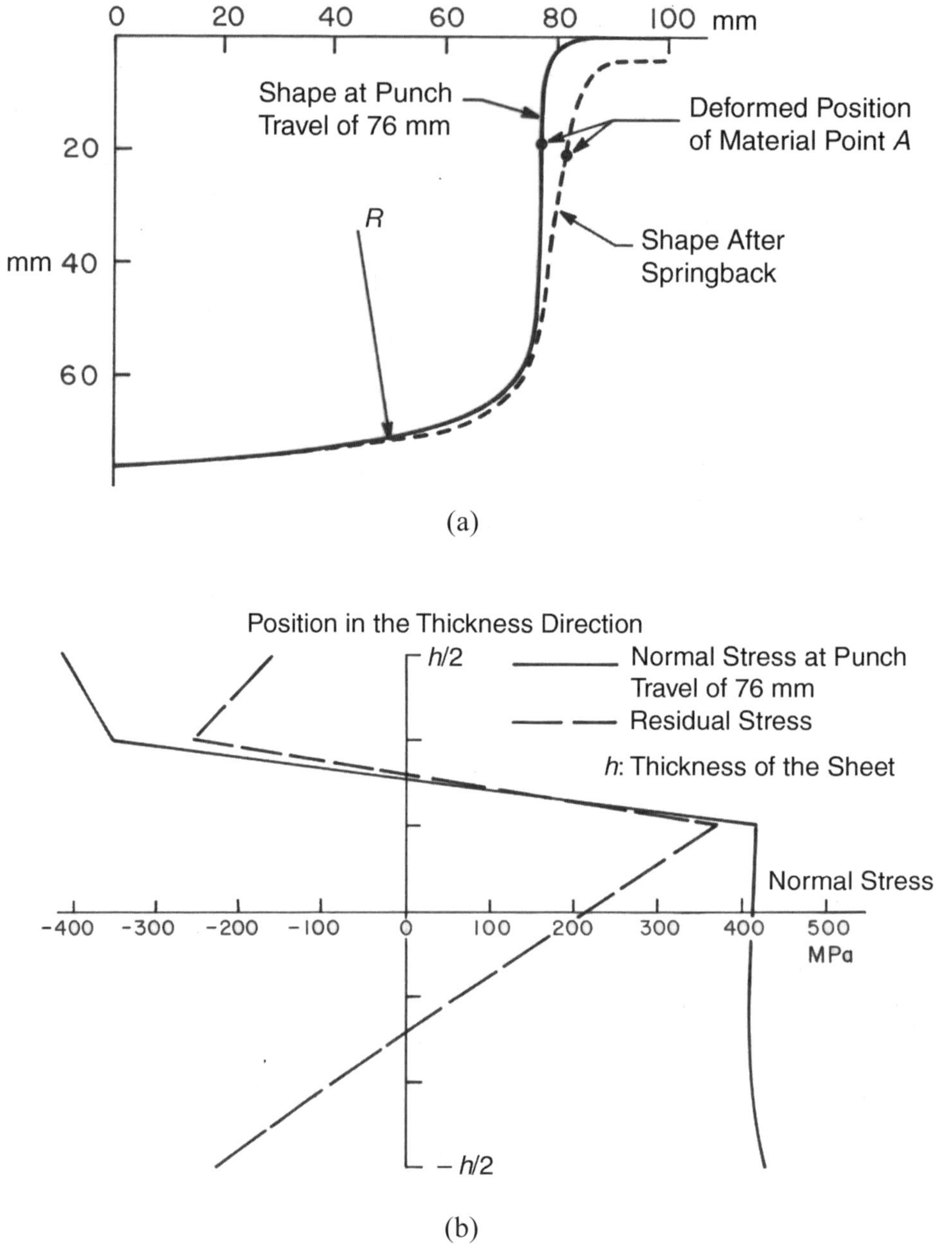

Figure 7.5 A plane strain draw with a curved bottom punch: (a) shapes of the sheet before and after springback, and (b) stress distributions before and after springback.

anisotropic hardening rule. There is significant difference between the forming stresses predicted by these two different hardening rules when the punch bottomed out at the punch travel of 20 mm. Note that the forming stress can affect the amount of springback after the sheet is released from the forming tools.

7.7 Remeshing and Adaptive Meshing

In simulations of bulk forming processes such as extruding or forging processes, the finite element mesh could be severely distorted because of large strains that occurred during the processes. To eliminate large aspect ratios in the distorted finite element

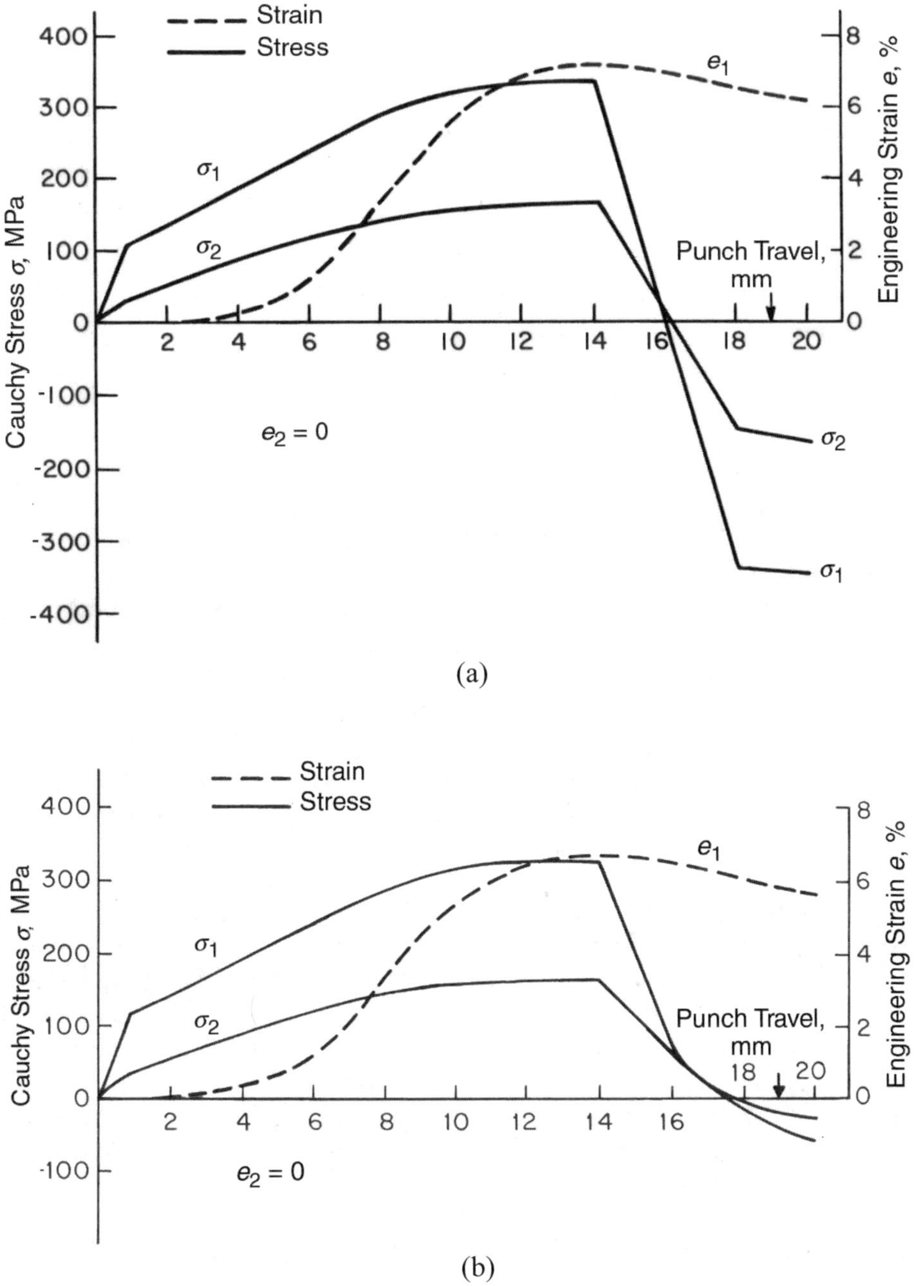

Figure 7.6 The stress and strain histories at the top surface of element 15, as shown in Figure 7.4: (a) results from the isotropic hardening rule, and (b) results from the Mroz hardening rule.

mesh, we must redo the mesh in the model (called remesh). Engineers usually refine or coarsen the existing finite element meshes in simulations of sheet metal forming processes where the strains in most areas are not as large as those for bulk forming processes. Therefore, the adaptive meshing technique is an excellent candidate for simulation of sheet metal forming processes to achieve a savings in computation cost without sacrificing accuracy.

There are basically two types of adaptive meshing techniques: p-adaptivity, and h-adaptivity. For p-adaptivity, a high-order polynomial for shape or interpolation function is used in elements to model areas to be refined. For simulations of sheet metal forming processes, the main load comes from the contact between a sheet and the tool surfaces. Therefore, it would not improve accuracy significantly for high-performance elements with a shape function of a higher-order polynomial in the contact areas. The most efficient way to improve accuracy and to save computing time is to add more elements to the model in the contacting tool surface with a sharp curvature. This is exactly the function of the h-adaptive technique, which refines the finite element mesh or adds more elements to the original mesh. For the contact of a flat tool surface or a surface with a mild curvature, the original coarse mesh may be restored to save computing cost without sacrificing accuracy.

As mentioned, corner radii of tools (dies and punches) generally are very tight (5 mm or less) for forming automotive sheet metal parts. Figure 7.7(a) shows a partial section of a draw die corner over which a sheet is drawn in a drawing process to form a sheet metal part. The original finite element model with nodes 1 through 5 is not fine enough. Because only nodes 3 and 4 contact the die corner, the deformed shape of the portion of the sheet, which is shown by the dashed line joining nodes 3 and 4, does not conform to the surface at the die corner. We should add new nodes between nodes 3 and 4 to have the deformed shape conform to the die surface. For a sheet contacting a die corner of nearly 90°, we should use at least seven nodes to model the section. Therefore, five new nodes *c* through *g* are added to the section of the existing finite element model between the original nodes 3 and 4, as shown in Figure 7.7(a). This addition of elements or nodes is called "fission" in adaptive meshing. To avoid replacing an inaccurate computation with the original coarse finite element model by the refined one in that portion and repeating the computation, the simulation software should detect in advance that the portion of the sheet possibly may contact a tool surface with a sharp curvature. Consequently, the software would refine that portion of the original mesh using the adaptive meshing technique before continuation of the finite element analysis.

Note that a sheet is moving during a drawing process. After the portion of the sheet goes through the die corner, this portion is drawn to a draw wall that is flat, as shown in Figure 7.7(b). Therefore, a coarse mesh to model this portion of the sheet on the draw wall is accurate enough in the simulation for saving computation cost. We may delete the added nodes *f* and *g*, as shown in Figure 7.7(b). This deletion of elements or nodes to restore the original mesh is called "fusion" in adaptive meshing. In the meantime, the original nodes 2 and 3 are drawn to contact the die corner. Therefore, we add two more new nodes, *a* and *b*, to refine this portion of the model (see Figure 7.7(b)).

7.7.1 Refinement and Restoration for Triangular Shell Elements

Figure 7.8(a) shows two curved triangular shell elements on the θ-plane (see Chapter 6). We want to refine one of the elements into four new triangular elements. To keep the aspect ratio of the added elements reasonable, we may pick the midpoints *a*, *b*, and *c* of the sides 1-2, 2-3, and 3-1, respectively, as the new nodes to create four new triangles 1*ac*, 2*ba*, 3*cb*, and *abc*. The coordinates of a new node, for example, for a new node *c*, are obtained by the linear interpolation as

$$\boldsymbol{\theta}_c = \frac{1}{2}\left(\boldsymbol{\theta}_3 + \boldsymbol{\theta}_1\right), \quad \boldsymbol{\theta}_i = \left(\theta_i^1, \theta_i^2\right) \quad (\mathrm{i} = 1,3,c) \tag{7.7.1}$$

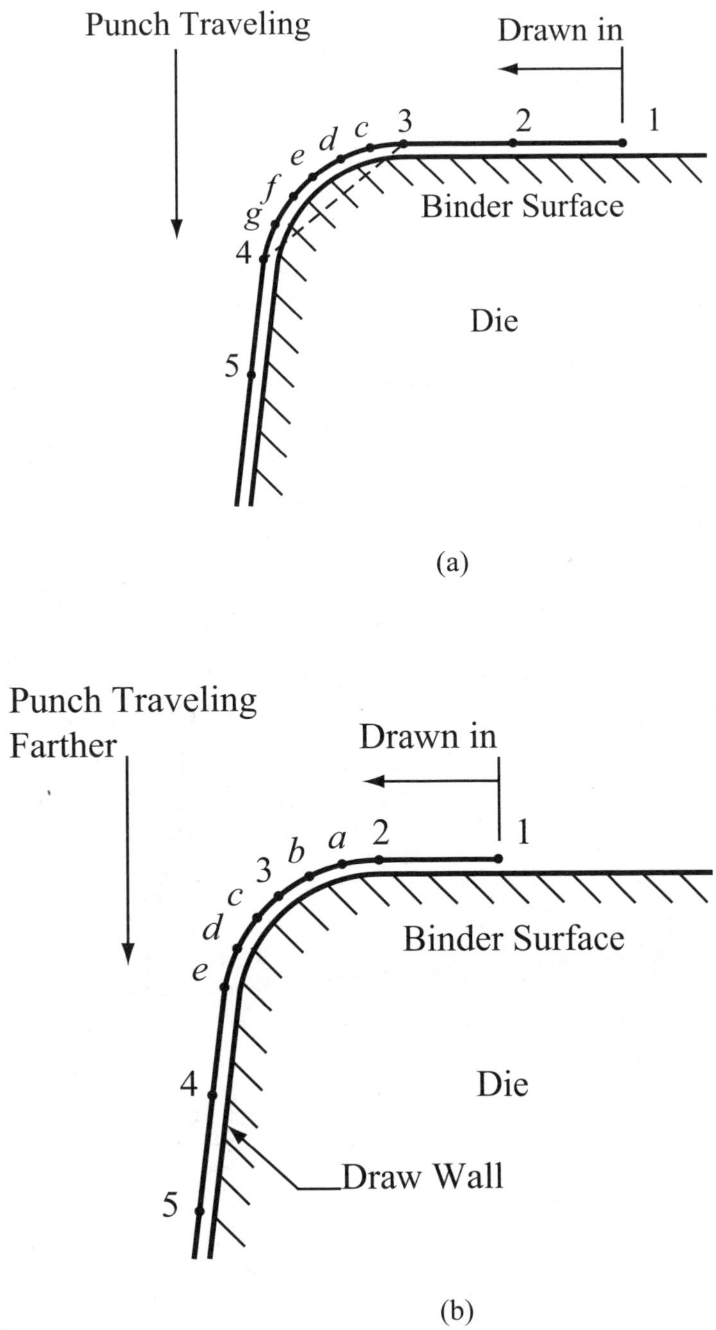

Figure 7.7 Drawing over a die corner: (a) adding new nodes, and (b) adding and deleting new nodes.

where the subscripts denote the node numbers. For the nodal displacement vectors, we may use the linear interpolation, for example, for new node c, as

$$\bar{\mathbf{U}}_c = \frac{1}{2}\left(\mathbf{U}_3 + \mathbf{U}_1\right) \tag{7.7.2}$$

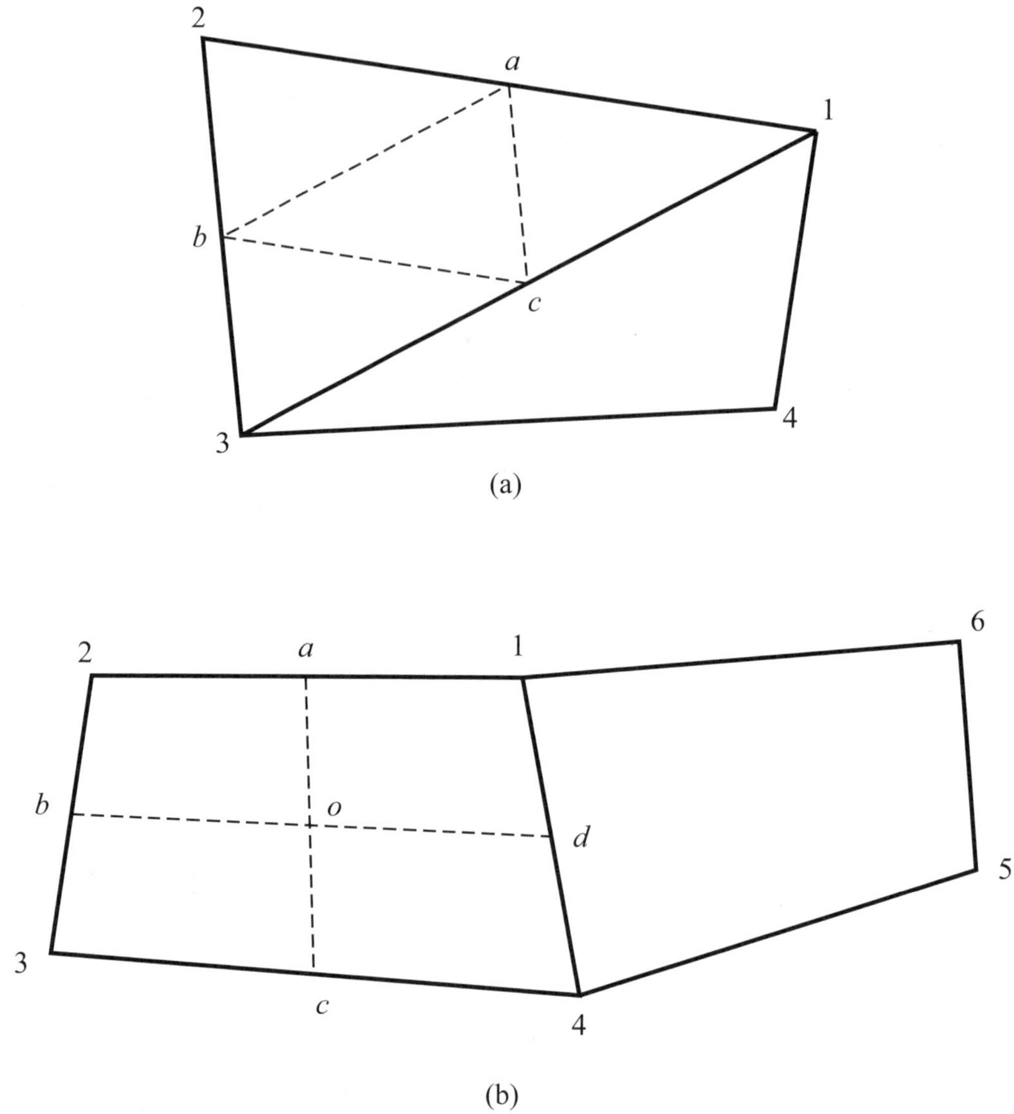

Figure 7.8 Refinement of a finite element mesh: (a) fission of a triangular element, and (b) fission of a quadrilateral element.

In the transition element 134, there is a side node *c* (called the slave node) connected to nodes 1, *a*, *b*, and 3 (called the master nodes). Because the shell element 134 (the transition element from fine to coarse mesh) cannot have a side node, we condense this slave node *c* by replacing the variables at this node by those at the master nodes. Using the linear interpolation, we may transform the linearly independent displacement vectors to the linearly dependent vectors as

$$\bar{\mathbf{U}} = \mathbf{\Phi U} \tag{7.7.3a}$$

where the bar denotes a vector, of which the elements are not linearly independent,

$$\bar{\mathbf{U}} = \begin{bmatrix} \bar{\mathbf{U}}_1 \\ \bar{\mathbf{U}}_3 \\ \bar{\mathbf{U}}_c \\ \bar{\mathbf{U}}_a \\ \bar{\mathbf{U}}_b \end{bmatrix} \qquad \mathbf{U} = \begin{bmatrix} \mathbf{U}_1 \\ \mathbf{U}_3 \\ \mathbf{U}_a \\ \mathbf{U}_b \end{bmatrix} \tag{7.7.3b}$$

and

$$\mathbf{\Phi} = \begin{bmatrix} \mathbf{I} & 0 & 0 & 0 \\ 0 & \mathbf{I} & 0 & 0 \\ \frac{1}{2}\mathbf{I} & \frac{1}{2}\mathbf{I} & 0 & 0 \\ 0 & 0 & \mathbf{I} & 0 \\ 0 & 0 & 0 & \mathbf{I} \end{bmatrix} \tag{7.7.3c}$$

and **I** is an identity matrix.

Now we transform the force vector including that at the slave node c to those at the master nodes only by application of the virtual work

$$\bar{\mathbf{F}}^T \delta \bar{\mathbf{U}} = \mathbf{F}^T \delta \mathbf{U} \tag{7.7.4a}$$

where

$$\bar{\mathbf{F}}^T = \begin{bmatrix} \bar{\mathbf{F}}_1 & \bar{\mathbf{F}}_3 & \bar{\mathbf{F}}_c & \bar{\mathbf{F}}_a & \bar{\mathbf{F}}_b \end{bmatrix} \tag{7.7.4b}$$

$$\mathbf{F}^T = \begin{bmatrix} \mathbf{F}_1 & \mathbf{F}_3 & \mathbf{F}_a & \mathbf{F}_b \end{bmatrix} \tag{7.7.4c}$$

Taking the variation of the displacement vectors in Eq. 7.7.3a and substituting $\delta\bar{\mathbf{U}}$ into Eq. 7.7.4a, we have

$$\mathbf{F} = \mathbf{\Phi}^T \bar{\mathbf{F}} \tag{7.7.5}$$

We may use this equation to condense the force vector at the slave node c to those master nodes 1, 3, a, and b. For the dynamic explicit integration method, the fission process is complete. However, we need to condense the sub-tangent stiffness matrix $\bar{\mathbf{K}}_c$ at the slave node c to those at all master nodes connected to c. For the slave node c, the sub-tangent stiffness matrix has the form

$$\bar{\mathbf{K}}_c \Delta\bar{\mathbf{U}} = \Delta\bar{\mathbf{F}} \tag{7.7.6}$$

where Δ denotes the increment. Substituting the increments in Eqs. 7.7.3a and 7.7.5 in the incremental form into Eq. 7.7.6 gives the condensed sub-tangent stiffness matrix as

$$\mathbf{K}_c \Delta\mathbf{U} = \Delta\mathbf{F} \tag{7.7.7a}$$

where

$$\mathbf{K}_c = \mathbf{\Phi}^T \bar{\mathbf{K}}_c \mathbf{\Phi} \tag{7.7.7b}$$

We may refine the elements at a few levels simultaneously. For the first level of refinement in adaptive meshing, we subdivide the triangle 123 into four triangles of 1ac, 2ba, 3cb, and abc, as shown in Figure 7.8(a). We subdivide each new triangle into another four for the second level of refinement in adaptive meshing. Therefore, the total number of new elements would be 16. Following this procedure, we may go for more levels of refinement in adaptive meshing. However, there are some restrictions, as shown in the LS-DYNA manual (Hallquist [1998]). First, the levels of adaptivity implemented in a mesh must be such that the level of difference between the adjacent elements is no more than one. Second, the total number of refined elements can be restricted by the available memory in a computer. Third, for the implicit method, too many levels of adaptivity may disturb equilibrium; consequently, the Newton-Raphson iteration may have convergence problems.

Another unsolved problem is how to obtain the stress and strain histories in new elements with multiple integration (Gaussian) points. The constitutive equations for elastic-plastic deformation require the histories of the stresses and strains at each integration point. We may obtain these values at the integration points in the added elements by interpolations from those in the existing elements or nodes. The interpolated values might have problems in consistency, with the constitutive equations or the interpolated stresses not necessarily satisfying the yield function. Also, the strain increments may not be compatible with the incremental nodal displacement vector. Some adjustments are required to minimize the discrepancy. Only one integration point used in some elements for the explicit method can avoid consistency and compatibility problems. In addition, there are no convergence problems for explicit integration. However, this is a challenge for the implicit method using finite elements with multiple integration points.

After the refined elements are drawn over the tool surface with a small curvature, we may delete the added elements for refinement and restore the original coarse mesh. This adaptivity is called "fusion." Although the principle is simple, development of software to detect the contacting tool surface with small curvatures in advance might be quite involved. Elimination of a new (added) node is nothing except transforming the force vector acting on the node to the original nodes connected to it. For the slave node c, we transform the force vector acting on it to the master nodes 1 and 3, as shown in Figure 7.8(a) via Eq. 7.7.5. Similarly, we tranform the force vector acting on the new node a to the original nodes 1 and 2, and the force vector acting on the new node b to the original nodes 2 and 3, respectively. Therefore, we can eliminate the new nodes a, b, and c and can restore the original triangle element 123. Following this procedure, all the added nodes can be eliminated so that the original coarse mesh is restored.

7.7.2 Refinement and Restoration for Quadrilateral Shell Elements

Figure 7.8(b) shows the reference surface of two original quadrilateral shell elements on the r–s plane (see Chapter 6). We refine one of the elements into four new quadrilateral elements. To keep the aspect ratio of the added elements reasonable, we may pick the midpoints a, b, c, and d of the sides 1-2, 2-3, 3-4, and 4-1, respectively, and a point o at the center of the element on the reference surface as new nodes to create four new quadrilaterals $1aod$, $2boa$, $3cob$, and $4doc$.

The adaptive technique for the triangular mesh in the previous section can be applied to the quadrilateral mesh as well. The only difference is that a center node o as shown in Figure 7.8(b) is being used to refine the quadrilateral element 1234. The coordinates of the new side nodes are obtained by interpolation using Eq. 7.7.1, while the coordinates of the new center node o are obtained by the average values of the four existing corner nodes as

$$\mathbf{R}_o = \frac{1}{4}(\mathbf{R}_1 + \mathbf{R}_2 + \mathbf{R}_3 + \mathbf{R}_4)$$
$$\mathbf{R}_i = (r_i, s_i) \quad (i = o,1,2,3,4) \qquad (7.7.8a)$$

In the Cartesian coordinates,

$$\mathbf{X}_o = \frac{1}{4}(\mathbf{X}_1 + \mathbf{X}_2 + \mathbf{X}_3 + \mathbf{X}_4) \qquad (7.7.8b)$$

The nodal displacement vector at node o also is the average of those at the four existing corner nodes as

$$\bar{\mathbf{U}}_o = \frac{1}{4}(\mathbf{U}_1 + \mathbf{U}_2 + \mathbf{U}_3 + \mathbf{U}_4) \qquad (7.7.9)$$

The remaining procedures for adaptive meshing are the same as those for the triangular meshes. Therefore, we will not repeat them here.

7.8 Numerical Examples of Various Forming Operations

In this section, we show several numerical examples for various forming operations obtained from the dynamic explicit integration method and the quasi-static implicit integration method with equilibrium check. (Later in this book, it is called the quasi-static method, for simplicity.) In the quasi-static method, Mroz's hardening rule for plastic deformation is used in most of these numerical examples for better simulation results.

7.8.1 Numerical Examples of Sheets During Binder Wrap

Tang [2000] predicted the deformation for the binder wrap of a floor pan made of mild steel by both dynamic explicit integration and quasi-static methods. Figure 7.9(a) shows the part shape, and Figures 7.9(b) and 7.9(c) show the results from the dynamic explicit integration and the quasi-static method, respectively. They are completely different. The dynamic solution has abrupt change immediately beyond the binder surface inside the punch opening. The abrupt change is caused by the inertia effect because there is no punch pressing the sheet inside the punch opening at the binder wrap stage. The quasi-static solution does not show any abrupt change, but the deformation is not symmetric.

Now we examine the deformation for the binder wrap of a mild steel fender, as discussed in a paper by Balun *et al.* [1993]. Figure 7.10(a) shows four sections and one profile taken for measurements. A comparison of the quasi-static solutions with the measurements is shown in Figure 7.10(b). Note that there are virtually no differences between the numerical predictions and the measurements because of the scale used for the figure. The maximum discrepancy between the numerical predictions and the measurements is 3.6 mm at the section of $x = -508$. The numerical solutions are accurate enough for practical use.

7.8.2 Numerical Examples of Sheets During Stretching or Drawing

A fender from the benchmark of NUMISHEET'93 (1993) is used as an example. NUMISHEET'93 is the abbreviation for "Numerical Simulation of 3-D Sheet Metal Forming Processes—Verification of Simulation with Experiment—in 1993." The conference proceedings is edited by Makinouchi *et al.* [1993]. Chapter 1 of this book briefly introduced this international conference. In NUMISHEET'93, there were three benchmark problems: square cup drawing, front fender stamping, and two-dimensional draw or bending.

The material of the fender was mild steel, and its original thickness of the sheet was 0.81 mm. Figure 7.11(a) shows the part shape of the fender, and Figures 7.11(b) through 7.11(d) show the major and minor strains and the thickness change, respectively. Based on the major and minor principal strain components shown in Figures 7.11(b) and 7.11(c), we can apply the forming limit diagram (FLD) mentioned in Chapter 3 of this book for the fender material to determine where fracture or splitting may occur. Figure 7.11(d) shows a spot (labeled A) where the deformed metal is too thin (thickness reduction by 39%) for an outer panel. Figure 7.11(e) displays the standard deviation of the predicted thinning from measurement. This measurement was taken at a section near the front of the fender marked "Sec C" in Figure 7.11(d). Figure 7.11(e) also shows the

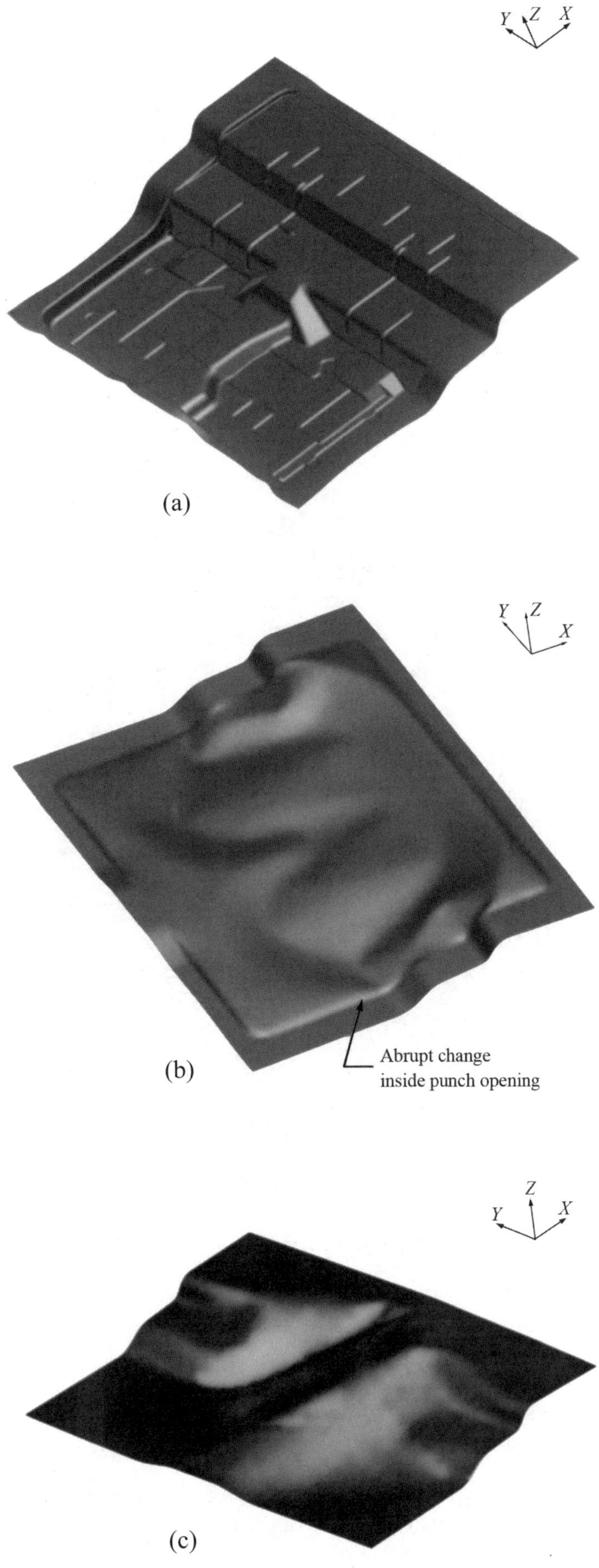

Figure 7.9 Binder wrap for a floor pan: (a) part shape, (b) results from the dynamic explicit integration method for binder wrap, and (c) asymmetrical deformation mode from the quasi-static implicit method.

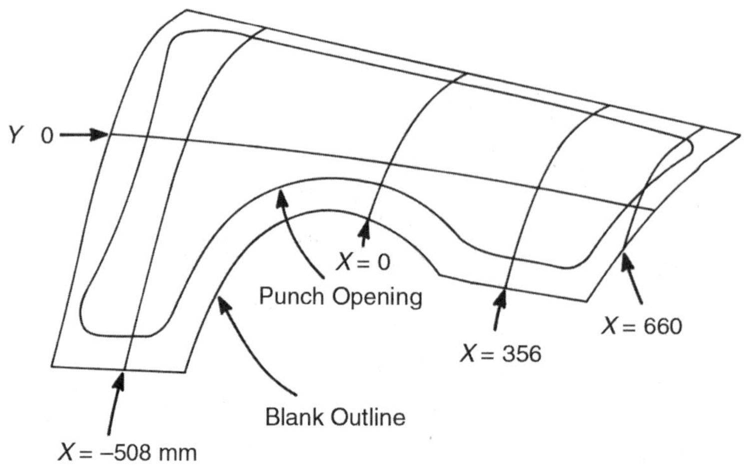

(a)

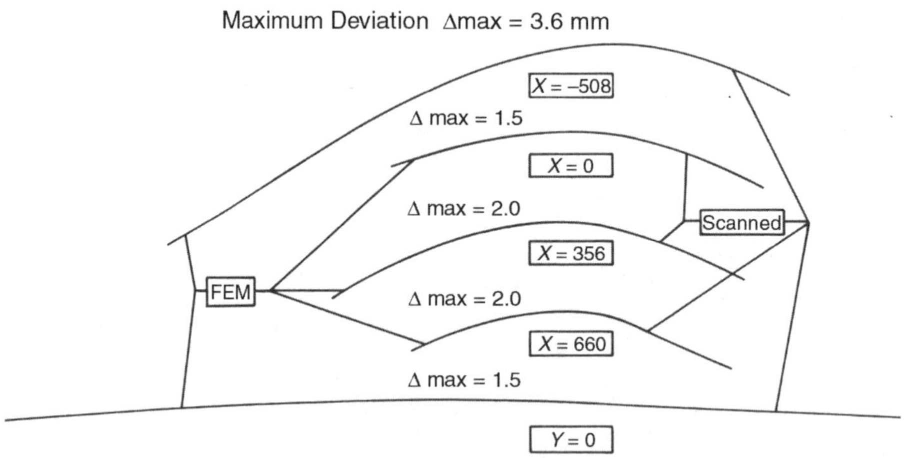

(b)

Figure 7.10 Binder wrap for a fender: (a) quasi-static solution, and (b) comparison of the numerical solution with actual measurements.

standard deviation of the forming results from participants from different institutes and the simulation software. In Figure 7.11(e), f50, f75, and f150 mean the draw-bead restraint force of 50, 75, and 100 N/mm, respectively, used for simulation. Also, "T" denotes triangular membrane elements, whereas "R" denotes rectangular membrane elements. Some results in Figure 7.11(e) demonstrate excellent prediction.

7.8.3 Numerical Examples of Springback After Various Forming Operations

In this section, we show a few simple examples of computing springback after releasing from drawing, trimming, and flanging.

The first example is a ring cut (trimmed) from an aluminum cylindrical cup, as shown in Figure 7.12(a). The original thickness of the cup material was 1.6 mm. Because the

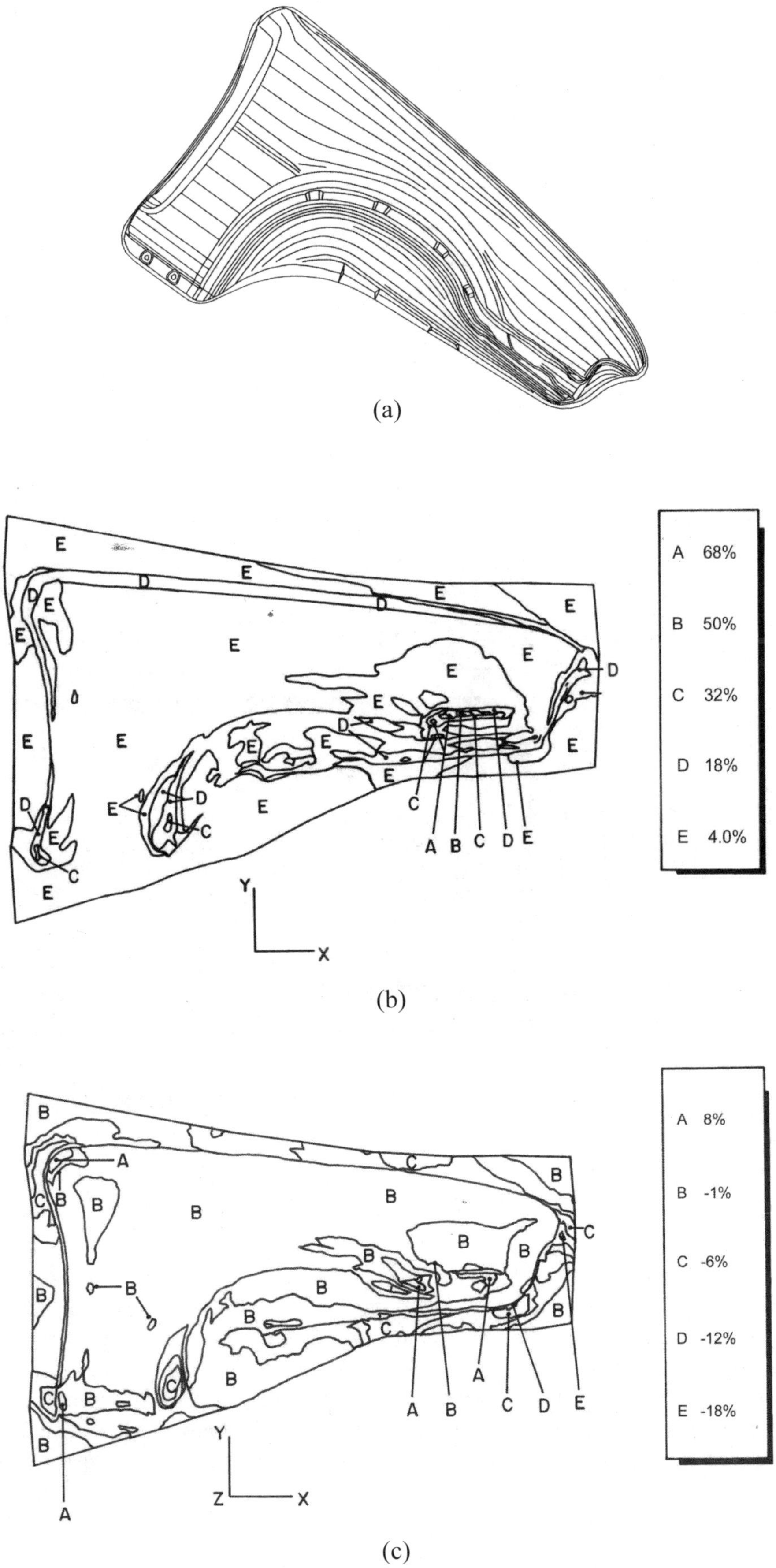

Figure 7.11 NUMISHEET'93 benchmark fender: (a) part shape, (b) major strain distribution, and (c) minor strain distribution.

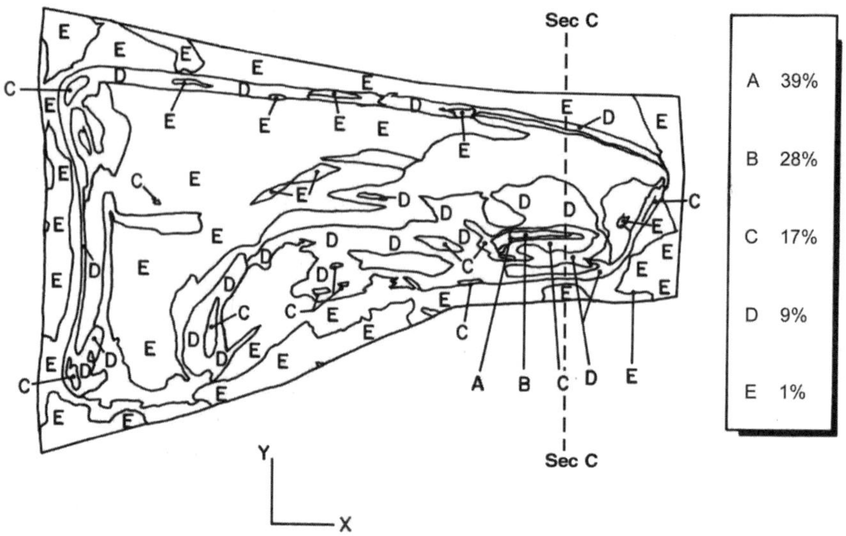

(d)

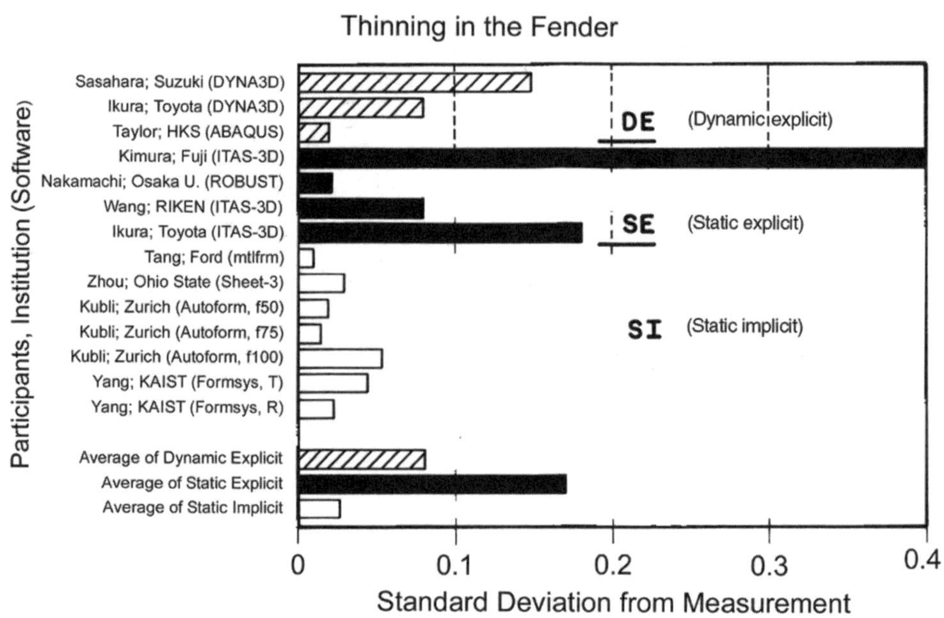

(e)

Figure 7.11 *(Continued)* (d) Thickness reduction, and (e) standard deviations of various numerical results from actual measurements.

ring was stretched mainly along the axis of the cup during the draw and forming process, there was little springback after it was cut from the cup. However, after a slit parallel to the axis of the cup was made, a wide gap was observed, as shown in Figure 7.12(b), because of springback from the bending residual stresses. The gap predicted by the quasi-static method was 48 mm versus that from the measurement of 50 mm. Because the amount of springback was huge, this is a good example to be used for checking simulation software.

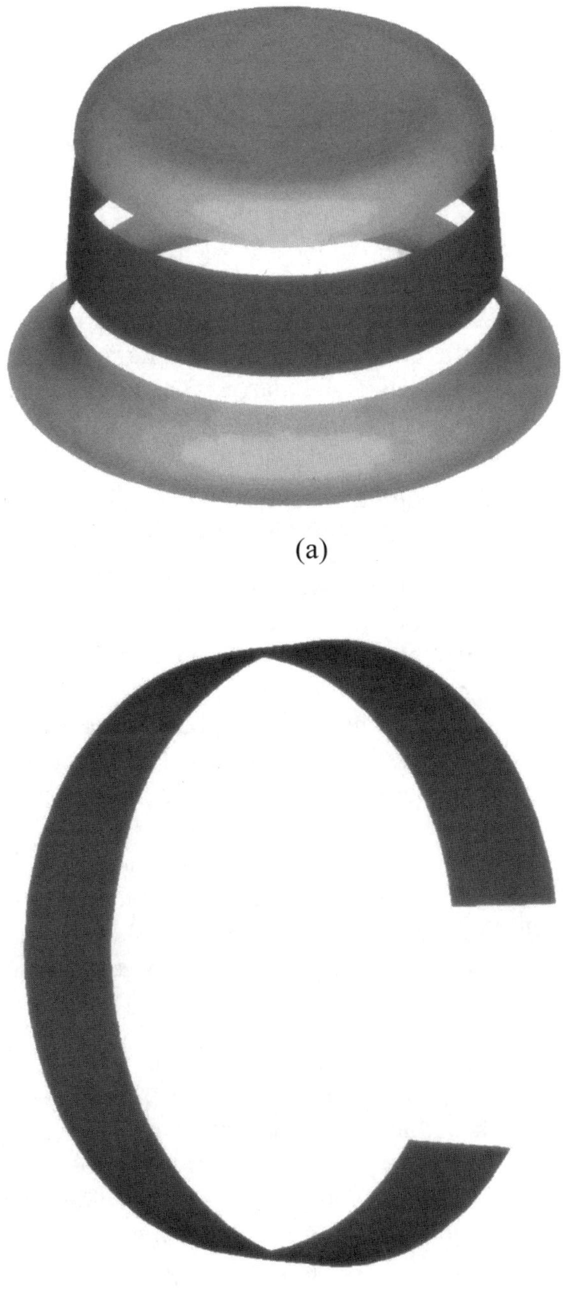

(a)

(b)

Figure 7.12 Springback of a ring cut from a cup: (a) ring trimmed from a cup, and (b) ring split with a wide opening.

The second example is the springback of the outer portion of a hood. The aluminum hood with the original thickness of 0.9 mm was formed and trimmed as shown in Figure 7.13(a). The reason for using trimmed parts to compute the springback was that we do not want to model the draw-beads after the part was released from the forming tools. Also, because the geometries of the draw-beads are rather complex, the draw-beads generally have tight corner radii that require three-dimensional solid elements for a more exact model. Therefore, engineers would skip the analysis of springback for parts with draw-beads immediately released from forming tools and go directly to trimmed parts. Taking advantage of symmetry, only one-half of the hood was modeled and analyzed for the draw and forming and for springback after trimming by both dynamic explicit integration and quasi-static methods. Measurements of springback of the trimmed hood also were made to verify the accuracy of prediction. Because the trimmed hood outer portion is not stiff, it was hung almost vertically to minimize the gravitational effect on deformation for measuring the amounts of springback. Figure 7.13(b) shows the lines (sections) where the measurements were made. Figure 7.13(c) shows the numerical results at section a-a (see Figure 7.13(b)) in comparison to the die shape that is the desired shape without any springback. Figure 7.13(c) shows two sets of results predicted by two different computer codes using the dynamic explicit integration method for the forming simulation with the quasi-static method and the dynamic relaxation method for springback computation. Figure 7.13(c) also shows one set of results predicted by the computer code using the quasi-static implicit method with equilibrium check for both forming and springback predictions. The maximum deviation for the three predictions from the measurements is less than 10 mm.

The third example is an aluminum hood inner with the original thickness of 0.9 mm, as shown in Figure 7.14(a). The dynamic explicit integration method was used for the forming simulation, and the quasi-static method was used to compute the springback of the trimmed hood inner portion. The other set of numerical results was from the quasi-static implicit method with equilibrium check for both the forming and springback predictions. Measurements of springback at several selected points also were made for validation of the computational results. The measurement points are shown in Figure 7.14(b). (Due to symmetry, only half of the panel is shown.) Figure 7.14(c) shows the results from the dynamic explicit integration method with the quasi-static method for springback computation, the results from the quasi-static implicit method with equilibrium check for both forming and springback predictions, and the measurements. The maximum deviation for the two predictions from the measurements is less than 2 mm.

The last example is the springback after a simple, straight flanging operation. The sheet material was aluminum. The ratio of the die gap Δ to the sheet thickness h was assumed to be 1.1. Numerical calculations were performed for the ratio of the die corner radius r to the sheet thickness h ranging from 1 to 5, which is realistic in forming automotive sheet metal parts. The purpose of the study by Xia and Tang [1999] was to understand the effects of the ratio r/h on springback angles and the accuracy of using thin shell element models. Figure 7.15(a) shows a mathematical model for three-dimensional solid elements. Two models with four and eight elements through the thickness of the sheet were used for the three-dimensional solid element model. Figure 7.15(b) shows a mathematical model for thin shell elements. Because we normally use the middle surface of a thin shell as the reference surface on which the forming tools contact, we modify the die corner radius to $r + h/2$ and the die gap to $\Delta - h$ as shown in Figure 7.15(b) to account for the thickness of the sheet metal when thin shell elements are used. For the shell model, seven and fifteen integration points were employed to integrate the stresses through the thickness of the shell elements. Figure 7.15(c) shows the computational results for the models with the three-dimensional solid elements and thin shell elements. The dashed lines in the figure are the results for the solid model with four elements through the thickness and for the shell model with seven integration points through the thickness, respectively. The results of the springback shown in the figure indicate that the springback decreases as r/h decreases. However, this study has some engineering significance. First, the difference of the springback angles predicted by both models is less than 5% for r/h greater than 4. Second, four elements through the thickness for

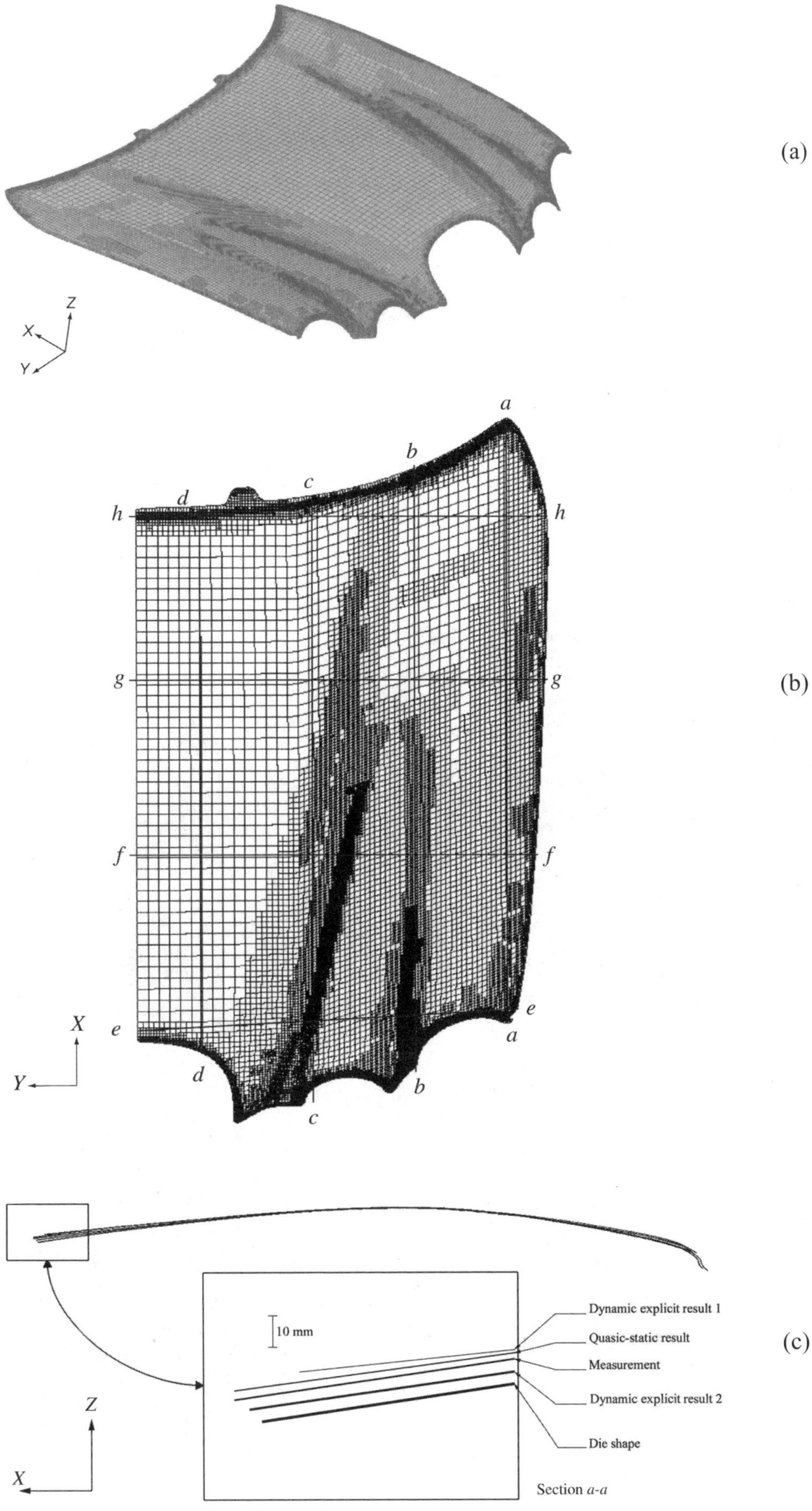

Figure 7.13 Springback of a trimmed hood outer: (a) part shape, (b) location of sections for measurements of springback, and (c) results from the dynamic explicit and quasi-static integration methods for section a-a.

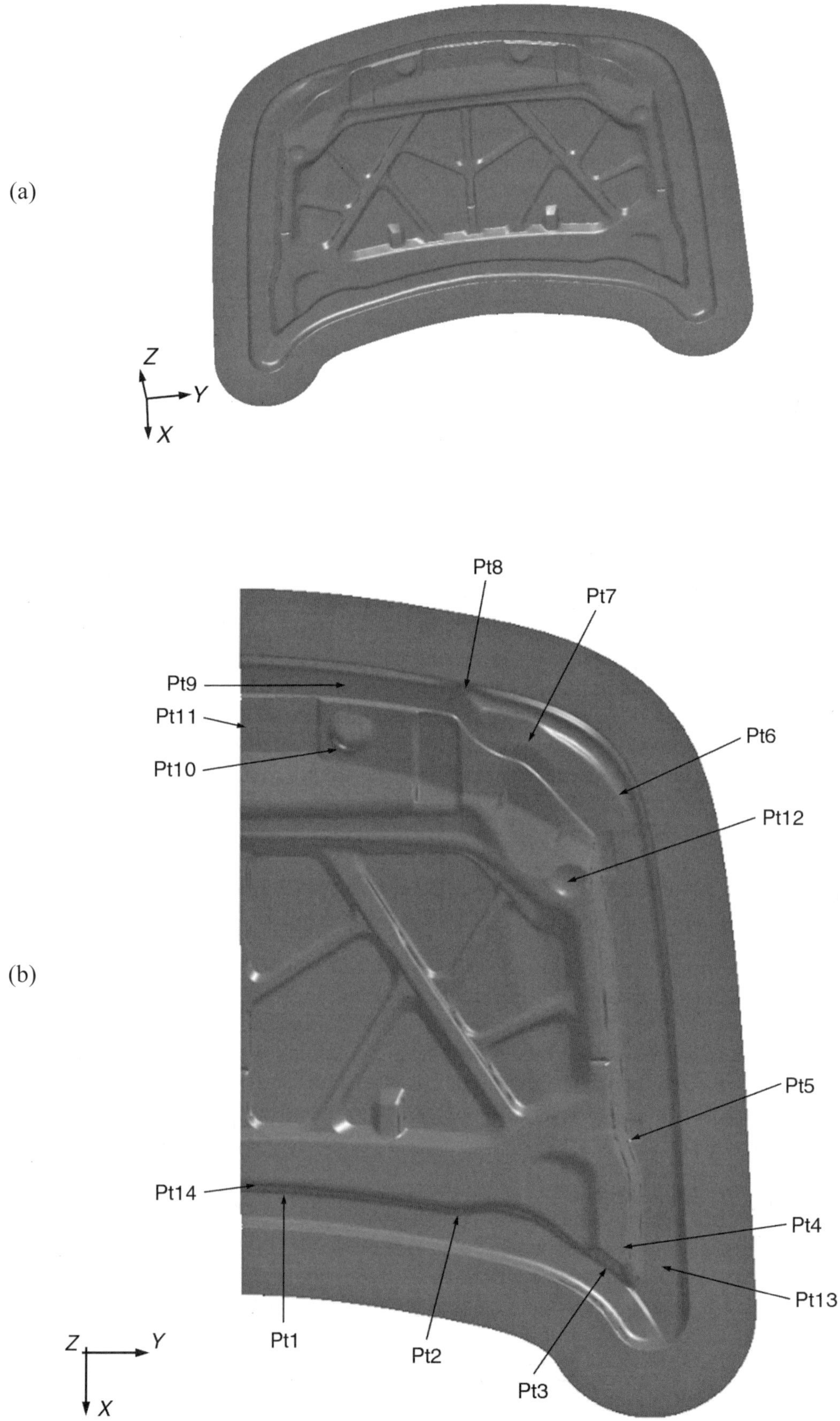

Figure 7.14 Springback of a trimmed hood inner: (a) part shape, and (b) locations of points for measurement of springback.

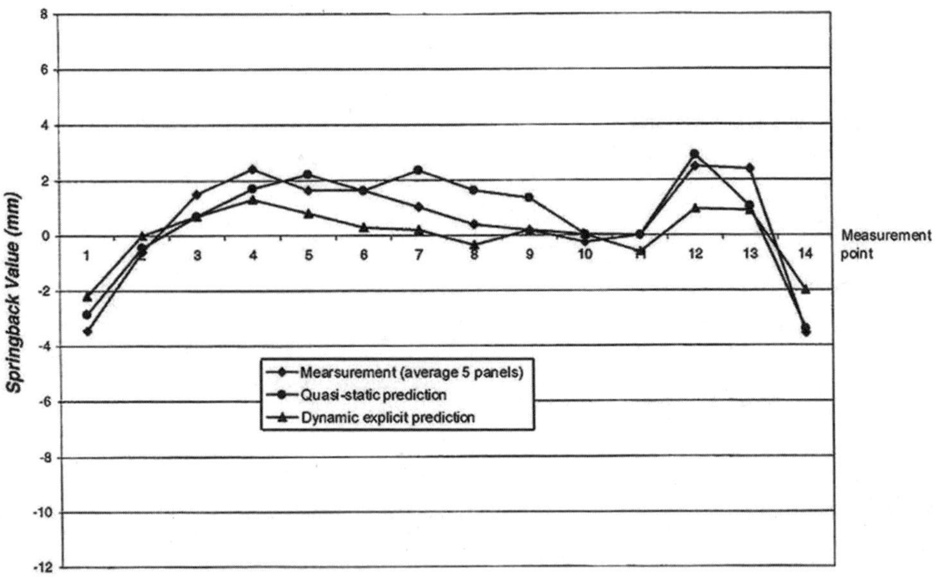

(c)

Figure 7.14 *(Continued)* (c) Results from the dynamic explicit and quasi-static methods.

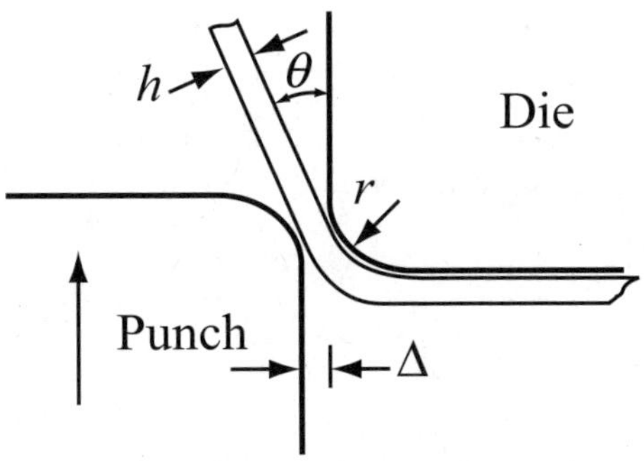

(a)

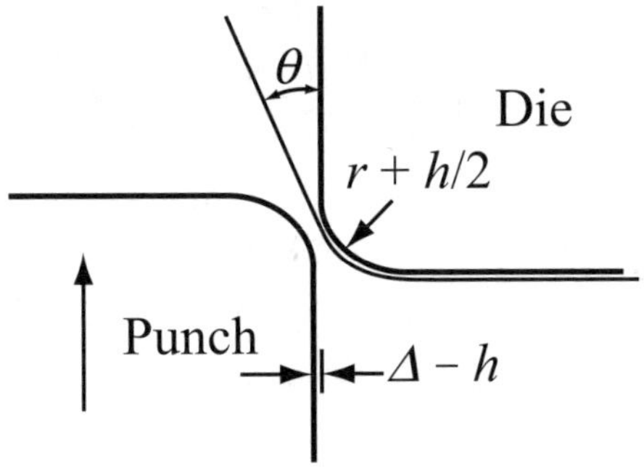

(b)

Figure 7.15 Springback after a flange operation: (a) model for three-dimensional solid finite elements, and (b) model for thin shell finite elements.

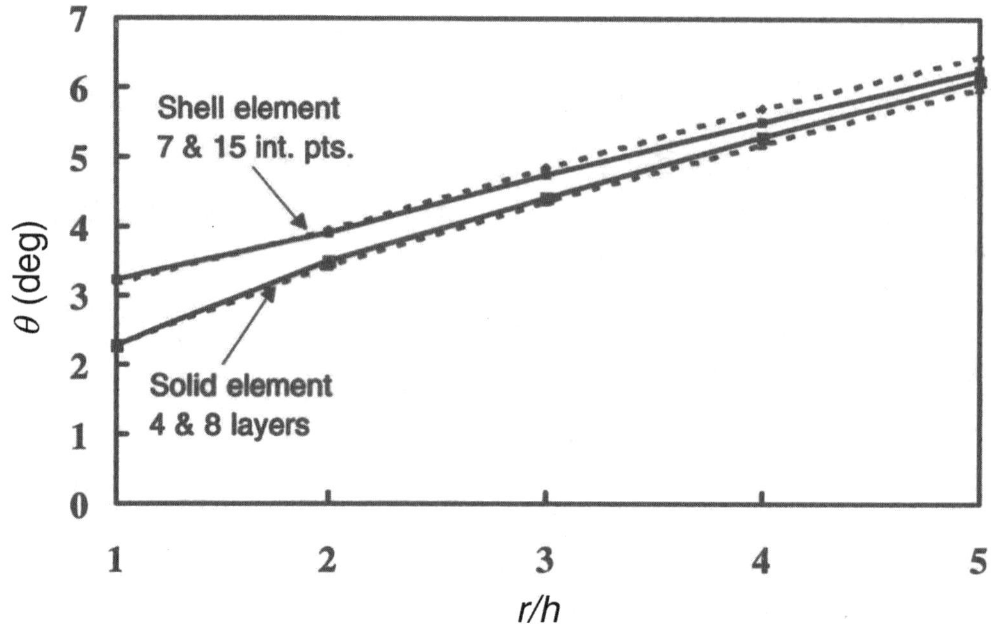

(c)

Figure 7.15 *(Continued)* (c) Predicted springback angles.

the solid element model or seven integration points through the thickness of the shell element model are accurate enough for practical applications.

7.9 References

Balun, T., Ling, P., Lou, M.M.K., and Tang, S.C., "Detection and Elimination of Wrinkles on Auto-Body Panels by the Binder Set Analysis," SAE Paper No. 930515, Society of Automotive Engineers, Warrendale, PA, 1993.

Bathe, K.J., *Finite Element Procedures*, Prentice Hall, Englewood Cliffs, NJ, 1996.

Belytschko, T., Liu, W.K., and Moran, B., *Nonlinear Finite Elements for Continua and Structures*, John Wiley, New York, 2000.

Chu, C.C., "A Three Dimensional Model of Anisotropic Hardening in Metals and Its Application to the Analysis of Sheet Metal Formability," *Journal of Mechanics and Physics of Solids*, Vol. 32, 1984, p. 197.

Gupta, A. and Kumar, V., "WSMP: A High-Performance Shared- and Distributed-Memory Parallel Sparse Linear Equation Solver," IBM Research Report, Computer Science/Mathematics, RC 22038 (98932), 2001.

Hallquist, J.O. (compiled), *LS-DYNA Theoretical Manual*, Livermore Software Technical Corp., Livermore, CA, 1998.

Hill, R., *The Mathematical Theory of Plasticity*, Clarendon Press, Oxford, 1950.

Hughes, T.J.R., *The Finite Element Method*, Prentice Hall, Englewood Cliffs, NJ, 1987.

Makinouchi, A., Nakamachi, E., Oñate, E., and Wagoner, R.H. (eds), Proceedings of the 2nd International Conference—NUMISHEET'93, Isehara, Japan, 1993.

Mroz, Z., "On the Description of Anisotropic Work-Hardening," *Journal of Mechanics and Physics of Solids*, Vol. 15, 1967, p. 163.

Newmark, N.M., "A Method of Computation of Structural Dynamics," *ASCE Journal of Engineering Mechanics Division*, Vol. 85, 1959, p. 67.

Tang, S.C., "Analysis of Springback in Sheet Forming Operation," *Advanced Technology of Plasticity*, Proceedings of the Second International Conference on the Technology of Plasticity (2ICTP), Vol. 1, Stuttgart, Germany, 1987, p. 193.

Tang, S.C., "An Anisotropic Hardening Rule for the Analysis of Sheet Metal Forming Operations," *Advanced Technology of Plasticity*, Proceedings of the Third International Conference on the Technology of Plasticity (3ICTP), Vol. 3, Kyoto, Japan, 1990, p. 1149.

Tang, S.C., "Trend on Simulation of Sheet Metal Forming," SAE Paper No. 2000-01-1108, Society of Automotive Engineers, Warrendale, PA, 2000.

Tang, S.C. and Carnes, J.C., "Method for Aiding Sheet Metal Forming Tooling Design," U.S. Patent No. 5,379,227, 1995.

Tang, S.C., Xia, Z.C., and Ren, F., "Application of the Radial Return Method to Compute Stress Increments from Mroz's Hardening Rule," *Journal of Engineering Materials and Technology*, American Society of Mechanical Engineers, Vol. 123, 2001, p. 398.

Wang, N.-M. and Budiansky, B., "Analysis of Sheet Metal Stamping by a Finite-Element Method," *Journal of Applied Mechanics*, American Society of Mechanical Engineers, Vol. 45, 1978, p. 73.

Xia, Z.C. and Tang, S.C., "Locking Phenomena in the Use of Solid Elements for Sheet Metal Forming Simulation," SAE Paper No. 1999-01-1005, Society of Automotive Engineers, Warrendale, PA, 1999.

CHAPTER

8

Buckling and Wrinkling Analyses

8.1 Introduction

Although sheet metal forming processes are mostly the stretching of thin sheets, compressive stresses still can be generated, such as in drawing. For the quasi-static formulation, the compressive stresses may cause structural instabilities of thin sheets. In the application of finite element analysis to thin shells, a global tangent stiffness matrix for the quasi-static method is not positive-definite for an unstable structure. At a point in the deformed state for a sheet, the determinant of a global tangent stiffness matrix changes signs from positive to negative, and, consequently, buckling or wrinkling occurs. This point is called the critical point. From a mechanics point of view, buckling or wrinkling occurs at the critical point where the global tangent stiffness matrix is singular or the determinant of the global tangent stiffness matrix is zero. A shell can deform in a long wavelength (buckling) mode and in a short wavelength (wrinkling) mode. In this book, we use the unified name "buckling" for both long- and short-wavelength deformation modes. In sheet metal forming processes, we are interested in not only the onset of buckling but also the post-buckling behavior, because the undesired buckled shapes during the early forming stage may be stretched in the final forming stage, and then the sheet metal part would not have any visible defects. That means the analysis of sheet metal forming processes must cover the entire range of deformation through the possible post-buckling stage.

In dynamic formulation, the equations of motion in Eq. 7.2.6 in Chapter 7 for the explicit integration always have a solution as long as the point mass at each node of a finite element model is not zero. We can solve the equations of motion in Eq. 7.3.31 for the implicit integration because of the mass and because the damping matrices are nonsingular. However, the tangent stiffness matrix for the implicit integration may become ill conditioned. Consequently, a much smaller time interval may be used to obtain an accurate numerical result if no special treatment is used.

In this book, we discuss only the buckling for the quasi-static formulation, which we believe is a better method for simulation. There are two types of buckling. Schematically, Figure 8.1(a) shows snap-through buckling, and Figure 8.1(b) shows bifurcation buckling. The applied load ρ is stationary at a critical point or limit point where the onset of snap-through buckling occurs. In press shops, a "bang" sound often is heard during snap-through buckling. The deformation mode changes at a bifurcation point during bifurcation buckling. The deformation mode is unstable along the primary path, but it is stable along the secondary path, as shown in Figure 8.1(b) for bifurcation buckling. A good example of bifurcation is the occurrence of wrinkling during a cup draw. The primary mode of deformation at the initial draw stage is axially symmetrical. However, the mode changes, and wrinkles appear in the flange during the early draw stage. Finally, the flange with wrinkles is drawn to the wall, as shown in Figure 8.2.

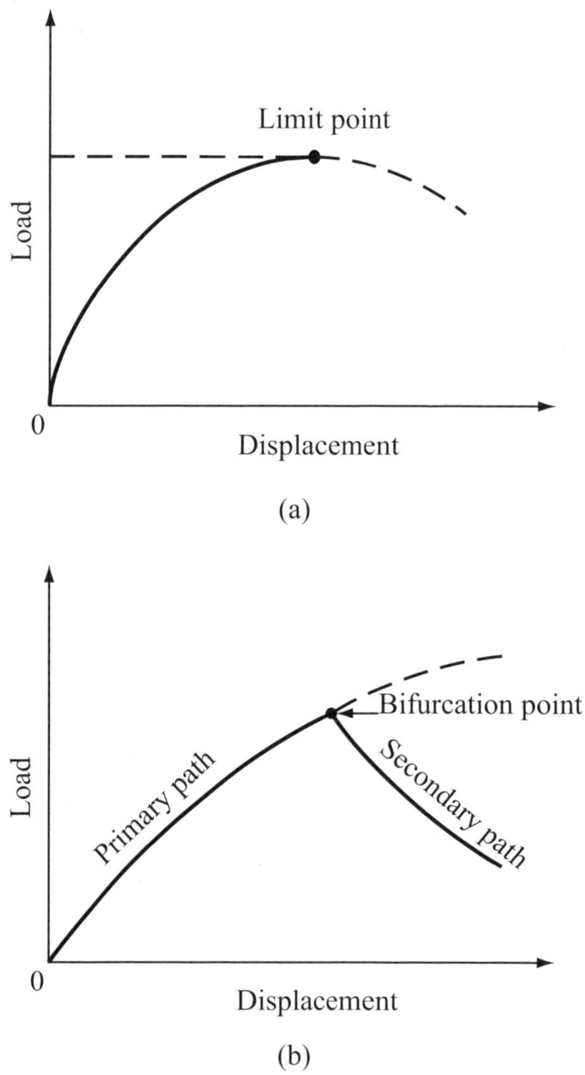

Figure 8.1 Load-displacement curves for (a) snap-through buckling and (b) bifurcation buckling.

Riks [1979] proposed a general method called the "arc length method" to compute the deformation mode near the critical point, where the onset of buckling initiates, and the post-buckling deformation. His method involves an external load parameter ρ. In analyses of sheet metal forming processes, we essentially solve a contact problem. However, the external load is not specified explicitly. Therefore, Riks' approach cannot be applied

Figure 8.2 Cup draw in the final draw stage, with flange wrinkles in the early draw stage.

directly. Nevertheless, his concept to analyze the onset of buckling and post-buckling deformation can be valid for analyses of sheet metal forming processes.

For snap-through buckling, the external load must be involved explicitly in Riks' constraint equation. If snap-through buckling occurs in a thin shell contacting a forming tool (where the external load is not specified explicitly), a different approach is needed and is discussed in this chapter. For a good shell element, the stable bifurcation mode would be triggered automatically due to small unbalanced forces that act as imperfections. The global tangent stiffness matrix in the neighborhood of a bifurcation point is ill-conditioned. Some special technique to overcome this difficulty in the numerical solution will be discussed at the end of this chapter.

8.2 Riks' Approach for Solution of Snap-Through and Bifurcation Buckling

At a critical point where the onset of buckling occurs, the global tangent stiffness matrix in the quasi-static analysis of thin shell structure is singular. Therefore, computation of a solution cannot be continued. Riks proposed the arc length method, making the solution continue to the post-buckling state. The arc length method works for both snap-through and bifurcation buckling problems.

If the dynamic terms are omitted, the principle of virtual work in Eq. 4.2.1 gives the equations of equilibrium as

$$f_i(\mathbf{x};\rho) = 0 \quad (i = 1,2,\dots,N) \tag{8.2.1}$$

where f_i contain the internal forces as well as the external force parameter ρ, $\mathbf{x}$ is a deformation or displacement vector for the deformed configuration of a shell, and N is the number of degrees of freedom in a finite element model to discretize a shell structure with properly assigned displacement boundary conditions. We note that the equations of equilibrium cannot be derived from the principle of minimum potential energy because a large amount of plastic deformation occurs during a forming process. Riks introduced the equilibrium paths of a shell in terms of a parameter s as

$$\begin{aligned} \rho &= \rho(s) \\ \mathbf{x} &= \mathbf{x}(s) \end{aligned} \tag{8.2.2}$$

with the initial conditions

$$\begin{aligned} \rho &= 0 \\ \mathbf{x} &= 0 \end{aligned} \tag{8.2.3}$$

at $s = 0$. With both ρ and $\mathbf{x}$, we can consider $[\rho;\mathbf{x}]$ as a point in an $N+1$ dimensional Euclidean space R_{N+1} with the orthonormal basis

$$\mathbf{e}_i = [\mathbf{e}_0, \mathbf{e}_1, \mathbf{e}_2, ..., \mathbf{e}_n] \tag{8.2.4}$$

The point $[\rho;\mathbf{x}]$ then can be expressed as

$$[\rho;\mathbf{x}] = \rho\mathbf{e}_0 + x_i\mathbf{e}_i \tag{8.2.5}$$

In addition to the equilibrium, Riks introduced the equation of

$$\left(\frac{d\rho}{ds}\right)^2 + \frac{dx_l}{ds}\cdot\frac{dx_l}{ds} = 1 \tag{8.2.6}$$

to constrain the load parameter ρ. The parameter s can be interpreted as the arc length of the curve in Eq. 8.2.2.

8.2.1 Critical Points

We differentiate the equations of equilibrium in Eq. 8.2.1 with respect to the parameter s once and twice, respectively, as

$$\dot{f}_i(\mathbf{x};\rho) = f_{i,j}\dot{x}_j + f_{i,0}\dot{\rho} = 0 \tag{8.2.7a}$$

$$\ddot{f}_i(\mathbf{x};\rho) = f_{i,j}\ddot{x}_j + \left(f_{i,0}\ddot{\rho} + \dot{f}_{i,j}\dot{x}_j + \dot{f}_{i,0}\dot{\rho}\right) = 0 \tag{8.2.7b}$$

where the single dot means the differentiation with respect to s once, the double dot means the differentiation with respect to s twice, and the subscript 0 means the partial differentiation with respect to ρ. The partial derivatives $f_{i,j}$ are the members of the global tangent stiffness matrix **K**, which is formed by assembling the element tangent stiffness matrices in Eq. 6.2.34 or Eq. 6.3.34 from Chapter 6 with specified displacement boundary conditions. We interpret Eq. 8.2.7a as the global tangent stiffness equation or equilibrium in the rate form.

At a critical point, the matrix $f_{i,j}$ is singular. Therefore, the smallest eigenvalue is zero because the tangent stiffness matrix before the critical point is positive-definite. The corresponding eigenvector (normalized) is $\mathbf{a}(1)$ with components $a_i(1)$. Here, the argument 1 represents the first mode corresponding to the smallest eigenvalue of zero. Later in this chapter, we deal with the smallest eigenvalue and the corresponding eigenvector. Therefore, we omit the argument 1 for simplicity. According to a theorem for ordinary differential equations, the nonhomogeneous parts of Eqs. 8.2.7a and 8.2.7b should satisfy the orthogonality conditions

$$\left[a_i f_{i,0} \dot{\rho}\right]_{s=s^*} = 0 \tag{8.2.8a}$$

$$\left[a_i \left(f_{i;0} \ddot{\rho} + \dot{f}_{i,j} \dot{x}_j + \dot{f}_{i,0} \dot{\rho}\right)\right]_{s=s^*} = 0 \tag{8.2.8b}$$

where the asterisk means that the values in the brackets are evaluated at the critical point $s*$.

It follows from the first requirement in Eq. 8.2.8a that there are two different cases:

$$\dot{\rho}\left(s^*\right) = 0 \tag{8.2.9a}$$

or

$$f_{i,0}\left(s^*\right) a_i = 0 \tag{8.2.9b}$$

The first equation indicates a critical point where the external load is stationary and snap-through occurs. The second equation indicates that bifurcation occurs or that the pre-buckling deformation path cannot continue at the critical point, while a different path of deformation can be generated after the critical point.

The conventional formulation of nonlinear shell structure problems is that either the load parameter ρ is specified or a component of deformation is specified. When ρ is specified, each point of solution is determined by the intersection of the surface $\rho = \eta$ and the equilibrium path governed by Eq. 8.2.1 as schematically shown in Figure 8.3(a). When a component x_k of the deformed coordinate (or a displacement component u_k) is specified, the solution is determined by the intersection of the surface $x_k = \eta$ and the equilibrium path governed by Eq. 8.2.1, as shown schematically in Figure 8.3(b).

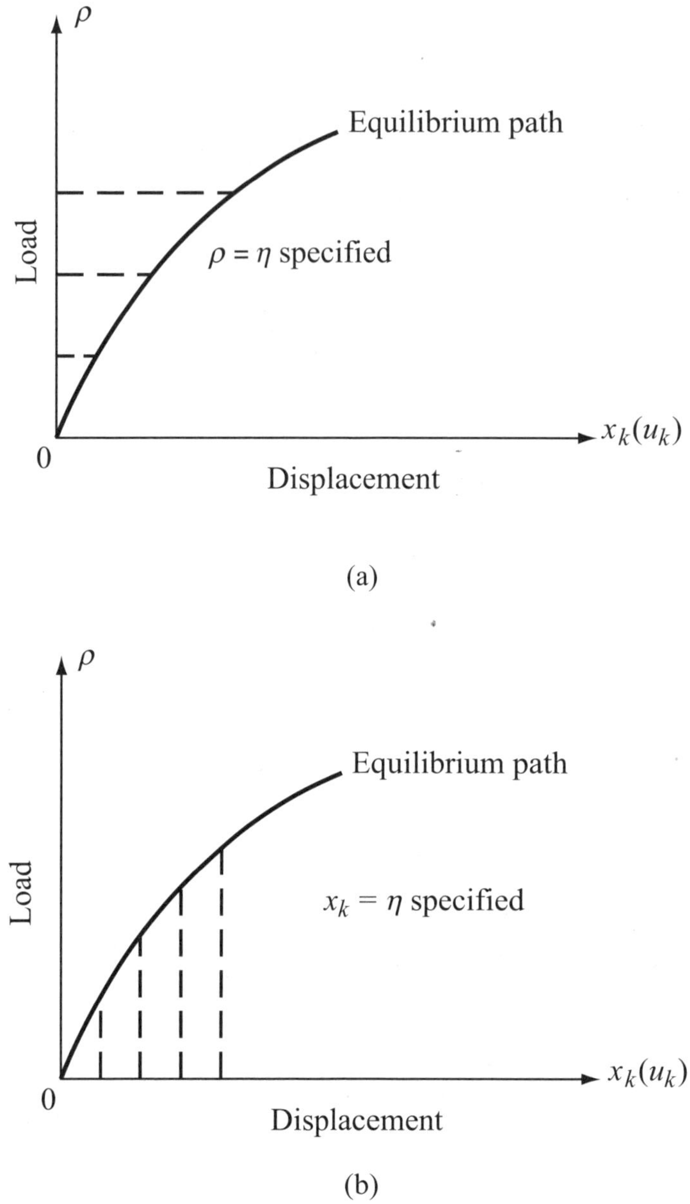

Figure 8.3 Solutions for (a) a specified load and (b) a specified displacement.

At the critical point,

$$\dot{\rho} = \frac{d\rho}{ds} = 0 \tag{8.2.10}$$

for $\rho = \eta^*$. Incrementing the value η beyond the limit value η^* leads to the loading surface

$$f_0 = \rho - \eta = 0 \tag{8.2.11}$$

that does not intersect with the equilibrium path, at least in the neighborhoods of the critical point defined by Eq. 8.2.10, as shown in Figure 8.4(a). This is called snap-through buckling. For the other case,

$$\dot{x}_k = \frac{dx_k}{ds} = 0 \tag{8.2.12}$$

for $x_k = \eta^{**}$. Incrementing the value η beyond the limit value η^{**} leads to the loading surface

$$f_0 = x_k - \eta = 0 \tag{8.2.13}$$

that does not intersect with the equilibrium path, at least in the neighborhoods of the critical point defined by Eq. 8.2.12, as shown in Figure 8.4(b). This is called snap-back. At these critical points, the conventional formulation for a specified load or deformation parameter fails to obtain a solution.

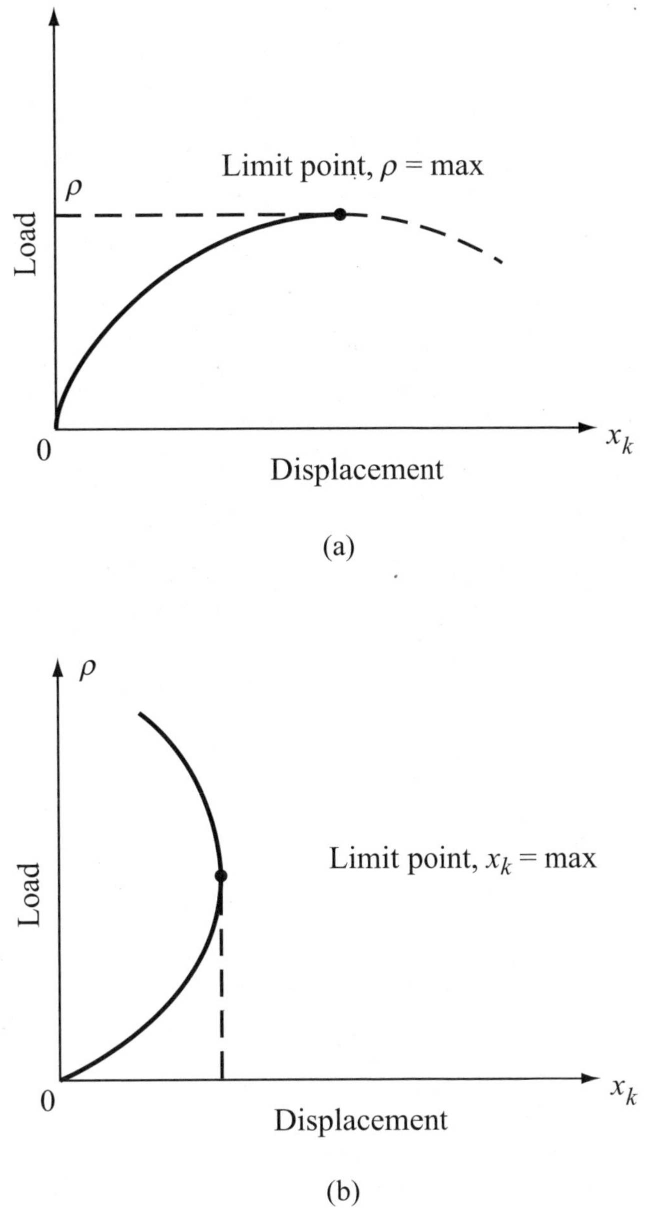

Figure 8.4 Load-displacement curves for (a) snap-through buckling and (b) snap-back.

Riks proposed the constraint condition in Eq. 8.2.6 so the loading surface includes those arguments ρ, $\mathbf{x}$, and s, such that

$$f_0(\rho, \mathbf{x}, s) = 0 \tag{8.2.14}$$

After we find the critical point, the smallest eigenvalue, and the corresponding eigenvector, we combine the equilibrium path in Eq. 8.2.1 and the loading surface in Eq. 8.2.14, and then we can calculate the post-buckling deformation, as discussed in the next section.

8.2.2 Establishment of Governing Equations in the $N+1$ Dimensional Space

Riks proposed Eq. 8.2.14 as a plane in the $N+1$ Euclidian space with the parameter s

$$f_0(\rho; \mathbf{x}; s) = \dot{\rho}_1(\rho - \rho_1) + \dot{\mathbf{x}}_1 \cdot (\mathbf{x} - \mathbf{x}_1) - (s - s_1) = 0 \tag{8.2.15}$$

where $[\rho_1, \mathbf{x}_1]$ and $[\dot{\rho}_1, \dot{\mathbf{x}}_1]$ denote a point on the path and its corresponding tangent (normalized) to the surface at the point, respectively, and s_1 denotes the arc length of the point $[\rho_1, \mathbf{x}_1]$ from the origin of the coordinate system. (This subscript 1 does not mean the first component of a vector.) Differentiating Eq. 8.2.15 with respect to s yields

$$\dot{\rho}(s_1)\dot{\rho}(s) + \dot{x}_l(s_1)\dot{x}_l(s) - 1 = 0$$

or

$$\dot{\rho}(s_1)\dot{\rho}(s) + \dot{x}_l(s_1)\dot{x}_l(s) = 1 \tag{8.2.16}$$

which is equivalent to Eq. 8.2.6 as s approaches s_1.

Combining Eqs. 8.2.15 and 8.2.1 generates a system of $N+1$ equations. We write them in the condensed form as

$$\begin{aligned} &f_L(\mathbf{x}; s) = 0 \\ &L = 0, 1, 2, \ldots, N \\ &x_0 = \rho \end{aligned} \tag{8.2.17}$$

We note that the range of the subscript L is from 0 to N. Using this system, we can solve $N+1$ unknowns x_L ($L = 0, 1, 2, \ldots, N$) for a specified arc length s. The equilibrium state expressed by Eq. 8.2.1 is nonlinear. Therefore, Eq. 8.2.17 is nonlinear despite Eq. 8.2.15 being linear. It can be solved incrementally with a step size Δs by the Newton-Raphson iteration.

Applying the Newton-Raphson iteration, we can compute the equilibrium path from the arc length zero to some specified value s_m with an incremental step size Δs. We assume one point is given together with its path tangent and that it denotes these vectors by

$$\begin{aligned} \mathbf{x}_1 &= \mathbf{x}_1(s_1) \\ \dot{\mathbf{x}}_1 &= \dot{\mathbf{x}}_1(s_1) \end{aligned} \tag{8.2.18}$$

An estimate of the next point of the path is then given by

$$\mathbf{x}^{(1)} = \mathbf{x}_1 + \dot{\mathbf{x}}_1 \Delta s$$

The approximation solution $\mathbf{x}^{(1)}$ can be improved by the Newton-Raphson iteration

$$\begin{aligned} & f_{L,M}\left(\mathbf{x}^{(i)}\right)\Delta x_M^{(i)} = -f_L\left(\mathbf{x}^{(i)};\ s_1 + \Delta s\right) \\ & (L, M = 0, 1, 2, \ldots, N) \\ & (i = 1, 2, 3, \ldots) \end{aligned} \tag{8.2.19a}$$

$$\mathbf{x}^{(i+1)} = \mathbf{x}^{(i)} + \Delta \mathbf{x}^{(i+1)} \tag{8.2.19b}$$

until the residue $\left\| f_L\left(\mathbf{x}^{(i+1)};\ s_1 + \Delta s\right) \right\|$ is sufficiently small.

For the computation of the next point on the path, we must compute the tangent $\dot{\mathbf{x}}(s)$. This can be done by solving the set of the following simultaneous equations

$$f_{L,M}(\mathbf{x})\, y_M = -f_{L,s} \quad (L, M = 0, 1, 2, \ldots, N) \tag{8.2.20a}$$

where

$$f_{L,s} = -\delta_{L0} \quad \text{(Kronecker delta)}$$

because only Eq. 8.2.15 explicitly includes the arc length s. Here, we normalize the tangent vector

$$\dot{\mathbf{x}}(s) = \frac{\mathbf{y}}{\|\mathbf{y}\|} \tag{8.2.20b}$$

We can solve both Eqs. 8.2.19 and 8.2.20 only if the coefficient matrix $f_{L,M}$ is non-singular. In the next section, we will investigate the properties of this matrix.

8.2.3 Characteristics of Governing Equations in the $N+1$ Dimensional Space

Let **J** be the coefficient matrix of the linear system in Eq. 8.2.19 or 8.2.20, and

$$\mathbf{J} = \begin{bmatrix} f_{0,0} & f_{0,1} & \cdot & \cdot & \cdot & f_{0,N} \\ f_{1,0} & & & & & f_{1,N} \\ \cdot & & & & & \cdot \\ \cdot & & f_{i,j}(i,j=1,2,\ldots,N) & & & \cdot \\ \cdot & & & & & \cdot \\ f_{N,0} & f_{N,1} & \cdot & \cdot & \cdot & f_{N,N} \end{bmatrix} \tag{8.2.21}$$

With some matrix manipulations, we can write the determinant J of the matrix **J** as

$$J = f_{0,0}K + f_{0,j}d_j \tag{8.2.22}$$

where K is the determinant of the global tangent stiffness matrix $f_{i,j}$ in Eq. 8.2.7, and d_j are defined as

$$d_j = -\frac{\partial K}{\partial f_{i,j}} f_{i,0} \tag{8.2.23}$$

We multiply both sides of the equations of equilibrium in Eq. 8.2.7a by $\dfrac{\partial K}{\partial f_{i,k}}$

$$\frac{\partial K}{\partial f_{i,k}} f_{i,j}\dot{x}_j + \frac{\partial K}{\partial f_{i,k}} f_{i,0}\dot{\rho} = 0 \tag{8.2.24}$$

where $\partial K / \partial f_{i,k}$ are the cofactors for the matrix **K**. Therefore,

$$\frac{\partial K}{\partial f_{i,k}} f_{i,j} = K\delta_{kj}$$

Using the definition of d_i in Eq. 8.2.23, Eq. 8.2.24 can be written as

$$K\dot{x}_i - d_i\dot{\rho} = 0 \tag{8.2.25}$$

At a critical point, $K = 0$. For snap-through buckling, $\dot{\rho} = 0$; however, d_i are not necessarily zero. Therefore, the determinant of the matrix **J** in Eq. 8.2.22 is not necessarily zero. However, for bifurcation buckling, $\dot{\rho} \neq 0$. Therefore, d_i must be zero because

$K = 0$ in Eq. 8.2.25, and Eq. 8.2.22 shows that $\mathbf{J}$ is singular because $J = 0$. The appendix of Riks' paper shows that $d_i = 0$ also can be derived from Eq. 8.2.9b.

8.2.4 Solution for Snap-Through Buckling

For this case, the coefficient matrix $\mathbf{J}$ of the linear system of governing equations in Eqs. 8.2.19 and 8.2.20 is not singular at the limit point. (Note that $\mathbf{K}$ is singular.) A Newton-Raphson iteration can be applied without any modification. To compute the incremental solution for a specified arc length increment Δs $(= s - s_1)$, the initial trial solution is based on the tangent vector from Eq. 8.2.20. For the subsequent iteration, we solve the system of Eq. 8.2.19

$$f_{L,M}\left(\mathbf{x}^{(i)}\right)\Delta x_M^{(i)} = -f_L\left(\mathbf{x}^{(i)};\ s_1 + \Delta s\right) \quad (L, M = 0, 1, 2, \ldots, N) \tag{8.2.26a}$$

for

$$\Delta x_M^{(i)} = -\left[f_{L,M}\left(x^{(i)}\right)\right]^{-1} f_L\left(\mathbf{x}^{(i)};\ s_1 + \Delta s\right) \tag{8.2.26b}$$

and update the coordinates

$$x_M^{(i+1)} = x_M^{(i)} + \Delta x_M^{(i)} \quad (i = 1, 2, \ldots) \tag{8.2.27}$$

until $\left\| f_L\left(\mathbf{x}^{(i)};\ s_1 + \Delta s\right)\right\|$ is sufficiently small. The solution is convergent. Figure 8.5 shows schematically the method that can be used to pass the critical point for snap-through buckling. Note that s_i in the figure represents s_1 as the starting point. Using Eq. 8.2.20, we compute the normalized tangent vector at the starting point $\mathbf{x}(s_i)$. We locate the approximate solution along the tangent and then use Eq. 8.2.19 to obtain the convergent solution $\mathbf{x}(s_{i+1})$, as shown in Figure 8.5.

8.2.5 Methods to Locate the Secondary Path for Bifurcation Buckling

We rewrite Eq. 8.2.25 here as

$$K\dot{x}_i - d_i\dot{p} = 0 \tag{8.2.28}$$

when $\dot{p}$ is not zero for bifurcation buckling at the critical (bifurcation) point. Therefore, d_i must be zero because $K = 0$. Equation 8.2.22 shows that the determinant of the coefficient matrix $\mathbf{J}$ is zero. Hence, we cannot continue the solution of the system in Eq. 8.2.19 for Δx_M along the primary path where the shell structure is unstable. We need to locate a secondary path at the critical point, and the solution can continue beyond the critical point on the secondary path where the matrix $\mathbf{J}$ is nonsingular.

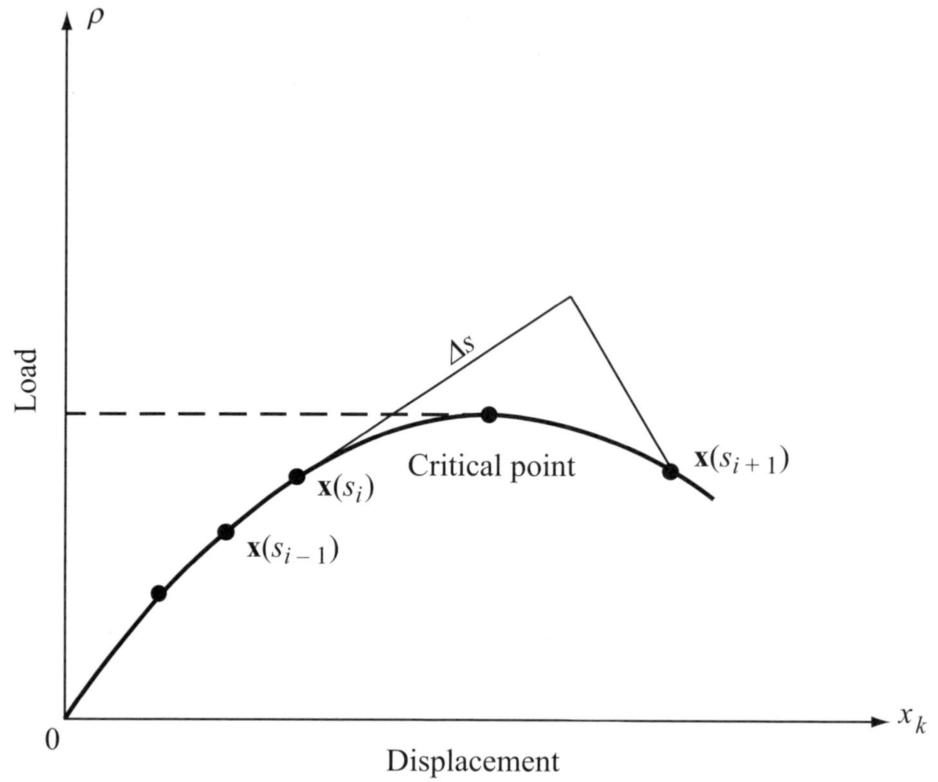

Figure 8.5 Schematic of a solution procedure to pass the critical point for snap-through buckling.

We assume the tangent to the secondary path at the critical point s^* has the form

$$[\dot{\rho}_{II}, \dot{\mathbf{x}}_{II}] = [\mu\dot{\rho}_I, \mu\dot{\mathbf{x}}_I + \mathbf{a}] \tag{8.2.29}$$

where the subscripts I and II denote the primary and the secondary paths, respectively, as shown schematically in Figure 8.6, $\mathbf{a} = [a_1, a_2, ..., a_N]$ is the eigenvector corresponding to the zero eigenvalue defined previously, and the factor μ is to be determined.

Combining Eqs. 8.2.8b and 8.2.9b yields

$$a_i\left[f_{i,j}\dot{x}_j + f_{i,0}\dot{\rho}\right]_{s=s^*} = 0 \tag{8.2.30}$$

Inserting the tangent vector $[\dot{\rho}_{II}, \dot{\mathbf{x}}_{II}]$ in Eqs. 8.2.29 and 8.2.30, we can solve for

$$\mu = -\frac{a_i f_{i,jk} a_j a_k}{2\left(a_i f_{i,jk} a_j \dot{x}_{Ik} + a_i f_{i,j0} a_j \dot{\rho}_I\right)} \tag{8.2.31}$$

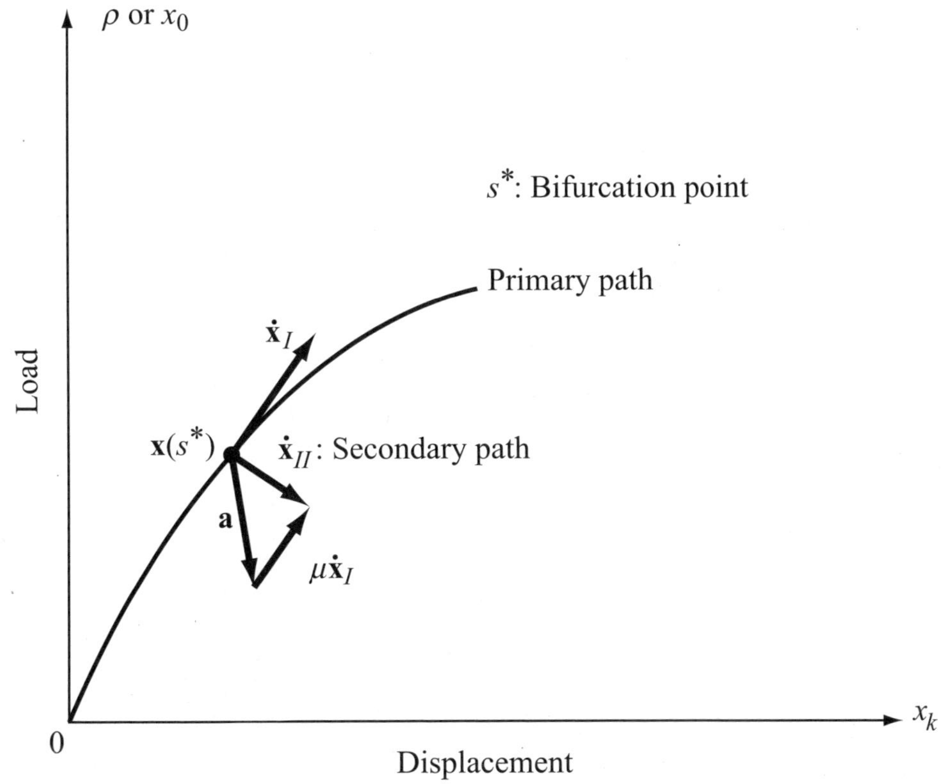

Figure 8.6 Schematic of the primary and secondary paths.

We can normalize the tangent vector to the secondary path at the critical point as

$$\dot{\mathbf{x}}_{II}\left(s^*\right) = \frac{\mathbf{a} + \mu\dot{\mathbf{x}}_I\left(s^*\right)}{\left\|\mathbf{a} + \mu\dot{\mathbf{x}}_I\left(s^*\right)\right\|} \tag{8.2.32}$$

This method requires the second derivatives $f_{i,jL}$ (in Eq. 8.2.31).

Once μ is determined, the first point on the secondary path can be estimated as

$$[\rho_{II}^{(1)}, \mathbf{x}_{II}^{(1)}] = \left[\rho_I\left(s^*\right) + \mu\dot{\rho}_I\Delta s,\ \mathbf{x}_I\left(s^*\right) + \left(\mu\dot{\mathbf{x}}_I + \mathbf{a}\right)\Delta s\right] \tag{8.2.33}$$

After we obtain the estimation in Eq. 8.2.33, we may improve the solution by the Newton-Raphson iteration in Eq. 8.2.19a as

$$f_{L,M}\left(\mathbf{x}^{(i)}\right)\Delta x_M^{(i)} = -f_L\left(\mathbf{x}^{(i)};\ s_1 + \Delta s\right)$$

$$(L, M = 0, 1, 2, \ldots, N) \tag{8.2.34a}$$

$$(i = 1, 2, 3 \ldots)$$

and

$$\mathbf{x}^{(i+1)} = \mathbf{x}^{(i)} + \Delta\mathbf{x}^{(i+1)} \tag{8.2.34b}$$

Because the first estimated point $x_L^{(1)}\left(s^* + \Delta s\right)$ is away from the critical point $x_L\left(s^*\right)$, the matrix $f_{L,M}$ is no longer singular. We can solve Eq. 8.2.34 to obtain the first point $\mathbf{x}_1$ after bifurcation on the secondary path. Therefore, the incremental solution can continue by the aforementioned solutions of Eqs. 8.2.20 and 8.2.19 or 8.2.34. The continuation of solution for points $\mathbf{x}_2$, $\mathbf{x}_3$, ..., $\mathbf{x}_m$ (up to another critical point, if any) on the secondary path is illustrated schematically in Figure 8.7. The disadvantage of this method is that computing the second derivatives is difficult.

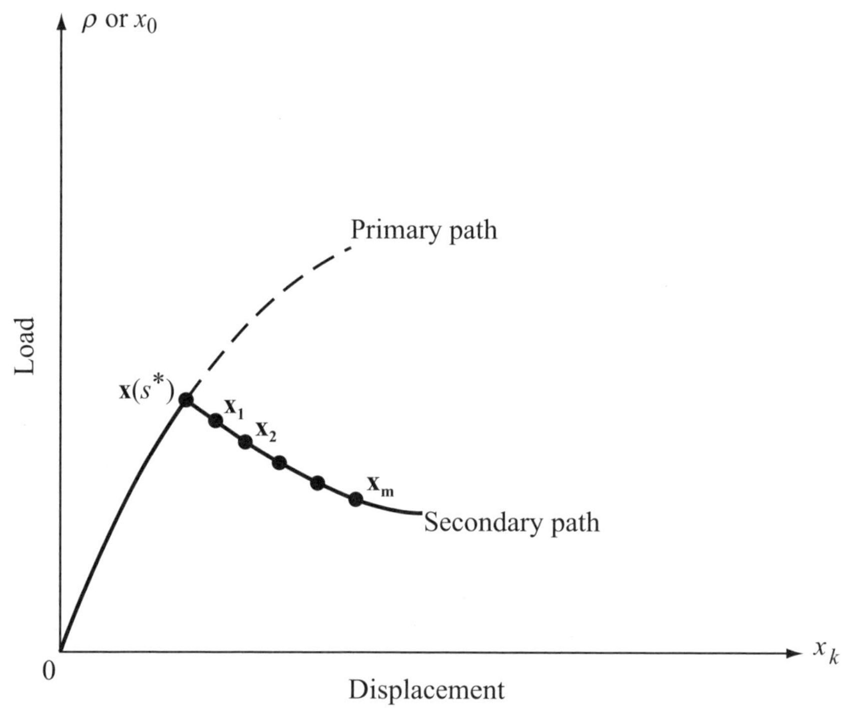

Figure 8.7 Schematic for continuation of the solution to the secondary path.

An alternative method to find the secondary path at the critical point is by assuming the normal of the plane in Eq. 8.2.15 at the critical point as

$$f_{0,L} = \frac{\dot{x}_{IL}\left(s^*\right) + \mu_1 a_L}{\left\| \dot{x}_{IL}\left(s^*\right) + \mu_1 a_L \right\|} \quad (L = 0, 1, 2, ..., N) \tag{8.2.35}$$

where $\dot{x}_{IL}\left(s^*\right)$ is the tangent to the primary path at the critical point s^*, and $f_{0,L}$ is orthogonal to $\dot{x}_{IL}\left(s^*\right)$. Note that we set $a_0 = 0$. Multiplying both sides of Eq. 8.2.35

by $\dot{x}_{IL}\left(s^*\right)$ yields

$$\mu_1 = -\frac{\dot{\mathbf{x}}_I \cdot \dot{\mathbf{x}}_I}{\dot{\mathbf{x}}_I \cdot \mathbf{a}} \tag{8.2.36a}$$

when $\dot{\mathbf{x}}_I \cdot \mathbf{a} \neq 0$. Otherwise,

$$f_{0,L} = \frac{a_L}{\left\|a_L\right\|} \tag{8.2.36b}$$

Using the normal $f_{0,\mathrm{L}}$, we estimate the first point onto the secondary path as

$$x_{IIL}^{(1)} = x_{IL}\left(s^*\right) + f_{0,L}\Delta s \tag{8.2.37}$$

Once the first point on the secondary path is estimated, the rest of the computation is to improve its accuracy on the secondary path by means of the Newton-Raphson iteration as specified in Eq. 8.2.19 or Eq. 8.2.34. We use Eq. 8.2.20 to compute the tangent vector and then the initial guess for the next step of the increment of the arc length s. We then apply Eq. 8.2.34 to improve its accuracy. Because the point is away from the critical point, the coefficient matrix **J** is no longer singular until the next critical point.

These two methods as discussed here require the tangent vector $\dot{\mathbf{x}}_{\mathbf{I}}\left(s^*\right)$ at the bifurcation point on the primary path. The following subsection will discuss a method to compute the tangent vector.

8.2.6 Method to Locate Critical Points and the Tangent Vector to the Primary Path for Bifurcation Buckling

For bifurcation buckling, not only the matrix **K** but also the matrix **J** is singular at the critical point. Therefore, we cannot use Eq. 8.2.20 to compute the tangent vector $\dot{\mathbf{x}}$ on the primary path. To determine whether the matrix is singular, we first must locate the critical point. Instead of handling a large linear system from the tangent stiffness matrix, we recommend locating a critical point at the arc length s^* by an approximate method of the finite difference. Following Riks [1979], we compute a sequence of the pre-critical points on the primary path, such as

$$\mathbf{x}\left(s_1\right), \mathbf{x}\left(s_2\right), \ldots, \mathbf{x}\left(s_i\right) \tag{8.2.38}$$

for $0 < s_1 < s_2 < \cdots < s_i$, where s_i is close to the critical point s^*. Using the position vectors $\mathbf{x}\left(s_{i-2}\right)$, $\mathbf{x}\left(s_{i-1}\right)$, and $\mathbf{x}\left(s_i\right)$, we compute the lowest eigenvalues $\omega\left(s_{i-1}\right)$ and $\omega\left(s_i\right)$, respectively, and then we obtain the derivative of the lowest eigenvalue with respect to the arc length s at s_i by means of the finite difference approximation as

$$\dot{\omega}(s_i) \approx \frac{\omega(s_i) - \omega(s_{i-1})}{s_i - s_{i-1}} \tag{8.2.39}$$

Here the dot denotes the differentiation with respect to the arc length s. Because the eigenvalue at the critical point is zero, Taylor's expansion gives

$$\Delta s^* = s^* - s_i \approx -\frac{\omega(s_i)}{\dot{\omega}(s_i)} \tag{8.2.40}$$

Therefore, the position vector on the primary path at the bifurcation is

$$\mathbf{x}(s^*) \approx \mathbf{x}(s_i) + \Delta s^* \dot{\mathbf{x}}(s_i) \tag{8.2.41a}$$

where the tangent vector is computed as

$$\dot{\mathbf{x}}(s_i) = \frac{\mathbf{x}(s_i) - \mathbf{x}(s_{i-1})}{\left|\mathbf{x}(s_i) - \mathbf{x}(s_{i-1})\right|} \tag{8.2.41b}$$

Note that this method also can be applied to locate a critical point for snap-through buckling.

8.3 Methods to Treat Snap-Through and Bifurcation Buckling in Forming Analyses

As mentioned in the previous sections, the external load must be explicitly involved in Riks' constraint equation. Except for the special case with a single point of contact, this condition is not satisfied for general contact problems of sheet metal forming analyses. Therefore, a different approach should be used. In addition, the global tangent stiffness matrix approaching a critical point is ill conditioned. Some special technique to overcome this difficulty in the numerical solution will be discussed in this section. For a good shell element, the stable bifurcation mode would be triggered automatically due to small unbalanced forces that act as imperfections.

8.3.1 Introduction of Artificial Springs at Selected Nodes

For a solution close to a critical point for snap-through buckling and bifurcation, the global tangent stiffness matrix is ill conditioned. The incremental displacements at certain points on a sheet increase rapidly for a step. We pick several points where the incremental displacements have extremely high values and impose stiff artificial springs at those points to remove the ill condition of the global tangent stiffness matrix.

We use a simplified model of the conventional draw operation as an example to demonstrate the computation of snap-through buckling. As shown in Figure 8.8, the sheet initially is convex, viewed from the outside of the lower die. As the punch moves

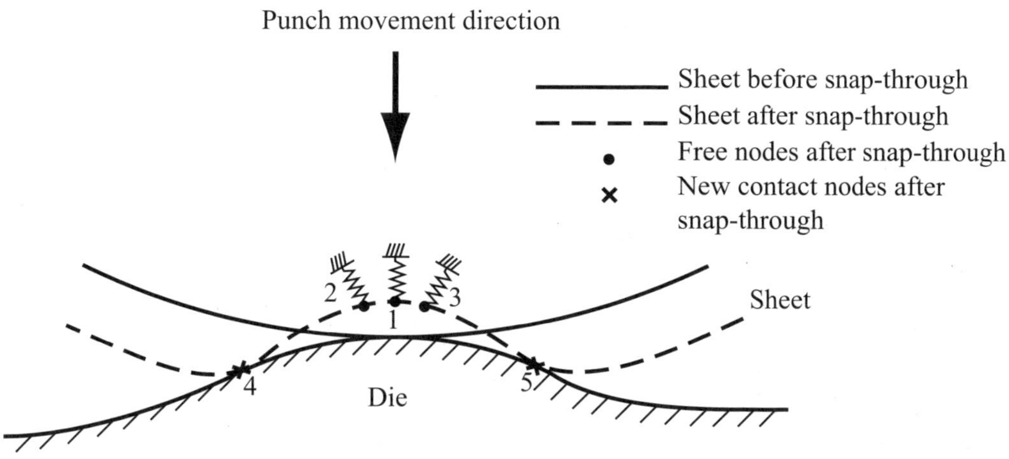

Figure 8.8 A sheet deformed from a convex shape to a concave shape at snap-through, and the use of artificial springs to obtain a stable solution.

downward, node 1 of the sheet contacts the lower die. As the punch moves farther downward, a snap-through occurs, and the sheet near the die changes from a convex shape to a concave shape. As the punch moves farther downward, new nodes 4 and 5 contact the die. During snap-through, node 1 and the neighboring nodes (say, nodes 2 and 3) have large values of displacement increments in the Newton-Raphson iteration. We ignore the results of this step. Instead, we impose artificial linear springs in the normal direction, as shown in the figure, and repeat the computation for this step. If we impose springs that are sufficiently stiff, the large incremental displacements can be avoided because the global tangent stiffness becomes well conditioned due to the addition of the stiffness from the artificial springs. Similarly, artificial springs also can be added in the neighborhood of a bifurcation point where the global tangent stiffness matrix is ill conditioned.

8.3.2 Forming Analyses of Snap-Through Buckling and Numerical Examples

Because the external loads are not explicitly specified for multiple-point contacts in sheet metal forming analyses, Riks' arc length method cannot be applied directly. We may apply the method by adding artificial springs at some selected nodes to avoid an ill-conditioned tangent stiffness matrix when the solution path is in the neighborhood of the critical point where a snap-through buckling occurs. We use two examples to demonstrate this method to obtain solutions after snap-through.

Here, we use a real production automotive body panel, a quarter inner portion, to show post-buckling deformation after snap-through during a binder wrap. Figure 8.9 shows the results of both quasi-static and dynamic explicit solutions. Because some area on the quarter panel inner portion is not supported and is not contacted by any tool surfaces during the binder wrap, snap-through buckling is prone to occur. Linear springs with sufficiently large stiffness are imposed on nodes with large displacement increments such that the global tangent stiffness close to the critical point is no longer ill conditioned. After the convergence of the Newton-Raphson iteration, all imposed springs are released, and another iterative process to counterbalance the spring forces is required. Because the deformation mode after snap-through is stable, there is no convergence problem for the iteration of releasing springs. The incremental solution by the Newton-Raphson iteration can go through the critical point; thus, the post-buckling solution is obtained.

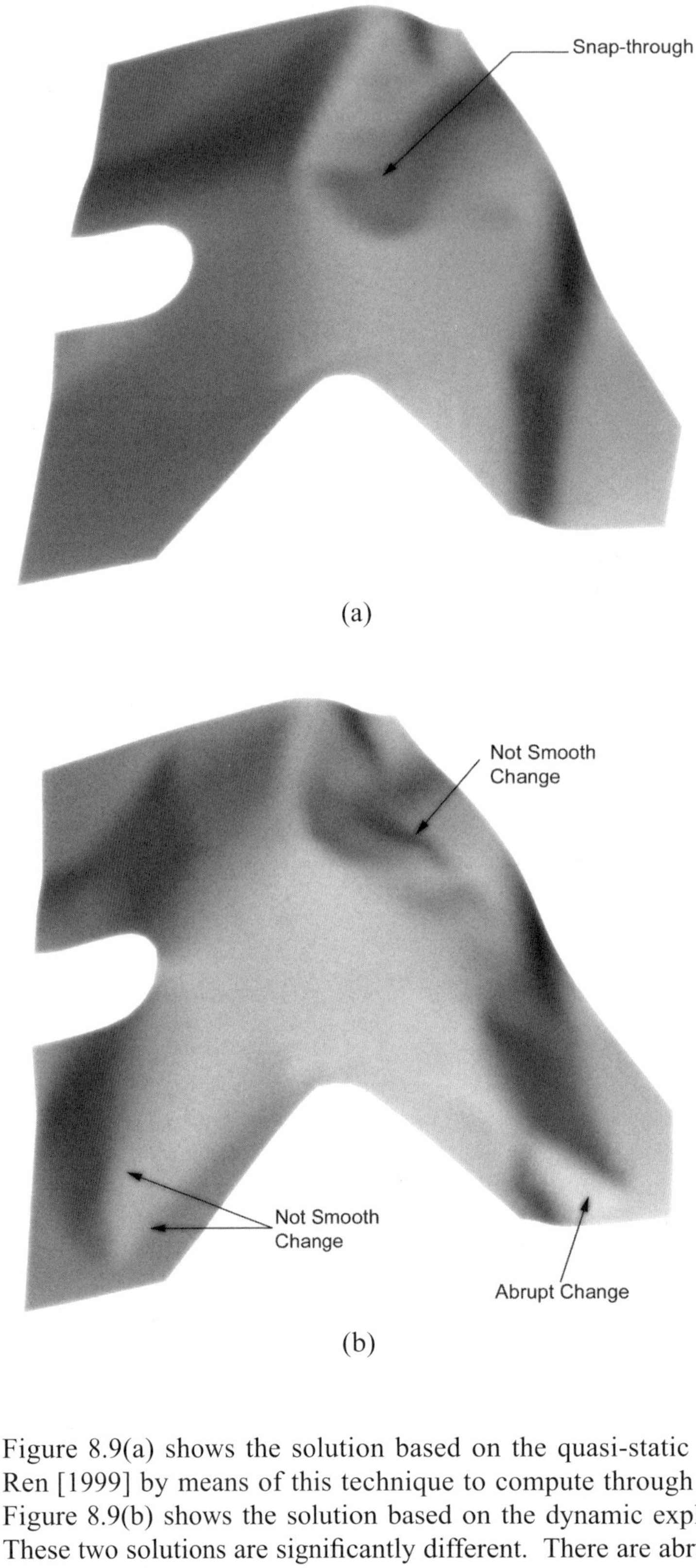

Figure 8.9 Binder wrap of a quarter panel: (a) quasi-static solution, and (b) dynamic explicit solution.

Figure 8.9(a) shows the solution based on the quasi-static method from Tang and Ren [1999] by means of this technique to compute through snap-through buckling. Figure 8.9(b) shows the solution based on the dynamic explicit integration method. These two solutions are significantly different. There are abrupt changes of deformation inside the punch opening from the dynamic explicit solution. This type of abrupt change also was reported by Tang [2000]. We do not think these abrupt changes of deformation would occur in a real forming process, based on the experience of practicing stamping engineers. The result from a dynamic code for the binder analysis of a fender discussed in a paper by Tang [2000] is not realistic when compared to that observed in the prototype shop. The simulation results of this binder wrap of the quarter inner had not been verified by the measurements in the shop. However, the simulation results based on the quasi-static implicit method and the field measurements for the binder wrap of a fender are presented in Chapter 7. The simulation results are in excellent agreement with the field measurements discussed in Chapter 7. For these reasons, we claim that solutions based on the quasi-static implicit method are more accurate.

For forming an automobile door inner panel, we often use an inverted draw operation with a three-piece tool, as shown in Figure 1.3 in Chapter 1. To avoid splitting during the forming process, the sheet during the binder wrap stage is given a convex shape viewed from the lower punch. A door inner with the binder wrap of this convex shape was used as an illustration in a patent awarded to Tang and Carnes [1995]. Figure 8.10(a) shows the part shape of the door inner portion. Figure 8.10(b) shows the convex shape of the sheet section after the binder wrap stage, and the concave shape of the sheet section

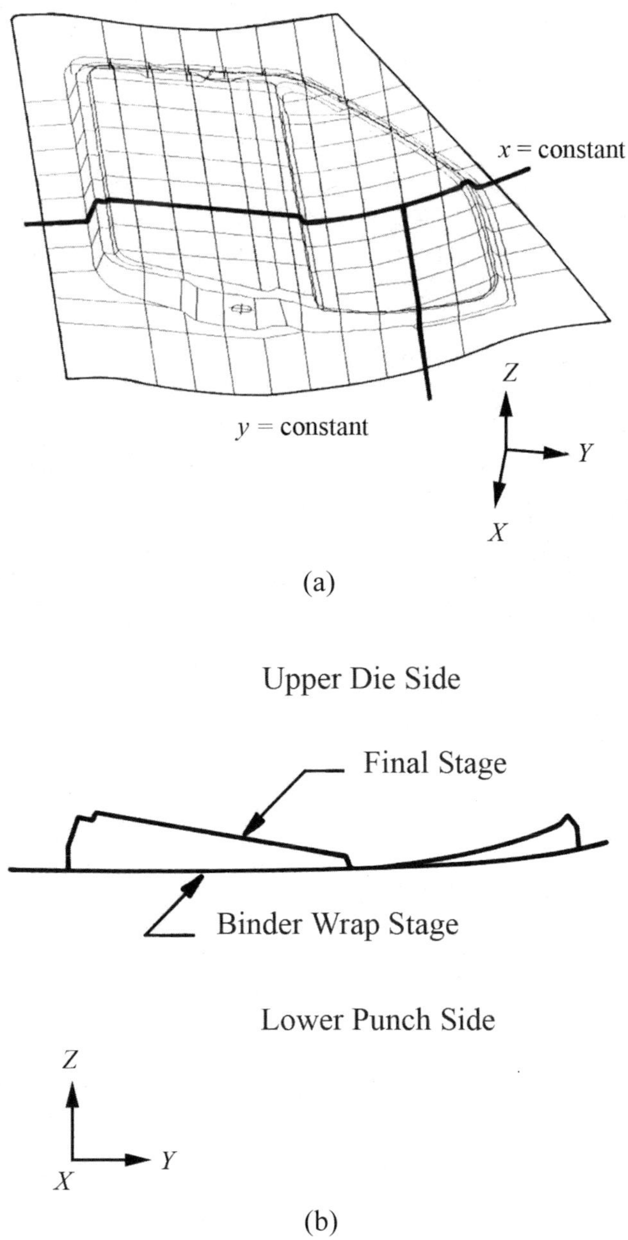

Figure 8.10 Forming a door inner panel: (a) part shape, and (b) the shape of the section (x = constant) changed from a convex shape after the binder wrap stage to a concave shape after the final stage.

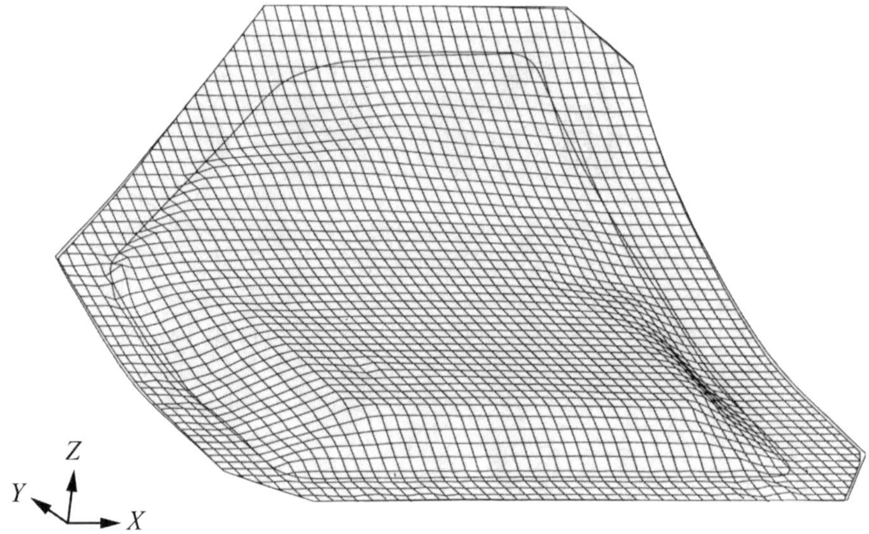

Figure 8.11 Shape of a door inner portion after snap-through buckling.

after the final die closure. Because the final shape of the part is concave, snap-through buckling occurs in the stretching and drawing process. We introduce a few artificial springs at the nodes where the incremental displacements are large for the load step close to the critical point. Figure 8.11 shows the formed shape of the inner panel at 40% of the travel of the upper die. For this stage, the panel already passes the snap-through buckling, and the forming process is stabilized.

8.3.3 Forming Analyses of Bifurcation Buckling and Numerical Examples

For bifurcation buckling, if we can determine the secondary path, then we can continue the incremental solution along the path until the next critical point, if any, because the coefficient matrix **J** in Eq. 8.2.21 is nonsingular beyond the critical point, and Eqs. 8.2.20 and 8.2.19 can be solved. In a forming analysis of solving the problem of surface contact with friction, the external load ρ is not specified explicitly. Therefore, Riks' arc length method must be modified.

The first method to determine the secondary path using Eqs. 8.2.29 to 8.2.32 requires $\dot{\rho}$, which is not involved explicitly in contact problems for sheet metal forming analyses. Therefore, the method cannot be applied to determine the secondary path. The alternate method using Eqs. 8.2.35 to 8.2.37 may be used to determine the path. We need to compute the first point on the secondary path x_i ($i = 1, 2, ..., N$) after the critical point. Therefore, we rewrite Eq. 8.2.35 for a normal vector orthogonal to the tangent to the primary path $\dot{x}_{Ij}$ at the critical (bifurcation) point

$$n_j = \frac{\dot{x}_{Ij}\left(s^*\right) + \mu_1 a_j}{\left\|\dot{x}_{Ij}\left(s^*\right) + \mu_1 a_j\right\|} \qquad (j = 1, 2, ..., N) \tag{8.3.1}$$

which does not contain $\dot{\rho}$. Because n_j is orthogonal to $\dot{x}_{Ij}$, multiplying both sides of Eq. 8.3.1 by $\dot{x}_{Ij}$ yields

$$\mu_1 = -\frac{\dot{\mathbf{x}}_I \cdot \dot{\mathbf{x}}_I}{\dot{\mathbf{x}}_I \cdot \mathbf{a}} \tag{8.3.2a}$$

when $\dot{\mathbf{x}}_I \cdot \mathbf{a} \neq 0$. Otherwise,

$$n_j = \frac{a_j}{\left\| a_j \right\|} \tag{8.3.2b}$$

Using the normal n_j, we estimate the first point onto the secondary path as

$$x_{IIj}^{(1)} = x_{Ij}\left(s^*\right) + n_j \Delta s \tag{8.3.3}$$

Once the first point on the secondary path is estimated, we can improve the solution on the secondary path using the Newton-Raphson iteration in Eq. 8.2.19 by omitting the first equation of the system and Δx_0 or $\Delta \rho$ as

$$f_{j,k}\left(\mathbf{x}^{(i)}\right)\Delta x_k^{(i)} = -f_j\left(\mathbf{x}^{(i)}\right)$$
$$(j,k = 1, 2, ..., N) \tag{8.3.4a}$$
$$(i = 1, 2, 3, ...)$$

and

$$\mathbf{x}^{(i+1)} = \mathbf{x}^{(i)} + \Delta \mathbf{x}^{(i+1)} \tag{8.3.4b}$$

After the first point on the secondary path is located accurately, we can apply Eq. 8.2.20 (omitting the first equation of the system and Δx_0 or $\Delta \rho$, while specifying the increment of the tool travel at the contact points with the moving tool surface) and Eq. 8.3.4a repeatedly to compute the subsequent points until the final result on the secondary path is reached. Because the first point is away from the critical point, the coefficient matrix **K** ($f_{j,k}$) in Eq. 8.3.4a or the modified Eq. 8.2.20 is no longer singular. Therefore, we can solve the equations until another critical point, if any, is encountered. If there is, we can repeat these methods for snap-through and bifurcation buckling. Note that the matrix **K** is a sub-matrix of **J** in Eq. 8.2.21. This computation is identical to that in Section 7.4.4 of Chapter 7, except for the contact forces being included. In Chapter 7, the method would fail if the coefficient matrix **K** is singular. The virtual time increment Δt has the same order of magnitude as the arc length increment Δs in this chapter.

Kim *et al.* [1999] used this method to analyze the wrinkling of a cylindrical cup during the deep drawing process. Figure 8.12 shows the wrinkling of a cylindrical cup under the blank holder force of 4 kN at the punch stroke of 30 mm (Kim *et al.* [1999]).

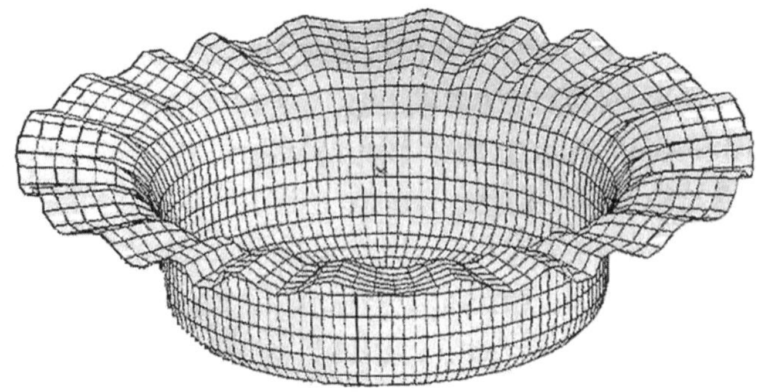

Figure 8.12 Wrinkling of a cylindrical cup.

For automotive panels, the method discussed in this section to compute deformation after bifurcation buckling usually is not necessarily needed. The quasi-static simulation program MTLFRM (Ford's in-house program) can automatically continue the solution to the region after bifurcation buckling. For the shell element used in MTLFRM, the stable bifurcation mode would be triggered automatically due to small unbalanced forces that act as imperfections. The dotted line in Figure 8.13 shows schematically the solution path with a small imperfection, which is approaching the primary path before the critical point and the secondary path after the critical point. Even with a small imperfection, the global tangent stiffness matrix may be ill conditioned when the solution closes to the critical point. Several artificial springs are required to impose at the nodes with large displacement increments. The artificial springs may be released after the solution is stabilized along the secondary path. Two examples of real automotive sheet metal parts are used in the following for illustration.

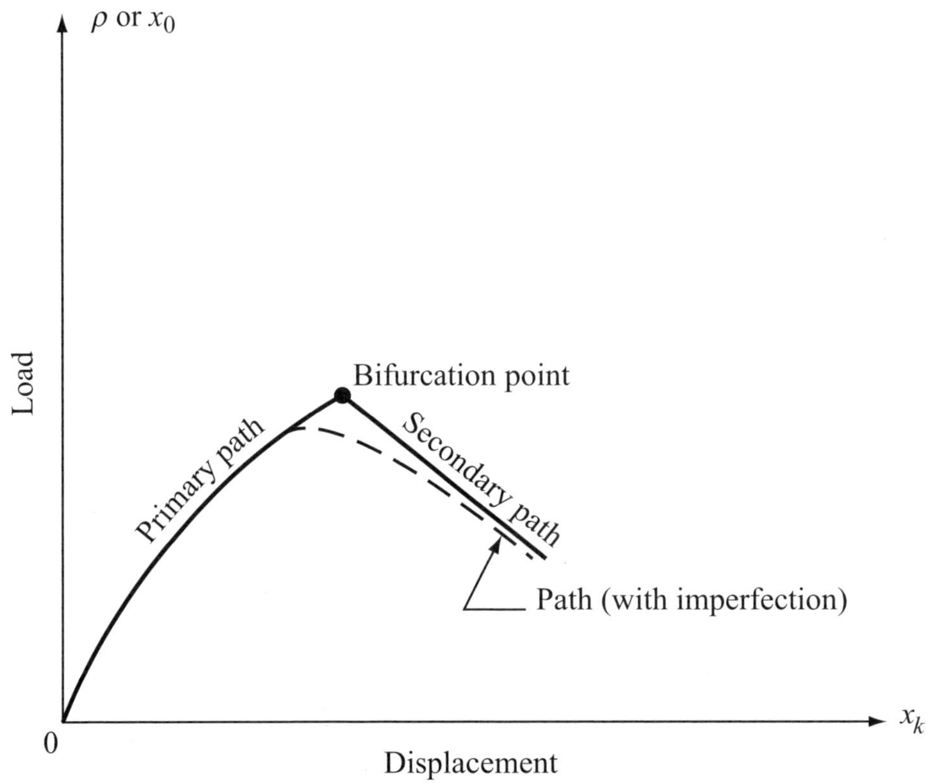

Figure 8.13 Schematic of the solution path with a small imperfection.

The first example involves the binder wrap of a floor pan, which was analyzed by both the dynamic explicit integration method and the quasi-static implicit integration method, as discussed in Chapter 7 of this book. The dynamic explicit integration method gives a symmetrical deformation mode, as shown in Figure 7.9(b). However, the quasi-static implicit integration method gives an asymmetrical buckling mode, as shown in Figure 7.9(c).

The second example involves an inverted draw process of a front fender with four-piece tools, as shown in Figure 1.4 of Chapter 7. Figure 8.14(a) shows the part shape. The material of the sheet is mild steel with a thickness of 0.7 mm. There was no apparent buckling in the simulation in the binder wrap stage by the quasi-static method. At 40% of the lower binder ring travel during the punch wrap stage, wrinkles started to generate along the hood line of the fender. Figure 8.14(b) shows the wrinkles at the end of the punch wrap stage. Because the wavelength of the wrinkles is long, only a relatively coarse element mesh with less than 1000 curved triangular thin shell elements is needed to show the wrinkles. This usually is the case for a large outer body panel with wrinkling modes of long wavelength during forming processes. A revised design of the tools was made by straightening the surface along the hood-line area and adding a draw bar in front of the fender. A draw bar can stretch loose metal that would cause wrinkling or buckling. The result from the simulation of the revised design is displayed in Figure 8.14(c), where the wrinkles had been eliminated. Figure 8.14(d) shows two pictures of the prototype fenders: the original design that has wrinkles along the hood line, and the revised design that is free of wrinkles. For a comparison of Figure 8.14(b) with the upper picture in Figure 8.14(d), not only wrinkling occurs in the same area but also the shapes are almost identical. For a comparison of Figure 8.14(c) with the lower picture in Figure 8.14(d), both shapes show freedom of any wrinkling because of adding a draw bar in the front and straightening the binder surface near the hood-line area.

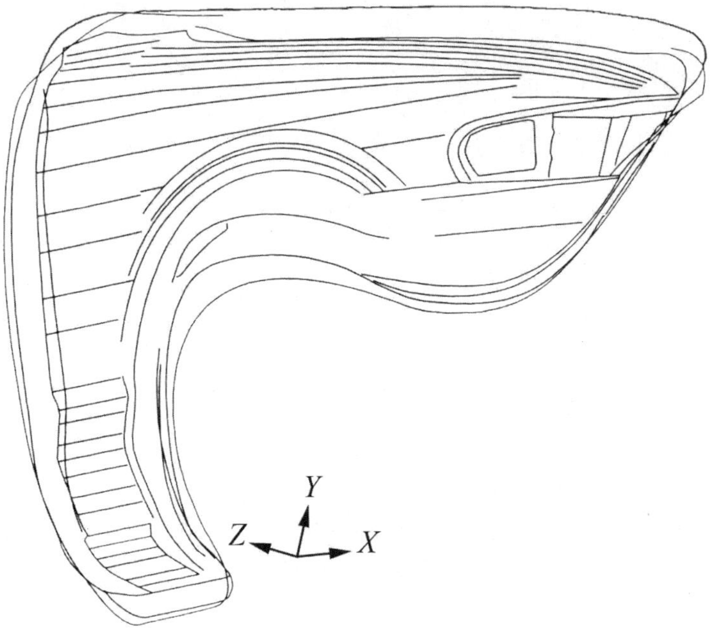

(a)

Figure 8.14 A front fender formed by an inverted draw process: (a) part shape.

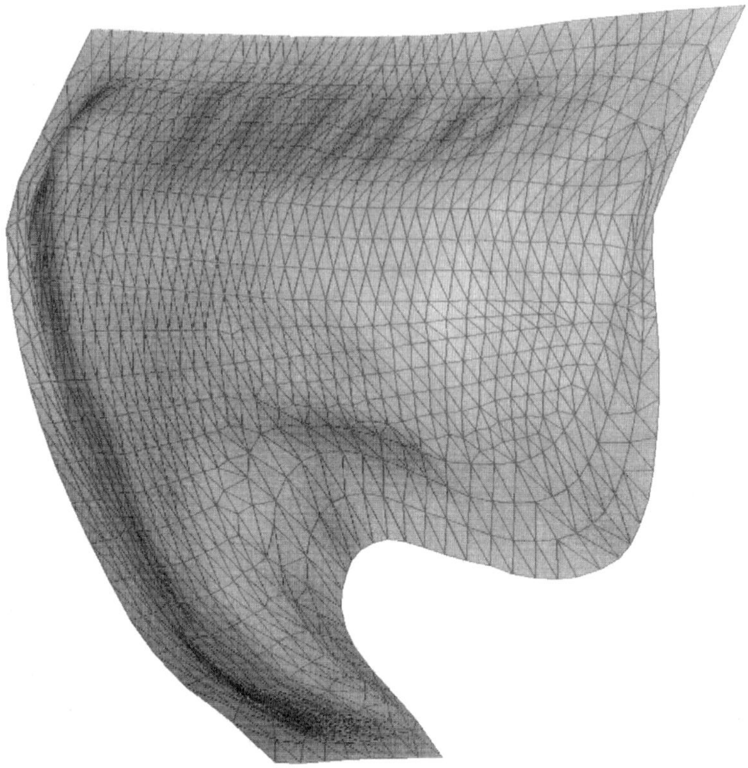

(b)

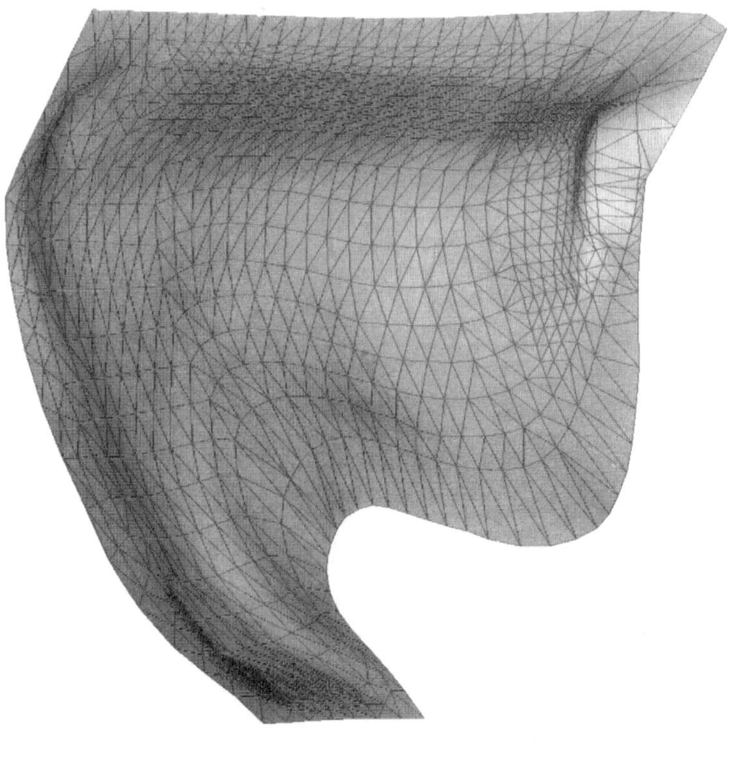

(c)

Figure 8.14 *(Continued)* (b) Solution with wrinkles based on the original design, and (c) solution without wrinkles based on a modified design.

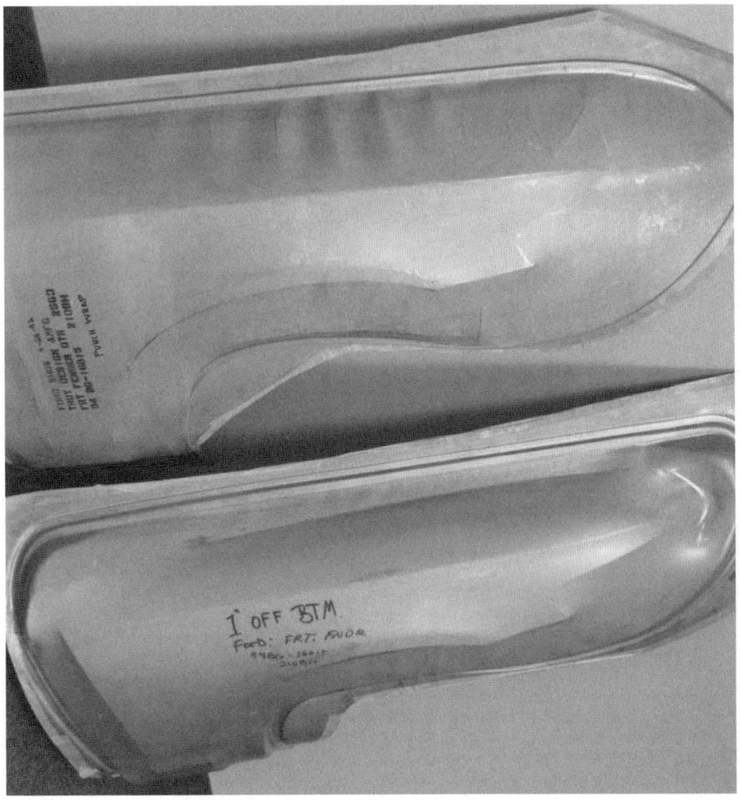

(d)

Figure 8.14 *(Continued)* (d) Formed fenders from the prototype shop.

8.4 References

Kim, J.-B., Yang, D.-Y., Yoon, J.-W., and Barlat, F., "An Analysis of Wrinkling Initiation and Growth of Anisotropic Sheet in Deep Drawing Process with Controlled Blank Holding Forces," Proceedings of the 4th International Conference—"NUMISHEET'99," Vol. 1, Besancon, France, 1999, p. 335.

Riks, E., "An Incremental Approach to the Solution of Snapping and Buckling Problems," *International Journal of Solids and Structures*, Vol. 15, 1979, p. 529.

Tang, S.C., "Trends on Simulation of Sheet Metal Forming Processes," SAE Paper No. 2000-01-1108, Society of Automotive Engineers, Warrendale, PA, 2000.

Tang, S.C. and Carnes, J.C., "Method and Apparatus for Predicting Post Buckling Deformation of Sheet Metal," U.S. Patent No. 5,390,127, 1995.

Tang, S.C. and Ren, F., "On the Solutions by Different Numerical Methods in Simulation of Sheet Metal Forming Processes," Proceedings of the Sixth International Conference in the Technology of Plasticity (6ICTP), Vol. III, Nuremberg, Germany, 1999, p. 2149.

Index

About the Authors

Dr. Sing C. Tang worked for the Research Laboratory of Ford Motor Company for 38 years. He was a Corporate Technical Specialist in the Research Laboratory before he retired in 2001. He received his B.S. in Civil Engineering from the National Taiwan University in 1956 and his Ph.D. in Structural Engineering from the University of Michigan in 1963. After joining Ford in 1963, Dr. Tang conducted extensive research on projects such as powertrain bending vibration, flow analysis in turbo machines, air bag modeling, elastic-plastic low thermal cycle stress analysis of a gas turbine compressor, estimation of damageability of bumpers, and vehicle crash analysis. During his last 15 years at Ford, his research concentrated on accurate simulations of sheet metal forming processes. As a leader for the Simulation Group at Ford, Dr. Tang always promoted the quasi-static implicit method for simulation of sheet metal forming processes based on theoretically sound mechanics principles. The quasi-static implicit method is particularly good for simulation of binder wrap and prediction of springback. Dr. Tang was awarded a Henry Ford Technological Award for his outstanding research accomplishments at Ford. He is a fellow of the American Society of Mechanical Engineers (ASME) and has edited one book and published more than 100 technical papers. Dr. Tang also holds six U.S. patents.

Dr. Jwo Pan is a Professor in the Department of Mechanical Engineering at the University of Michigan in Ann Arbor, Michigan. He received his B.S. in Mechanical Engineering from the National Taiwan University in 1974 and his Sc.M. and Ph.D. in Solid and Structural Mechanics from Brown University in 1978 and 1981, respectively. Dr. Pan has worked in the area of sheet metal forming, yielding and fracture of plastics and rubber, weld residual stress and failure, fracture, fatigue, plasticity theories, spot weld, and material modeling for crash simulations. He also has worked closely with engineers in the automotive industry. Dr. Pan has served as Director of the Center for Automotive Structural Durability Simulation funded by Ford Motor Company and as Director for the Center for Advanced Polymer Engineering Research at the University of Michigan. He is a Fellow of the American Society of Mechanical Engineers (ASME) and a Fellow of the Society of Automotive Engineers (SAE). Dr. Pan is on the editorial boards of the *International Journal of Fatigue* and the *International Journal of Damage Mechanics*. He has edited 20 books and special journal volumes. Dr. Pan also has served as an associate editor of the ASME *Journal of Pressure Vessel Technology*. He has written 110 technical papers published in refereed journals, as well as 130 technical conference papers and presentations.